ORKNEY

HIGHLAND

TAYSIDE

*NORTH
SEA*

LOTHIAN

DUMFRIES AND
GALLOWAY

DURHAM

NORTH
YORKSHIRE

*IRISH
SEA*

S.
YORKSHIRE

CLWYD

NOTTINGHAM
SHIRE

SHROPSHIRE

NORFOLK

NORTHAMPTON
SHIRE

HEREFORD
AND WORCESTER

DYFED

BUCKINGHAM
SHIRE

GWENT

ESSEX

BERKSHIRE

HAMPSHIRE

KENT

DEVON

DORSET

E. SUSSEX

CORNWALL

*ENGLISH
CHANNEL*

Ireland

SMITHSONIAN NATURAL HISTORY SERIES

John C. Kricher, Series Editor

Books in this series explore the diverse plants, animals, people, geology, and ecosystems of the world's most interesting environments, presented in an accessible style by world-renowned experts.

Ireland

Michael Viney

A Smithsonian Natural History

SMITHSONIAN
BOOKS

WASHINGTON
& LONDON

For Ethna and Michele
with love and deep appreciation

Copy editor: Debbie K. Hardin
Production editor: Robert A. Poarch
Designer: Brian Barth

Library of Congress Cataloging-in-Publication Data
Viney, Michael, 1933–
    Ireland : A Smithsonian Natural History/Michael Viney.
        p.   cm. — (Smithsonian natural history series)
    Includes bibliographical references (p.   ).
    ISBN 1-58834-110-0 (alk. paper)
    1. Natural history—Ireland. I. Title. II. Series.
QH143.V555 2003
    508.415—dc21   2002070687

British Library Cataloguing-in-Publication Data is available
Manufactured in the United States of America
10 09 08 07 06 05 04 03   5 4 3 2 1

Color signature printed in China

♾ The paper used in this publication meets the minimum requirements of the American National
Standard for Information Sciences—Permanence of Paper for Printed Library Materials ANSI
Z39.48-1984.

Contents

Editor's Note

It is often called the "Emerald Isle," the land of "forty shades of green." An island surrounded by the North Atlantic Ocean and the Irish Sea, it is a land of peat bogs, coastal marshes, high cliffs, soft green pastures defined by neat stone walls, and fields ablaze with wildflowers. It is a lush place, where clear skies repeatedly yield to gentle rain. It is Ireland.

This is a region whose countryside rarely provides anything other than the most splendid of vistas. Emerging from between the clouds as the Aer Lingus plane entered final approach to Shannon Airport on my first visit to Ireland in August of 1978, I can still vividly recall the verdant agricultural landscape, a patchwork of rolling pasture populated by sheep and donkeys, bordered by narrow roads and generous hedgerows. Our first stop was the Burren, in County Clare, a region unique for its geology and amazing wildflower display, acres of limestone outcroppings, where trees are absent but where the moist, mild climate permits cattle to forage in the pastures every month of the year. The Cliffs of Moher, near Ennistymon, rise up from the sea, rugged and imposing. High winds greeted me as I walked to the edge of the sheer cliffs to look out on colonies of seabirds, common murres (called guillemots in Ireland), kittiwakes, fulmars, and Atlantic puffins. Gannets were plunging into the sea in pursuit of fish. Soaring on the updrafts were many rooks and jack-daws, members of the Corvidae, or crow family, along with a few choughs (pronounced "chuff"), also a corvid and one of Ireland's rarest bird species. Choughs, recognized by their long, red bill, look otherwise like crows, uniformly black. They appear to take thorough pleasure in soaring to great heights and diving, wings folded back, rolling and turning, as though the bird is imagining its own roller coaster in the sky. This was my introduction to Ireland.

I have had the good fortune to travel widely in Ireland and to experience first-hand some of its diverse natural history. In County Wicklow, south of Dublin, I have walked long distances over soft, boggy peat in search of such species as ring ouzel and dipper. I have marveled at flocks of diverse shore-birds and waterfowl in the marshes at Wexford Slobs, and enjoyed the masses of seabirds that can be found off the coast of Donegal, nestled on the north-western side of Ireland. I have come to learn something of the extraordinary natural and human history of this fascinating land.

When searching for titles for the Smithsonian Natural History Series it seemed obvious to me that Ireland must be among them. I called my cousin, Bruce Carrick, whose home is in County Westmeath, in the Irish midlands, and asked for his advice about who might author such a volume. Without hesitation, he told me I should contact Michael Viney.

Michael was born in Brighton, England, in 1933, but moved to Ireland in 1961. He had a long career in journalism and television that took a radical turn in 1977 when, along with his wife and small daughter, he abandoned the urban environment of Dublin and moved to rural County Mayo. There he began writing a weekly column for the *Irish Times*, titled "Another Life." The column chronicled Michael's new self-sufficient lifestyle of living off of and in concert with the land, but it soon grew to be the story of his immersion in nature. The column has appeared without interruption for twenty-five years.

Michael has been a member of two scientific expeditions to northeast Greenland (in 1983 and 1987) to study the summer breeding biology of the barnacle goose, a migrant between the Arctic and the west of Ireland. His popular book, *A Year's Turning* (1996), inspired an Irish television series that he and his wife filmed and their now-grown daughter edited. The series documented the Vineys' lifestyle and observations of nature over the course of a year.

I am very pleased that Michael Viney has chosen to share his understanding and perspectives on the natural history of Ireland. My cousin was right. You could not ask for a better guide to introduce you to nature in this wonderful land that is Ireland.

John C. Kricher

Preface

My house on a hillside on Mayo's rugged west coast looks out to islands, sand dunes, mountains and bogs, and to farmhouses scattered among small, rough fields. Each window seems to hold a different part of Ireland's story.

At morning the sun glancing down from the ridge picks out boulders spilled across the hillside by an ice-age glacier. They share the fields with other curious corrugations: broad ridges under the grass, parallel as the wales of corduroy. These are — or were — potato ridges, dug before the Famine of the 1840s, when this hillside was dense with people and peat smoke. After the blight's catastrophe, the ridges were abandoned in the great emptying-out of rural Ireland by emigration in the late nineteenth century. Today they are grazed by sheep and mountain hares, and as the last, spiky gables of the old thatched cabins melt down into mere rock heaps, the strata of geological and human history lie welded in the landscape.

My study looks out to the Atlantic, where morning rainbows arch above Inishturk, a jagged, Ordovician island with speckles of gold in its veins. Its buckled profile was prefigured 400 million years ago, when the ancient Iapetus Ocean slammed shut (at that time this coast was in Labrador's backyard). The rainbows speak for the intense humidity and processional showers that nourish the mosses of Ireland's peatlands, the luxuriance of stream-side ferns, the greenness of endless hedged pastures. The ocean's influence is mostly benign, yet in geological time the story of weathering and erosion has been a fierce one, carving Ireland's initial relief from a huge, uplifted block of sedimentary stone. In today's winters, storms of 80 miles per hour (128 kph) and more batter at the high western cliffs of the islands and send breakers clawing at the sandy fields and dunes below me. As global warming raises the sea level and notches up the power of the wind, images of salt marsh and tundra wake from a suddenly recent—merely postglacial—past.

My seat beside the wood stove is angled to watch Mweelrea, the highest mountain in Connacht, springboard of the winter sun and enormous autumn moons. The mountain rises sheer from Killary Harbour (a glacially deepened fjord), and at its peak one clatters over flakes of sandstone shattered in the ice age cold. Yet Mweelrea's summit was left a bare dome above the ice—a nunatak—and thus speaks for all the odd corners of the island that might have served as what science calls "refugia." From the tops of mountains to strippedback seabeds far offshore, there might have been tolerable sanctuaries for this or that plant or animal: how otherwise explain their presence at improbable locations today? Were there really land bridges from Britain and Europe as the ice melted, and if so where were they, and when? The questions find geologists and biologists strongly at odds, and the quirks of Ireland's postglacial colonization are a recurring theme of this book.

Beyond the dark cleft of Killary Harbour the mountain cluster of the Twelve Bens gleams with granite and quartzite (to travel a mere 30 miles (50 km) in Ireland is often to change the bedrock entirely), and the moorland bogs of Connemara step back in a series of horizons, brownish-gray or blue. At this innocent distance, they seem to flow on unbroken, part of a difficult wilderness long stripped away from the western margins of continental Europe. In Ireland, the mossy fabric of the peatlands, often many feet deep, seem immune to change, part not merely of a heritage but of a culture. Now the great raised bogs of the midlands are down to mere shreds and remnants, and the rolling bog of the uplands and western seaboard is increasingly a motheaten blanket, drained or scooped away for a host of different and profitable ends. Yet the peatlands are still, in many places, magical in their strangeness, and provide, in fossil pollens and timbers and buried artifacts, an often astounding archive of the past.

On the map, the corner of the coast where I live is a peaceful cul-de-sac: the road thins to a tendril, winds round the mountain and stops at the sea. Yet to dwell here is to become aware of the great traffic of migratory birds. In spring and autumn, seabirds and waders flow around Ireland like water around a rock, part of a great arc of movement between the Arctic and Africa. Some, like the gannets, shearwaters, and petrels, come to nest on uninhabited islands; others, like the whimbrels of Iceland, pause only briefly on passage to and from the north. Working outdoors on my acre, I mark the more intimate migrations of swallows, wheatears, warblers, and, in autumn, listen and watch for the first skein of Icelandic whooper swans, flying low along the dunes to the lake.

The awakening to such rhythms has drawn natural history to the forefront of my life. Some twenty years ago, a sudden, mid-career resolve inspired my wife and me, with a small daughter, to move from Dublin to remote rural Mayo. The rewards were to be those of a low-income "alternative" lifestyle in which we grew much of our own food, or fished for it, in a landscape of rugged beauty. We were media people, with the habit of curiosity, used to digging out facts and making sense of them. As we settled into the endless experiments of a more self-reliant existence, these instincts were turned with increasing fascination on plants, birds, insects, and sea creatures. Almost daily beachcombers, we discovered dead whales and turtles and collected tropical drift seeds among the seaweed at high-water mark. Our windowsills filled up with a naturalist's bric-a-brac and our shelves with books about the natural world.

In time, I found myself drawn into expeditions to Greenland, organized to study the summer ecology of the barnacle goose *(Branta leucopsis)*, which winters along the western shores of Ireland. My role among the biologists was chiefly that of goose herd and net carrier, but the months spent in the rock-and-tundra wilderness of North Germania Land left me with a permanently liberating sense of Ireland's place in the western Palearctic: of the continuum of habitats and ocean zones that link up with our small green island.

They also led to a number of wildlife films for television and enriched the weekly column that I write for *The Irish Times*. "Another Life" began as a chronicle of our adventures with goats, hens, and bees, but the life it celebrates now is almost exclusively wild. The column brings me a steady flow of readers' observations and queries, and along with worries about wasps' nests or bats in the attic come meticulous descriptions of strange birds and insects and of glimpses and encounters of a revelatory kind.

The next decade or two could be a difficult time for the Irish landscape and its wildlife, as pressures flow out from a totally unprecedented economic boom, a rising population, and even more tourists seeking the "unspoiled" corners of the island. The impetus for conservation, substantially nourished by Ireland's membership in the European Union, has yet to find a full endorsement in the Republic's political soul. In this book, I have tried to take the ecologist's view of what is special about Ireland's natural fabric, in the hope that it may help to keep the best of it more or less intact. It is not in any way an inventory of the island's flora and fauna, but it is more concerned with the habitats and eco-systems, much influenced by humans, that have brought their lives together.

Michael Viney

Acknowledgments

Any modern natural history of Ireland must reach back to the work of two great polymath naturalists and popular communicators of the twentieth century, Robert Lloyd Praeger and Frank Mitchell. They set a standard of enthusiasm and authority in interpreting the Irish landscape that any lay writer about nature can only follow on tiptoe. Their modern successors in science have watched my newspaper writings wander through their territories with great forbearance and encouragement, from time to time sliding a relevant paper under my nose or pointing out a journal I should consult. Thus, over the years, they have helped me to build the overview and perspective essential to a book of this ambition. In that tutorial process, I would mention with special appreciation Pete Coxon, James Fairley, John Feehan, Ferdia Marnell, Kate McAney, Dan Minchin, Paul Mohr, Charles Nelson, James P. O'Connor, Patrick Sleeman, Chris Smal, and the late Tony Whilde. For indelible and formative experiences in the field, I thank David Cabot, who, with Roger Goodwillie, swept me off to the Arctic on a three-month expedition that helped me to set Ireland in the wider world of bird migration.

For specific assistance in research, I must express my gratitude to the following: Trevor Beebee, Simon Berrow, Andrew Bleasdale, Tom Curtis,

Karina Dingerkus, Tom Egan, Ralph Forbes, Howard Fox, Paul Giller, Valerie Hall, Daniel Kelly, Andrew Kitchener, Neil Lockhart, Gerald Mills, Fraser Mitchell, Nigel Monaghan, Evelyn Moorkens, Michael O'Connell, John O'Halloran, Donncha O Teangana, Matthew Parkes, Julian Reynolds, Michael Sheridan, Martin Speight, and William P. Warren. They, along with others, have guided me to a wide range of primary research, much of which I have sought to acknowledge in the text. For the general reader, however, I have closed each chapter with a brief list of selected references to the more readily available books.

Finally, I am deeply indebted to the efficiency and patience of Mary Glavin, of Mayo County Library, and of my wife, Ethna. The first has secured a seamless flow of source material to my remote rural address. The second has been a supportive and stimulating working partner and a skillful live-in editor with meticulous attention to process, without whom, indeed, the project might have foundered.

— *Chapter 1* —

Rediscovering the Wild

A feeling for nature, if not the actual study of nature, goes back a long way in Ireland, yet the current appreciation of wildlife, as a popular pastime, was slow to mature. For much of modern history a colonial occupation stood between the native Irish and the hobbies and education of an alien governing class. Then, in the impoverished early decades of an independent Ireland, the popular view of nature was urgently utilitarian and land-hungry.

Today's birders and plant lovers have thus rediscovered a kind of spontaneous affection for nature not much in fashion among the mainstream Irish for a long time. It recalls—perhaps even connects with—the delight in nature shown by Ireland's medieval monastic poets, brothers to the hermit monks who decorated manuscripts with animated birds and animals. Between the eighth and twelfth centuries, they celebrated what can seem a Celtic Walden, with a sound track of blackbirds and wind in the trees. "It is extraordinary," wrote Frank O'Connor in 1962 in *Kings, Lords and Commons*, "how clear and bright the landscape of early Irish poetry is, as though some mediaeval painter had illustrated it, with its little oratories hung with linen, its woodlands and birds, its fierce winters and gay springs." Some of the medieval Irish poems, praising the diversity and bounty of nature, become a virtual catalog of spe-

ATLANTIC
OCEAN

Tory Island

Malin Head

Rathlin Island

Derry
LONDONDERRY
Larne
ANTRIM

DONEGAL

U L S T E R

Donegal

Lough
Neagh
Belfast

Donegal Bay

Omagh
TYRONE

DOWN

Enniskillen

Armagh
ARMAGH

Downpatrick

Sligo
LEITRIM

FERMANAGH

Clones

Newry

Ballina

SLIGO

MONAGHAN

Cavan

Achill
Island

MAYO

Carrick-on-
Shannon

CAVAN

LOUTH

I R I S H
S E A

Clew Bay

Castlebar

ROSCOMMON

Longford

Kells

Drogheda

Westport

Roscommon

LONGFORD

MEATH

C O N N A C H T

WESTMEATH

Athlone

DUBLIN

GALWAY

Tullamore

KILDARE

Dublin

Connemara

Galway

OFFALY

Naas

WICKLOW

Aran Islands

Burren

L E I N S T E R

CLARE

LAOIS

Ennis

River
Shannon

TIPPERARY

Kilkenny

Carlow

CARLOW

Limerick

Thurles

KILKENNY

LIMERICK

Tipperary

WEXFORD

Blasket
Islands

Tralee

Wexford

Waterford

MUNSTER

WATERFORD

Killarney

Mallow

Saltee Islands

Dingle Bay

KERRY

CORK

Cork

Bantry

Cape Clear Island

Bill Nelson

Ireland

cies, conveying the sort of facts that might be gleaned from modern pollen-cores but with a great deal more color and feeling. "Bushy, leafy oak, you are high above every tree," wrote Suibne Geilt ("Mad Sweeney") as he wandered the woods, "little hazel, branchy one, coffer for hazel-nuts. Alder, you are not hostile; beautifully do you gleam. . . ." (quoted in *Early Irish Lyrics*, by Gerard Murphy). Even so many centuries before the great forest clearances in the fifteenth and sixteenth centuries under the Tudors, the human use of trees, plants, and wildlife was fiercely competitive and needed regulation. A botanical legacy has been the list of twenty-eight trees and shrubs ranked by value in *Bretha Comaithchesa*, the main law text on farming in pre-Norman Ireland. The Old Irish tree list, as it is commonly called, sorts the species by economic importance, with the oak (*Quercus* spp.) at the head of the seven "nobles of the wood," followed, unexpectedly perhaps, by hazel *(Corylus avellana)*, holly *(Ilex aquifolium)*, yew *(Taxus baccata)*, ash *(Fraxinum excelsior)*, Scots pine *(Pinus sylvestris)*, and wild apples *(Malus sylvestris)*. Next come the seven "commoners," such as alder *(Alnus glutinosa)* and willow (*Salix* spp.); then the "lower" species, such as blackthorn *(Prunus spinosa)*; and finally the mere bushes, which include bracken *(Pteridium aquilinum)* and gorse *(Ulex europaeus)*. The status of each depended on a fine calculation of use (oaks for acorns to feed to pigs; hazel for winter nuts and wattles; holly for cattle fodder and chariot shafts) and was reflected in the size of the fines for damaging them. Set against the precision of a legal tract, an Old Irish saga of events that may or may not have occurred in the reign of King Cormac Mac Airt (A.D. 227–266) and that was not set down until the eleventh century, may seem a dubious source of species data. Nonetheless, the list of wild creatures set out in the early *Book of Lecain* as a ransom for Fionn MacCumhaill did purport to be an inventory of the island's fauna: "Two ravens from Fiodh da Bheann, two wild ducks from Lough Sillane, two foxes from Slieve Cuilinn, two stags from the Burren. . . ." and so on through some fifty-five species, from porpoises to wrens, otters, bats, and wolves. For all its clear imperfections ("two peacocks from Magh Mell"—are you sure?) the list was charming for its own sake and grist to the mill of Victorian antiquarians poring over fossil bones in the collections of the Royal Irish Academy.

In so far as natural history suggests the spirit of inquiry, the Irish literature reaches back to A.D. 655, when an Irishman named Augustin (not the saint) was actively speculating on the means by which the mammals could have reached islands after Noah's flood in *Liber de Mirabilibus Sanctae Scripturae*, where he asked, "*Quis enim, verbi gratia, lupos, cervos et sylvaticos porcos, et vulpes, taxones et lepusculos et sesquivolos in Hiberniam deveheret?*" ("Who could have brought wolves, deer, forest pigs, foxes, badgers, little hares, and squirrels to Ireland?")

More than 500 years later, what is now called island biogeography was continuing to concern a visiting scholar, Giraldus Cambrensis, who completed his *Topographia Hiberniae* in 1188. This Welsh ecclesiastic belonged to one of the Anglo Norman warlord families involved in the invasion of Ireland, and his academic career was entwined with an exuberant social life at the court of Henry II. He meant his work to entertain, and along with sharp personal observation of wildlife he indulged an instinct for fantasy and myth (a floating island, a fish with three gold teeth) that seemed to swell with each revision of his work. It was also shaped by the medieval tradition of the bestiary, in which real observations of nature were made to serve fables of Christian morality. The result, in modern translation, is received by Irish scientists with varying degrees of patience. "An amalgam of fact, fibs and fantasy, and much of it . . . patently absurd," huffs mammalogist James Fairley, but to Christopher Moriarty, fisheries biologist and a student of the early naturalists, it is "a delightful, maddening work of genius."

If Giraldus was susceptible to myths about the natural world, some of them were widely shared and remarkably durable. He encouraged, but did not invent, the idea that the migratory barnacle geese appearing so mysteriously on Ireland's winter shores were hatched from the stalked barnacles growing on drift timber: he had seen a thousand of the "small, bird-like creatures" hanging from a single log! In the late nineteenth century this belief was still strongly and widely held along the Irish coast, and *Lepus anatifera*, with its chalk-white shell and feathery feeding appendages, is today irreversibly known as the goose barnacle.

But in among the fairy tales of the *Topographia* are real glimpses of a very different ecosystem—forests full of falcons (*Falco* spp.) and capercaillie *(Tetrao urogallos)*, marshes full of cranes—and some significantly accurate local intelligence on species present in Britain but absent from the Ireland of the time: not only serpents, but polecats *(Mustela putorious)*, beavers *(Castor fiber)*, hedgehogs *(Erinaceus europaeus)*, moles *(Talpa europaea)*, and purely freshwater species of fish. Among the curiosities Giraldus seems to have noted for himself was the surprising absence of the all-black carrion crow *(Corvus corone corone)* and its replacement by the black and gray form of the species, the hooded crow *(Corvus corone corvix)*. He included an observation of the sort repeated as novel even among modern birders: "These birds bring up small shellfish into the air, and let them fall again so that they may be able to break, by collision with a stone after a long fall, the shell which they cannot break with their beak."

Five centuries pass between Giraldus and the next record of a comparable observer. Roderic O'Flaherty, Irish and Catholic, was born in a castle in County Galway in 1629 and was heir to a huge tract of land to the west of Lough Corrib. He might have been a scholar and writer in some comfort, but Oliver Cromwell and his land confiscations intervened, and O'Flaherty died in poverty in 1718. He knew the landscape and wildlife of the western province, and his list of mammals added hedgehog and rabbit *(Oryctolagus cuniculus)* to the record (both seem to have been introduced to Ireland as food), together with descriptions of the frequent hunting of "fat deere"—that is, the now greatly diminished wild, red deer *(Cervus elaphus)*—in the rugged hills of Joyce country, in the northwest of Galway, which once belonged to his family. From the Connemara coast, he offered a glimpse of a seabird, the gannet *(Sula bassana)*, in which he wove observation with colorful report: "Here the ganet soares high into the sky to espy his prey in the sea under him, at which he casts himself headlong into the sea, and swallows up whole herrings in a morsell. This bird flys through the ship's sailes, piercing them with his beak." O'Flaherty's *A Chorographical Description of West or H-Iar Connaught* did not actually reach print until 1846 and, though written in English, stands as a rare product of scholarly inquiry into nature from within the native Gaelic culture. The very name O'Flaherty stands out from the stream of English and Scottish surnames that were to dominate the literature of Irish natural history for some three centuries afterward.

Arthur Stringer, who was published in 1714, just four years before O'Flaherty's death, might seem a case in point, but he was altogether exceptional in other ways. His social status fell far short of that of physician, clergyman, or academic—as was often the case—and his daily experience of nature was intimate and often bloody.

Stringer was huntsman to Lord Edward Conway, member of a Scottish family settled on lands beside Lough Neagh by grant of James I. Stringer led packs of hounds and horsemen in pursuit of deer, hare *(Lepus timidus hibern-*

icus), fox *(Vulpes vulpes)*, otter *(Lutra lutra)*, and pine marten *(Martes martes)*, but his absorption in wildlife behavior and biology went far beyond the kill, into painstaking work and observation—even to the scrutiny of stomach contents and droppings. After some thirty years of note-taking, he felt equipped to publish *The Experienced Huntsman*, a vigorous treatise on hunting methods and on the principal prey species of the gentry. Few copies survived into modern times, but the book was rescued from obscurity by Fairley and reprinted in 1977. It was, Fairley judged, "the earliest reliable book on the Irish fauna and . . . the first serious work to be devoted to the wild mammals of these islands."

The hunters Stringer served were a colonial elite, galloping through deer parks of the kind first introduced to Ireland by the Anglo Norman lords. Within estate walls this elite inhabited a well-husbanded, parallel world of nature in which fallow deer *(Dama dama)* were bred from imported stock and wild duck were lured out of the sky by decoys on artificial ponds. Later, the same class, with their associated merchants and professionals, were to produce the more adventurous shooter–naturalists whose stuffed specimens enriched museum collections and whose records and observations, in Ireland as elsewhere, became the raw material of nineteenth-century ornithology. Sir Ralph Payne-Gallwey, author of *The Fowler in Ireland* (1882), was a prodigious killer with a punt-gun (an oversized fowling piece mounted in a small, flat-bottomed boat) toppling 1,500 duck in one winter in the estuaries and bays of western Ireland.

William Thompson, whose *Natural History of Ireland* (1849–1851) was the pioneering work in Irish bird study, spent many nights shooting in the marshes of Belfast Lough and learned to tell the species overhead from the sound of their wings alone. Thompson was the son of a wealthy Belfast linen merchant, and the acknowledgments to contributors in his and two later standard bird books (Ussher and Warren's *Birds of Ireland* [1900], and Kennedy, Ruttledge, and Scroope's *Birds of Ireland* [1954]) reflect the enthusiasm of amateur hobbyists among a largely Anglo Irish and Protestant middle-class. Industrial, ship-building Belfast, with fortunes built on linen and engineering, became a particular locus in Ireland of the avid Victorian rediscovery of nature's diversity and beauty. In both islands, amateurs of natural history, primed with new and cheap books, took to the railways to collect ferns, butterflies, and seashells, sometimes with careless rapacity. City field clubs for mass expeditions and picnics were both fashionable and, quite often, strikingly egalitarian: Belfast's included an amateur geologist from the shipyards who helped rivet the *Titanic*. In the 1870s this field club also attracted an eleven-year-old Robert Lloyd Praeger, another son of a linen merchant, who was to become Ireland's greatest field botanist and perhaps the best-known figure in twentieth-century Irish natural history.

Praeger's popular reputation rests on his "topographical autobiography," *The Way That I Went*, written in his seventies and published originally in 1937. His life was shaped by a somewhat rarified scientific tradition, and he eventually became president of the Royal Irish Academy, founded in 1785 for the study of the sciences, "polite literature," and antiquities. The charm of his popular writing and its vigorous celebration of Ireland's landscape and natural wonders had a wide appeal in the young Irish Free State, still caught up in the novelty and pride of independence. Praeger's was also a story of driving energy and omnivorous curiosity that has influenced generations of Irish naturalists. Swimming through a flooded cavern with a candle stuck in his hat, lying out all night in island heather to time the comings and goings of nesting seabirds, crawling on hands and knees into the cattle-trampled prehistoric tombs of the Boyne River Valley, Praeger exemplified the essentially nineteenth-century tradition of passionate and polymath inquiry, becoming, as Britain's distinguished ecologist David Bellamy wrote in a preface to *Floreat Hibernia,* Timothy Collins's biography of Praeger, "botanist, geologist, zoologist, archaeologist, Irish Naturalist *Optimus Omnium. . . .*"

Praeger's botanical reputation was founded on a survey of plant distribution called the *Irish Topographical Botany*, published in 1901. This was a remarkable overhaul of the *Cybele Hibernica* (1866) compiled by Alexander Goodman More and David Moore, an inventory with many blank patches in the less accessible or less scenically attractive corners of the island. Praeger got out maps and railway timetables to plan five years of weekend walking—4,960 miles (8,000 km) across every kind of terrain: bog, esker, mountain, marsh, woodland, one dull midland field after another, often in the worst of weather. Faced with a watery barrier, a common enough hazard, he was as likely to walk on through as waste time on finding a crossing.

Toughness of this sort was quite common among the late-Victorian naturalists. Richard Barrington, the ornithologist, told of a day out above the Powerscourt waterfall in County Wicklow with Henry Chichester Hart, a

Robert Lloyd Praeger with his wife, Hedwig, standing on Rostellan Cromlech, a megalithic tomb, during a field club conference in 1907.

tireless botanist of mountains. The rain fell in torrents, but neither man referred to it and both were soon wet to the skin. Hart deliberately pushed through the briars, scrub, and long grass close to the edge of the stream to discourage his companion. To show equal nonchalance, Barrington walked into the water, sat down on a submerged rock and began to eat his sandwiches. Hart, without a word, did the same, whereupon "all rivalry ceased."

There were also the gentler sort, whose endurance lay in long and patient observation. A Dublin newspaperman, Charles Moffat (1859–1935) was considered both by Praeger and the modern mammalogist Fairley as probably the best field naturalist that Ireland has ever produced, with the patience to watch and wait for events to take place. His editorial colleagues on the *Daily Express* enjoyed what they saw as his bird-like demeanor: one said he would not have been surprised "if Moffat flew in and lighted on the gas bracket." Among his contributions (to the *Irish Naturalist* in 1903) was the conclusion, then original in Ireland, that the chief purpose of bird song was to advertise the territorial claims of the cock. His best mammal work was on bat behavior, and a long paper, "The Mammals of Ireland," for the Royal Irish Academy in 1938 is still essential to student mammalogists.

Although Moffat, like me, wrote a nature column for a newspaper, his dedication and accomplishment as a field naturalist rules out any affinity I might claim. I am in awe, also, of the field amateurs of the past whose work helped to build up collections for museums and herbariums and whose patient, precise drawings and paintings became the basis of so much of today's convenient knowledge.

A good many of these early naturalists were women—many more, indeed, than were given credit at the time. In the early nineteenth century, even serious scientists among Europe's "lady naturalists" would publish their work under male pseudonyms or let it be attributed to husbands or brothers, such were the social assumptions. In fieldwork for botany, women would often take on the more modest sorts of plant: mosses and seaweeds.

A century later, some Irish women naturalists certainly had recognition: Matilda Knowles, for example, as the island's leading lichenologist; Cynthia Longfield as a dragonfly expert of international reputation. There was also a special breed of women drawn from the spirited but often lonely culture of the rural Anglo Irish. My own favorite among these is Maude Jane Delap (1866–1953) of Valencia Island, County Kerry, not least because, in our beachcombing half a century apart we each found two True's beaked whales *(Mesoplodon mirus)*, still one of the Atlantic's least-known mammals. She rotted one of hers down in her asparagus bed and retrieved the whole skeleton for the Natural History Museum in Dublin.

Delap's story is a cameo of her kind. She was the seventh of ten children of a Church of Ireland rector and naturalist: there was no money, so the boys got the education and three of the daughters remained unmarried. Delap made a marine biologist of herself and reared generations of jellyfish in bubbling bell jars in the rectory, patiently matching medusas to their hydroids. She made genuine discoveries, and sent a succession of rare species up to Dublin: fish, sharks, and turtles, as well as whales; a sea anemone *(Edwardsia delapiae)* was named after her. In *Stars, Shells, and Bluebells*, Anne Byrne gives a vivid image of Delap and her sister Constance rowing a boat in rough seas below the cliffs, towing a trawl for specimens.

A whole culture away from these two committed sisters was the world of Séamas Mac an Iomaire, born in 1891 in the Irish-speaking community of fishermen and seaweed harvesters at Mweenish Island off the Connemara coast. His narrative of folk life and marine observation, *The Shores of Conne-*

mara, written in Irish in the 1920s when he was an emigrant in New York, and published in English only in 2000, stands unique in Irish natural history. Its direct, practitioner's relationship with nature has a warm sense of the wonderful, whether in the burrowing shellfish that could be better caught by whistling to hold their attention or the "flashing skate" that shone light from its eyes to show it the way through the kelp forest.

In Dublin, meanwhile, the period from the 1880s to World War I became a golden age for natural history, with Praeger at his most active and the field clubs booming. It was part of a burgeoning intellectual excitement about Ireland's cultural identity, a "revival" that underpinned the surge of nationalism. Natural history fitted in with a sense of exploration (of the west coast, in particular, once the railways from Dublin made it possible) to assemble the indigenous materials of "Irishness." Praeger's contribution included *A Tourist's Flora of the West of Ireland* (1907) and his leadership of the growing field club movement.

One modern biographer, Sean Lysaght, puts Praeger alongside William Butler Yeats, John Millington Synge, Lady Augusta Gregory, and Douglas Hyde as one of "the culture givers of the [nationalist] revival." Just as others had turned to archaeology and folklore to rediscover the roots of culture, Praeger helped in "the establishment of Ireland as a biological and geographical territory with an identity of its own."

This was also his driving impulse, Lysaght argued, in organizing the remarkable Clare Island Survey of 1909 to 1911, the world's first major inventory of nature in a single geographical location. A few years earlier, he had taken a score of naturalists to the small, privately owned island of Lambay, off the Dublin coast. They surveyed its organisms with the minute and methodical attention of a forensic police search and found five species new to science (three worms, a mite, and a bristletail) and seventeen new to the British Isles; they also added almost ninety species to the Irish flora and fauna. It was an unexpected triumph. Praeger was already stimulated by the post-Darwin focus on island biogeography and endemic species. Could it be that a comparably intensive survey of a much bigger, more remote, island in the far west would discover native Irish forms of species that, if not quite of Galápagos uniqueness, might certainly be a bit different from anything Britain could offer?

Clare Island, 3 miles (5 km) out in Clew Bay off the Mayo coast and populated by tenant farmers and fishermen, was chosen by Praeger and his colleagues. Over three years, more than one hundred scientists were deployed there, from Ireland, Britain, Denmark, Germany, and Switzerland. Professionals such as Robert Scharff, keeper of the National Museum, and Grenville Cole, director of the Geological Survey, worked alongside gifted field-club amateurs to catalog every species and biotic community on Clare and its adjoining islands and littoral, from the fish, mollusks, and seaweeds of the rocky coves to lichens, mosses, and insects at the summit of Knockmore Mountain. Among nearly nine thousand organisms, they found 109 animals and eleven plants new to science. But of endemic species, adapted by long isolation, there were none—not even a special mouse to match the island races of *Mus musculus domesticus,* the house mouse, which evolved on the Scottish islands of St. Kilda and the Outer Hebrides. The Clare Island Survey was the ultimate mobilization of Ireland's Victorian naturalists and produced, so to speak, the ultimate list in sixty separate biological reports. Eighty years later, in 1991, the Royal Irish Academy launched a complete resurvey, using the original as a baseline. Praeger had bequeathed a tool unique in Europe for measuring and analyzing change in a once-pristine ocean environment—analysis now becoming crucial in the unpredictable progress of global warming.

Ireland's own upheavals in the first quarter of the twentieth century—world war, the War of Independence, the founding of the Free State, and civil war—shattered the social comfort in which field clubs and collaborative natural history had thrived. The countryside suddenly lost much of its innocence. Big houses burned and gamekeepers fled; the rural Anglo Irish melted away or laid low. In Dublin, "royal" institutions of learning and Trinity College were displaced or marginalized. In the primary schools of the new state, "rural science," the study of plants and animals, was discarded to give more time for the teaching of the Irish language.

Praeger was his own kind of patriot (expressed in the emphatic subtitle of *The Way That I Went, "An Irishman in Ireland"*). In old age he began to urge protection of the landscape and he became the founding president of An Taisce, the Irish National Trust, in 1948. For decades after his death, however, and sometimes even today, this remained a marginal, mainly urban voice, deeply misunderstood and often resented in rural Ireland as raising obstacles to development and employment.

Until a popular interest in nature awoke in the late twentieth century, largely a product of television and the burgeoning environmental movement, most amateur naturalists found themselves pursuing somewhat solitary passions. In his garden in Fermanagh, the county surveyor J. P. Burkitt put metal bands on the legs of robins in the early 1920s, the better to keep track of individual birds, and thus invented the basic tool of ornithology. In Connacht, Robert F. Ruttledge spent a night on every rugged island off Galway and Mayo in his search for Leach's petrel *(Oceanodroma leucorhoa)*. He became the father figure of modern Irish birding, his hundredth birthday in 1999 celebrated by a movement now vigorously installed in the mainstream of national life.

As science of all kinds became specialized with the advance of the twentieth century and the technology of research soared beyond the amateur's grasp, the occupation of "naturalist" came to seem almost defiantly unprofessional. Yet the broad, polymath approach to the landscape and its natural history has been a cultural inheritance greatly prized by Ireland's environmental scientists. It was carried through from Praeger's time in the charismatic person of Frank Mitchell (1912–1998), Professor of Quaternary Studies at Trinity College, Dublin, geologist, botanist, archaeologist, geographer, ornithologist— honest naturalist, above all. His last book, *The Way That I Followed* (1990), an autobiographical tour of Ireland, was offered as companion to Praeger's *The Way That I Went*: two men in thorn-proof tweed jackets, striding the island half a century apart in a spirit of passionate inquiry.

The contrast between the world of science in Praeger's day and the second half of the twentieth century was part of what prompted Mitchell to his story. One development, in particular, had opened the history of the Irish landscape with "a flood of light" that Praeger had foreseen but did not live long enough to enjoy. As president of the Royal Irish Academy in the 1930s, he and his fellow field scientists were impatient to unravel the story of the bogs and the virgin postglacial ground they covered. They recruited a Danish botany professor, Knud Jessen, a leading expert in the new techniques for drilling into bogs and lake muds and studying the cores for the sequences of leaves, seeds, and pollens preserved in them. Mitchell, fresh out of college, was Jessen's courier and, later, collaborator.

A scene in Chapter 7 of *The Way That I Went* has become one of the most memorable snapshots of science in Ireland. Praeger, Jessen, Mitchell, and others are conferring in the middle of a great wet bog behind the small town of Roundstone in Connemara on a bad August day in 1935:

We stood in a ring in that shelterless expanse, while discussion raged on the application of the terms soligenous, topogenous and ombrogenous; the rain and the wind, like the discussion, waxed in intensity, and under the unusual superincumbent weight, whether of mere flesh and bone or intellect, the floating surface of the bog slowly sank till we were all half-way up to our knees in water. The only pause in the flow of the argument was when Jessen or Osvald, in an endeavor to solve the question of the origin of the peat, would chew some of the mud brought up by the boring tool from the bottom of the bog. . . .

In those early days of bog-coring, Praeger could feel pleased with his fore-sight in collecting drawers full of seeds for the National Museum with which fossil seeds, shed many thousands of years ago, could be compared. The po-tential of fossil pollen as a key to the landscape's history was almost unimag-inable. At one time, the surface of all pollen grains looked smooth under magnification; the modern electron microscope finds patterns that can iden-tify many pollens even to their subspecies. In Mitchell's later career, tech-niques of radiocarbon dating were refined as dramatically. In 1980, for ex-ample, he needed to sieve out five thousand fossilized leaves of arctic willow from a deposit of silt on Achill Island to get a big enough sample for a dating: a decade later, five leaves would have been plenty.

From palaeobotany to plate tectonics to glaciological computer modeling, Mitchell "the naturalist" absorbed technology fearlessly as he edged toward old age. Successive editions of his widely popular *Reading the Irish Landscape* kept pace with quaternary research. He reshaped, but never abandoned, his insistence on land bridges from Britain and continental Europe as the means by which "organized oakwoods," complete with flora and fauna, advanced to recolonize Ireland after the last Ice Age (see Chapter 4).

Mitchell died on the eve of the twenty-first century and before the explo-sion of economic change that would so profoundly affect the landscape of the republic. He would have been seriously concerned at the pace of construction and pressures on land and at the perilous balance between development and conservation.

In a Republic transformed by, at times, full urban employment and with emigration apparently at an end, an increasingly affluent and sophisticated society is in retreat from its agrarian past. Even before the eruption of commuter housing, almost half of all country dwellers had no direct contact with the land, and only about 145,000 people in Ireland called themselves farmers in 2002. Yet the level of interest in nature and general awareness of ecology have never been higher, and more and more townspeople now consider themselves stakeholders in the countryside and the welfare of its wildlife. Ireland's membership in the European Union has largely dictated the pace of conservation of habitats and species and has helped to fund a great deal of research. The process of deciding what is special in the Irish biota and how its habitats fit in with those of other European landscapes has been revelatory, even if controls on land use that flow from it are not always widely appreciated or understood.

SELECTED REFERENCES

Fairley, J. 1984. *An Irish Beast Book.* 2d ed. Belfast: Blackstaff.

Foster, J. W., and H. C. G. Chesney. 1997. *Nature in Ireland: A Scientific and Cultural History.* Dublin: Lilliput.

Lysaght, S. 1998. *Robert Lloyd Praeger: The Life of a Naturalist.* Dublin: Four Courts Press.

Mitchell G. F. 1990. *The Way That I Followed.* Dublin: Country House.

———. 1997. *Reading the Irish Landscape.* Dublin: Town House and Country House.

Praeger, R. L. 1937. *The Way That I Went.* Dublin: Hodges, Figgis. Reprinted 1997. Cork: Collins.

Stars, Shells, and Bluebells. 1997. Dublin: Women in Technology and Science.

— *Chapter 2* —

Fire and Ice

It was an Irish theologian who set the date of creation at 4004 B.C., "upon the entrance of the night preceding Sunday 23 October." James Ussher was Protestant primate of Ireland in the seventeenth century and also Archbishop of Armagh, where his cathedral crowned a drumlin, a whale-backed mound of boulder clay shaped beneath the ice sheet at Ireland's last glaciation.

Ussher arrived at his date and time from scriptural data and metaphor—"a work," wrote Stephen Jay Gould, reproving any retrospective ridicule, "within the generous and liberal tradition of humanistic scholarship." The "true" origins of Ireland, explored today with the help of radiometric dating, might seem scarcely less marvelous. The prodigious journey northward from the southern polar region, through ice cap to jungle to desert and back to ice again; the successive inundations by, and emergences from, the sea; the mountain-building crumplings and eruptions, the vast wearing-down, the shearing off from North America, and the spread of a new Atlantic Ocean—all this seems more than a match for any theological fable.

Such incredible change has resulted in an island originating in two separate land masses but planted for the moment between latitude 51.5°N and 55.5°N, on a level with the Canadian prairies, the Aleutian Islands, and

15

Moscow and at a longitude 5.5° to 10.5°W. Its 32,595-square-mile mass (84,420 km^2—comfortably driven across, Dublin to Galway, in four hours) is saucer-shaped. A rim of modestly uplifted hills (the highest peak at 3,414 feet, or 1041 m) surrounds central lowlands that undulate, for the most part, above a carpet of glacial debris left by the retreat of the last Ice Age, ending finally 10,000 years ago.

The oldest rocks rear up most dramatically from the crystalline "basement" of Ireland in cliffs and mountains at the island's rim. At opposite edges of the island, for example, are outcrops of Precambrian gneiss, mostly metamorphosed granite, tied to the deepest sinews that link Ireland with Scotland, continental Europe, and—until the Atlantic Ocean broke the connection—Newfoundland. The gneisses that form Inishtrahull, off Malin Head, and parts of Erris Head in County Mayo are 1,800 and 1,900 million years old—rocks that surface again in the islands of the Hebrides off the west of Scotland. At Kilmore Quay, on Ireland's southeast coast, the gneiss sloping out beneath the long Atlantic waves is possibly as old as one billion years. It was formed half a world away from the gneisses of Donegal and Mayo, and drifted to join them in the later and mighty collision of continents.

Invertebrate animals with calcareous shells first appeared in Earth's oceans some 550 million years ago. It was 100 million years earlier that a new sea had opened, and what are now Connacht and Ulster lay along its margin. Thick deposits of sand became hardened into the quartzites raised up in the immense cliffs of Achill Island, or crowning peaks in Donegal and Connemara with a gleam of snow-white scree. Deposits of lime hardened into marble—in Connemara, a beautiful jade-green rock, now cut and polished into fireplaces and tourists' ashtrays. No traces of living organisms have survived the metamorphism (compression and baking) of these ancient sediments. However, on the eastern side of the island fossils in Cambrian slates around Dublin, delicate and frondlike, trace the radial burrows of marine creatures living some 500 million years ago.

At that time, the two small blocks of land that would ultimately be bonded into Ireland—the northwest half into the southeast half—were set within separate continents at the bottom of the planet, facing across an ocean called the Iapetus, precursor to the young Atlantic. Northwestern Ireland was traveling (in company with Scotland, Norway, and the Canadian Maritimes) in the continent Laurentia, later to become North America. Southern Ireland, together with England, Wales, and Brittany, was part of the small continent called Avalonia, away to the southeast. The two halves were inching toward each other in the great convergence of Palaeozoic continents that would produce, for a few hundred million years, the single, universal land mass of Pangaea. Thus the little island of Ireland was to be welded together, as if in a press, deep within the biggest island Earth has ever known.

In its collisions and subductions, its stretchings and sunderings, plate tectonics and the internal heat engine that drives it has shaped the relief of the whole planet. As the Iapetus Ocean narrowed, about four hundred and fifty million years ago, the previously distinctive marine fauna of its Laurentian and Avalonian margins intermingled. After the two continents began to touch, the crumpling of the American margin created huge thrusts and folds of rock to make a mountain range running across from what is now Scandinavia, down through Scotland and northwest Ireland, to Newfoundland and Appalachia. The excavated and eroded mountains are called by geologists Caledonides, after Caledonia, the old name for Scotland.

The Caledonian upheavals had also released molten rock, fueling active volcanoes and also deep upwellings of magma that crystalized underground into separate masses of granite. The uplands of modern Wicklow are granite foundations of the far higher Caledonian Mountains, originally roofed by Ordovician slate and lava; and in south Connemara, Donegal, and Armagh, intrusions of granite have been smoothed into hills below a profile of tougher quartzite peaks or scoured by ice into undulating barrens, pierced and fretted by the sea.

The final squeezing together and elimination of the Iapetus Ocean about 400 million years ago married the two halves of Ireland along a line from the estuary of the River Shannon on the west coast to Dundalk Bay on the east. There are differences in the older marine fossils to either side of the line, showing them to have come from opposite shores of the Iapetus. The identity of fossils in Connemara and Newfoundland, on the other hand, is witness to their ancient connection along a common North American shoreline.

The union of the Irish land mass, which occurred at the latitude where South Africa is today, also lifted up the Caledonian Mountains into eroding rain and wind as Pangaea steadily moved northward. Torrential rivers in a raw and broken landscape carried off great quantities of sand and gravel into an alluvial plain, later to harden to the old red sandstone of Munster.

FOSSIL EVIDENCE

About 340 million years ago, as the Irish sector of Pangaea edged through the Equatorial zone, a new but gentle tectonic disturbance brought a warm sea flooding in across much of what is now northwest Europe. A few parts of Ireland, the mountains of Wicklow and Donegal among them, were left as islands in a shallow sea, but most of the region was submerged by lime-rich water, nourishing abundant corals, algal reefs, and other tropical marine life. Its debris of dead shells built up to a great thickness of sediment on the sea floor, and the Carboniferous limestone formed by its compaction is now the bedrock, studded with fossils, of nearly half the modern land area of Ireland. Later in the Carboniferous, sand and clay carried out from rivers consolidated into sandstone and shale and the fossil-patterned flagstone now quarried at Liscannor above the Cliffs of Moher. The shallow margins of the sea became swamps and lagoons and then forests of the steamy, waterlogged sort, tangled with tree ferns and giant horsetails and rustling with amphibians and huge insects. The abundant vegetation, buried and compressed under additional layers of sand and silt, became the coal that gives the Carboniferous period its name. In many parts of Western Europe the coal beds have survived as a rich resource, but in Ireland they were to be eroded away almost completely.

Near the end of Carboniferous time, about 300 million years ago, came another major phase of mountain-building. Tectonic collision between the European and African plates thrust up the Harz Mountains in central Germany and sent a surge of crustal pressure toward Ireland. As it approached what is now the south of the country, it squeezed and crumpled up the sea-floor sediments against the resistant backbone of older rocks to the north. The mountain range formed in this way (known as the Hercynides) was so high that it drew down the moisture from the prevailing south winds and made desert of the land in its lee. Today this ridge has been eroded down to stumps, surviving in Ireland as gentle hills of Old Red Sandstone, folded east to west through Counties Waterford, Cork, and Kerry. Much of western Europe, indeed, was worn down to a low relief in the arid, sand-swept conditions of the vast and doomed Pangaean continent.

Pangaea began to break up late in the Triassic period, about 200 million years ago, slowly disintegrating along much the same continental seams that had joined it together in the old Ordovician time. The elevation and spreading of submarine tectonic ridges reduced the storage capacity of the ocean

Marine fossils in carboniferous limestone.

basins, and a resulting worldwide rise in sea level drowned about half of North America and flooded most of Europe. It was a gentle, lapping inundation, warmed by the currents of the great Tethyan seaway opened up between Eurasia and an Africa still tied to India and Australia. During this Cretaceous period, the shallow reaches of the shelf seas nourished especially the microscopic, floating algae known as coccoliths, whose calcareous plates drifted down to form a white, fine-textured ooze.

This became the chalk of England's White Cliffs of Dover—indeed, of much of the southern coast and the matching cliffs of France. Most of Ireland, too, was blanketed in chalk, to a thickness of 300 feet (100 m) or more, but erosion has stripped it away so completely that, in Frank Mitchell's words, "to accept that it was once there almost requires an act of geological faith."

So profound and severe has this glacial and river erosion been that only the northeastern corner of Ireland can offer any tangible record of the Mesozoic and Tertiary history. Elsewhere on the island, the record is virtually blank for 250 million years of Irish sedimentation, from the accumulation of the coal measures to the dumping of till and gravels by the first Ice Age glaciers. Those missing sediments were scoured and washed away into the young Atlantic

Ocean, born in Jurassic time and widening ever since. That some Cretaceous chalk does survive as gleaming layers in the cliffs of Antrim is an accident of the next dramatic shift in Ireland's tectonic history.

About 55 million years ago the European and American plates began to be pushed apart by a rising plume of very hot rock in the creation of the Norwegian Sea. As Greenland, part of the American plate, began to drift west, fractures at either side of the newly rifted ocean poured out floods of white-hot lava. In the northeast of Ireland, fissures opened up, and molten basalt surged out across a landscape of chalk already weathered and pitted by erosion. Successive flows of lava hardened into a pile of layers some hundreds of feet thick, forming a basalt plateau whose worn edge today halts at the rim of a coastal escarpment stretching from Belfast Lough and around Fair Head to Lough Foyle.

The surface of each lava flow weathered into soil, and subtropical vegetation rooted into it, only to be buried by additional flows. Soil trapped in this way is sticky and red, like the laterites of sub-Saharan Africa, and sandwiched within it are thin layers of lignite, compressed from peat and lake mud. Small-scale mining for iron ore and bauxite in the nineteenth century and World War I uncovered fossilized tree stumps and logs, some of them charred by heat but identifiable as pine *(Pinus)*, cypress *(Cupressus)*, and monkey-puzzle *(Araucaria)*.

Strata of black basalt, red clays, and white chalk above a foundation of red Triassic sandstone are the basic signature of much dramatic scenery along the northeastern coasts, but there are also striking examples of particular volcanic effects. At Fair Head, facing across the North Channel to Scotland, precipitous cliffs are fissured vertically by great columns of dolerite, crystallized below the surface in a cavern of magma intruded into Carboniferous sandstone layers. Further west along the Antrim coast the Giant's Causeway steps up through seven triple tiers of hexagonal columns that cooled from a deep pond of lava. This dramatically weathered cliffscape, an early wonder to the human eye, is now a World Heritage Site and Ireland's leading geological attraction. (See also the appendix list of nature reserves.)

The volcanism that flared and smoldered through the region for two million years sent magma seeping far into underground faults and fissures. In the northwest of Ireland and on the west coast, plugs and dikes of dolerite are legacies of the fringes of North Atlantic igneous activity. Down the east coast, volcanoes at Slieve Gullion and Carlingford created giant craters and ring-dikes, discernible today in the curved pattern of heathery hills, and central peaks built of dolerite and granite, two contrasting igneous rocks. Close by, the clustered domes of the Mountains of Mourne tell a different story. Their pale granite crystalized from five separate surges of molten rock under a roof of Silurian sandstone. Today, with much of the sandstone eroded away, the granite arches up through the boulder-strewn slopes of Ireland's youngest mountains.

The scale and pace of erosion of Ireland's rocks has been exceptional, even by the often drastic norms of geomorphology, and has prompted some arresting conclusions. On the island's western rim, for example, many of the rocks are of exactly the same age and type as those that, in Scotland, form substantial highlands and upland plateaus. Yet around the fjord of Killary Harbour the mountains are glacially excavated into ten steep-sided blocks, divided by valleys and lowlands lying within 300 feet (100 m) of sea level. In some of these valleys, today's rainfall averages 98 inches (250 cm) a year, encouraging an early conjecture that Ireland's high humidity may have helped to wear the rock away faster than elsewhere in the British Isles.

The gap in the record of Ireland's Mesozoic and Tertiary rocks has taxed the imagination of geomorphologists. But one simple, potent image guided generations of thought about the shaping of the modern island. This saw Ireland emerging from the sea around 70 million years ago as a high-standing block, topped with a layer of chalk and standing level with the summits of today's tallest mountains. What happened after that, in this hypothesis, was a steady grinding-down to the basic shape of Ireland's modern topography.

The idea first gained consensus in the mid-nineteenth century from efforts to explain how several large Irish rivers, including the Shannon, had carved valleys through barriers of tough old Paleozoic rocks on their way to the sea. In the original theory, the emergent block was carpeted with Carboniferous sandstone, its surface beveled smooth by wave action. Rivers on this surface followed gentle slopes southward, and, when the sandstone layers were worn away, kept to their original pattern across the grain of the uncovered older rocks. In later revisions of the theory, the emergence of the block was postponed to the close of the Mesozoic, and the first drainage patterns were carved through a mantle of Cretaceous chalk. Even at the middle of the twentieth century, geomorphologists were still speculating on the probable elevation of the base of the vanished chalk—hardly lower, it was thought, than the summit of Lugnaquillia, Leinster's highest mountain, 3,039 feet (926 m) high.

But then came dramatic discoveries of chalk and Tertiary sediment, with plant fossils, surviving in rare pockets a mere 200 or 300 feet (60 to 90 m) above sea level in Counties Tipperary and Kerry. The pockets, hidden under young gravel and soil, filled fissures in the Carboniferous limestone bedrock. The fissuring of freshly exposed limestone through solutional, karstic action, so obvious in the Burren Hills of County Clare, made the story of superimposed river systems, engraving their way for eons through a massive, flat-topped block, increasingly unlikely. A closer examination of Ireland's rivers finds glacial or tectonic explanations for the geological puzzles in their courses to the sea.

UPHEAVALS AND EROSIONS

The recognition of plate tectonics in the mid-twentieth century revised the whole picture of an island in an eroding, structurally quiescent state once the early Tertiary volcanoes had become extinct. Now we view a land surface drastically faulted by crustal movements during the later Tertiary epochs. Uplifted areas were denuded, while down-warped areas were largely protected from weathering. The Irish midlands, a plain of Carboniferous limestone, is ringed by uplands, mostly of much older granite and metamorphic rocks that extend under the midlands' limestone. Tertiary tectonic movements that sank the midlands and lifted up the island's outer rim might explain, for example, the steep linear walls of some of the mountains around Killary Harbour. They could also explain why, over much of the midlands, the pre-Carboniferous rocks lie far below sea level, hidden by the mantle of limestone, while from Kerry to Donegal and elsewhere around the rim they rise up into mountain peaks and plateaus. Some researchers think that the Maumtrasna massif, for example, directly east of Killary Harbour, may have been raised up as recently as the Pliocene, less than six million years ago. In the Erriff River at the head of the Killary fjord the Aasleagh Falls are one of several rock sills in estuaries that encourage speculation on a tectonic "tilt" to Ireland, northwest to southeast, some time in the early Tertiary.

But none of this can minimize the prodigious erosion in periods leading up to the Ice Age. The great basalt-capped plateau of the northeast has been worn back on all sides, and many of the higher, younger lava flows have vanished, as hundreds of feet of rock have been stripped away. On some western mountains, notably Slieve League on the coast of Donegal and on Maumtrasna in Mayo, strata of Carboniferous sandstone have been reduced to tiny patches on the summits. Vast erosion of the Mournes was completed by the Quaternary ice sheets, which carried off millions of granite boulders and dumped them on the plains. Further south, on the Dublin and Wicklow Mountains, surviving tors of granite sometimes 20 feet (6 m) high speak of the great mantle of rock, shattered by freeze-thaw cycles, that was peeled away from the summits.

THE LIMESTONE LAYER

The central limestone lowlands have been chemically corroded through many hundreds of feet and over many millions of years. Irish people sometimes complain that the midlands are "monotonous." In fact, their flatness is considerably relieved by the contours of thick glacial drift, sometimes heaped into drumlins and eskers (see page 27), and by the remnants of peat bogs raised above postglacial lakes. The plain has all the appearance of a spread of glacial debris, yet the bedrock beneath is an unevenly planed surface of limestone, stretching almost 125 miles (200 km) from east to west. It is thought to be the product of perhaps forty million years of karstic denudation—the solution and honeycombing of limestone by naturally acidified water.

The erosive potency of dissolved carbon dioxide is demonstrated where nonlimestone boulders—of granite, say, or gneiss—have been carried on to limestone and left there by glaciers. These "erratics," relatively insoluble, have protected the limestone directly beneath them and are now left perched on pedestals, like mushrooms on stalks. The height of the pedestal is a straightforward measure of erosion of the limestone platform since the last glaciation: in the Fergus Basin of County Clare, for example, about 6 inches (15 cm) of limestone have been dissolved, or about a quarter of an inch (50 mm) every thousand years. Corrosion at that rate could remove more than 980 feet (300 m) of limestone in seven million years, which would have utterly destroyed the sediment surface. In reality, the rock was protected by residual clays and soils, which held back corrosion over most of the plain. Solution of limestone would have continued laterally, however, at the margins of the lowlands, extending the limestone plain as covering shales were stripped back.

At northern and western edges of the plain, so far unconsumed in this expansion, are Carboniferous uplands of atmospheric, often moodily beautiful scenery. Near the west coast, the pillared scarp of Ben Bulben (1,730 ft; 527 m) makes a striking landform outside Sligo Town. Geographer J. B. Whittow, in *Geology and Scenery in Ireland*, refers to "its implacable presence in the landscape, like the prow of a battleship thrusting westwards towards the Atlantic." The scarp, undercut by the passage of ice, forms one corner of a group of limestone plateaus that stretch away through Leitrim to the karstic potholes and caves of the mountains bordering Fermanagh. The best known of these karstic-solution systems, Marble Arch on the terraced flank of Cuilcagh Mountain (2,188 ft; 667 m), carves out some 4 miles (6.5 km) of underground river passages and lofty caverns that have been known for a century or more. Edouard Martel, the celebrated French speleologist, investigated them with a folding canoe in 1895; indeed, explorations of Fermanagh's caves are on record since 1796. A long cave at Cuilcagh's western end provides the source of the River Shannon. In open bogland to the north, the potholes of Tullybrack lead into great horizontal systems, one of which, the Reyfad Pot, also twists down through almost 590 feet (180 m) of limestone—the deepest system so far known in Ireland.

The island's five big cave complexes, spaced out through uplands from Fermanagh to Kerry (see particularly the Burren in Chapter 8), are the clearest evidence for the action of water that has riddled the island's limestone with cavities. Scientific interest in the caves goes back almost two centuries, notably in search of fossil animal bones (see Chapter 3). Yet the implicit geological message was somehow overlooked in the earlier scenarios for Ireland's missing strata. The idea of primal river systems carving their courses through limestone as if engraving an intaglio was a drainage theory that, in the most literal sense, could not hold water. Almost to the present day, the extent of water's underground action in the island's massive limestone strata, turning them to a spongelike mass, has been slow to dawn on many geologists.

THE ICE AGE
The role of ice in shaping the surface of Ireland arrived as an even more profound revelation. Errisbeg Mountain, "little promontory," rises to some 987 feet (301 m) behind the village of Roundstone on the fretted, jigsaw-puzzle coast of Connemara. Undramatic in itself, its rock-ribbed gabbro (igneous rock) summit commands a view landward that is uncommonly alive to an in-

visible presence: the ice of Ireland's Quaternary period. A maze of tiny lakes in the russet bog below fits the hollows in a sheet of glacial till and conjures a landscape with all the vacant innocence of arctic tundra. A distant frieze of mountains, the Twelve Bens, is scalloped with cirques (steep-walled basins) and U-shaped valleys, carved by glaciers inching southward to the sea. Some of the rocks they carried from Mayo and Galway can be found today as large erratics on the beaches of Kerry, some 125 miles (200 km) away.

From massive mountain sculpture to the merest scratches on rock, from the great heaps of moraines, drumlins, and eskers to stone festoons and polygons, the marks of ice and frost are everywhere in Ireland. As the mountaineer Joss Lynam has put it, glaciation rejuvenated the island's mountain landscape, carving sharp peaks and steep-sided valleys from the high plateau whose remnants can still be found in the flat summits of Devil's Mother ("Devil's Testicles" in Irish) in Connemara, Mweelrea, in County Mayo, and the Comeragh Mountains in Waterford. In the Comeraghs, ice clawed out Coumshingaun, "the glen of the ants," where cliffs of 120 feet (365 m) now shadow its lake and a moraine of massive boulders. This picture is repeated in mountain groups all over Ireland, yet when Louis Agassiz, the Swiss glacial theorist, visited Ireland in 1840, his demonstrations of the evidence for glaciation were greeted with skepticism by the geological establishment, in Ireland as elsewhere in Europe. Winding ridges of rocks and gravel, polished and scratched rock surfaces, erratic boulders far from their source—the phenomena he credited to the retreat of ice were still widely seen as the legacy of marine inundation. Within three decades, however, the Dublin-born Maxwell Close (a Protestant clergyman, like many naturalists and Earth scientists of his day) had produced a classic paper that introduced the Irish terms *droimnin* (drumlin) and *eiscir* (esker) to the scientific literature and persuaded his geological colleagues of the fact of glaciation.

The history of ocean currents and global climate in the Pleistocene epoch has grown more certain with today's advances in the study of oxygen isotopes applied to deep-sea sediments. A peak of mild climate around 18 to 17 million years ago was followed by fluctuating but mostly declining temperatures and establishment of the polar ice caps. The growth of the Antarctic ice cap, in particular, and the northward circulation of its cooling waters, helped to create a new and narrower range between temperate and glacial conditions in the Earth's mid-latitudes. Ice repeatedly advanced and retreated across some

30 percent of Earth's land surface, responding to 10,000-year cyclic variations in the planet's orbit and tilt of its axis.

Like the rest of Europe, Ireland has been covered by ice several times in the past two million years. Eight glacial cycles have been traditionally recognized in Britain and northern Europe, but the evidence of the deep-sea record suggests that there were more. In Ireland, each surge of ice swept away much of the evidence for the preceding glaciation, together with the record of the interglacial periods when a temperate climate returned and woodland reestablished itself.

The island's glacial deposits have traditionally been assigned to two final stages of the Quaternary period: the Munsterian and Midlandian—both names taken from the apparent southern limits of the ice sheets. In the Munsterian Cold Stage, 300,000 to 150,000 years ago, most of the island seems to have been buried at some time, right to the south coast and beyond. The first ice of this cold stage formed in the center of Ireland. It was joined by an immense ice mass that, building up in Scotland, spilled out across the north of Ireland and thrust southward down the basin of the Irish Sea. It ground along the coast of Leinster and pushed across the corner of Munster as far as Cork Harbour. Ice masses formed in the Irish mountains, too, flowing out east and south from Connemara and carrying boulders of Galway granite as far as Counties Kerry and Cork. Connemara ice reached the uplands of Leinster, pushing up their slopes to a height of more than 985 feet (300 m). The radiation of these great lobes of ice, separate in time as well as origin, left intersecting and overlapping deposits of till, some far-traveled, some quite local, to confuse Quaternary students. The lack of pollen in deposits from the last interglacial age has added to the problems of dating key events in Ireland.

During the Midlandian Cold Stage, which began about 80,000 years ago, the Irish midlands were deeply buried in ice—hence the simple name for a phase of the Quaternary that was, in fact, far from solidly frozen. Study of plant fossils and other Quaternary evidence, notably by Pete Coxon, and radiocarbon dating of mammal fossils in the Quaternary Fauna Project (see Chapter 3), now offers a far more complex picture of climate change.

Between an early development of ice and the Last Glacial Maximum, about 20,000 years ago, there were many thousands of years in which Ireland was warm enough for plants to flower. These years included a return to temperate conditions, with conifer woodland and beetles, at around 48,000 years ago (this from fossil evidence in lignite in County Antrim) and subsequent phases of cold, tundralike conditions, still warm enough for woolly mammoths *(Mammuthus primogenius)* and many other mammals, before the main advance of the Midlandian ice from about 25,000 years ago.

In the traditional view of the Midlandian, a great oval ice cap reached down across the island to a line roughly from the Shannon Estuary to Wicklow, where a glacial moraine divided relatively fresh and less weathered drift to the north from older and smoother material to the south. Beyond this, there was an isolated ice cap over the Cork and Kerry highlands but also a band of ice-free tundra across Munster and headlands on the northwest coasts.

Later revisions showed the ice overrunning this line in many parts of the midlands and southeast, and in one influential model William Warren of the Irish Geological Survey has proposed three major ice domes on Ireland, merging with the Irish Sea ice lobe to cover the island completely. The maximum thickness, heaped to the north of the island, is thought to have been about 1,970 feet (600 m).

When the ice began to melt and retreat, it left the surface of Ireland scarred and molded like a rough clay armature of the landscape seen today. Behind the hummocks of the moraines appeared the high, sinuous gravel ridges of eskers and then, further north in Ireland, the myriad rounded hillocks of the drumlins. Eskers, in particular, were to puzzle geologists grievously: as late as the 1890s, many still believed they were relics of the biblical flood. More significantly, they were also taken as showing lines of sea currents over a flooded Ireland and Britain: submersion of these islands under a Pleistocene sea was widely accepted into the latter part of the nineteenth century.

The typical esker is a sinuous ridge of sand and gravel, deposited by a meltwater stream in a tunnel beneath the ice: a cross-section often shows the horizontal bedding of the stones rolled along in the powerful flow of rushing water. Near Tullamore, in County Offaly, one of a group of three eskers meanders for 20 miles (32 km) across the midland plain and rises more than 50 feet (15 m) above it; the ridge of another is 100 feet (31 m) high in places.

The drumlin, on the other hand, is a streamlined land form, smooth and oval as half an egg, and usually shaped from boulder clay carried along during a major advance of ice. The axis of the drumlin lies parallel with the thrust of the ice sheet, so it can offer geologists an excellent guide to the final course of the ice.

Drumlins seldom lie singly in the landscape but instead nestle together in a "field" or a "swarm." They have shaped large stretches of the northern country-side into an intimate maze of little hills: in one study of ice movements in County Down, an area of 620 square miles (1,600 km²) was found to hold some 3,900 drumlins. On the west coast, ice, pushing rapidly out from its in-land ice domes into Donegal Bay and Clew Bay in County Mayo, molded the underlying till into drumlins that later became islands when the seabed rose again. They have taken on a distinctive shape, the seaward end sharply cliffed by wave erosion, and many have vanished except for a platform base.

The retreat of the ice produced some dramatic changes in the relative levels of sea and land, as one rose from the conversion of ice into water and the other rebounded, over time, from the former weight of ice. In the north of Ireland, where the burden of ice was heaviest and lasted longest, these changes are marked by beaches raised up in the profile of the shore. On the rocky prom-ontory of Malin Head, in County Donegal, for example, the village of Bally-hillin sits on a long gravel ridge about 65 feet (20 m) above the sea. Below it, small, unfenced strip fields run out to a low cliff and then continue again, as if on a terrace, to the brink of the bay's white sands. This succession of shore-lines was formed in intervals when the rebound of the land and the rise of the sea happened to be in synchrony. At Cushendun, County Antrim, a rising shoreline lifted sea caves above high-water mark.

The pattern and timing of changes in relative sea level—and of possible land bridges between Ireland and Britain and Continental Europe—have been crucial to Irish natural history. They are discussed in Chapter 4, which deals with the postglacial colonization of the island by plants and animals. But the Ice Age has a final drama with which to close this geological narrative.

The Greenland Ice Core Project of the 1990s has revealed a complex pat-tern of temperature cycles over the 100,000 years of the last cold stage, linked to the instability of ice caps and feedback loops in atmospheric and oceanic circulation. The warming trend that began in the North Atlantic about 20,000 years ago was delayed or reversed for 2,000 to 3,000 years by the release of meltwater and icebergs, and this type of negative feedback was repeated even after the warming that allowed plants and animals back to Ireland. Such events support uneasy speculation today about the outcome of global warm-ing for the climate of northwest Europe.

The abrupt thermal rise about 13,000 years ago boosted sea-surface temperatures in the North Atlantic by about 9°F (5°C) in just seven years. It followed a major ice collapse that allowed warmer currents back into the ocean and pushed the polar front northward. This brought conditions as warm as today's. But they did not last and deteriorated through a set of warm-cold cycles as glacial meltwater and cold Greenland currents pushed the polar front rapidly south again, sending icebergs to press around the north and west coasts of Ireland. The island was plunged back into a brief Ice Age for six to eight centuries, ending at around 10,000 years ago with another rapid warming. This "cold snap" in Ireland is called the Nahanagan Stadial, after a mountain lake south of Dublin that nurtured a glacier in its cirque (or, in some contexts, the Younger Dryas, from the Irish fossil record of the arctic plant mountain avens, *Dryas octopetala*).

The stabilization of climate to today's somewhat fragile-seeming equilibrium was helped by an upswing in the solar radiation during summer. This helped to melt the icebergs and let the moderating warmth of the Gulf Stream finally prevail in the northeast Atlantic. But if the upswing continues, promoted by the human contribution to global warming, the Gulf Stream could stop, with harsh consequences for the countries of western Europe (see Chapter 5).

Meanwhile, the story of Ireland's environment and wildlife during the final oscillations of the last Ice Age has been elaborated by revelations from the fossil record. Ireland may not have a single dinosaur to offer, but its fossil menagerie has its own ecological dramas.

SELECTED REFERENCES

Davies, G. L., and N. Stephens. 1978. *Ireland, a Geomorphology*. London: Methuen.

Edwards, K. J., and W. P. Warren. 1985. *The Quaternian History of Ireland*. London: Academic Press.

Holland, C. H., ed. 2001. *A Geology of Ireland*. Edinburgh: Scottish Academic Press.

Mitchell, F. 1990. *The Way That I Followed*. Dublin: Country House.

Mitchell, F., and M. Ryan. 1997. *Reading the Irish Landscape*. Dublin: Town House.

Whittow, J. B. 1974. *Geology and Scenery in Ireland*. London: Penguin.

Williams, M., and D. Harper. 1999. *The Making of Ireland: Landscapes in Geology*. London: Immel.

—— Chapter 3 ——

Footprints and Fossil Caves

More than 385 million years ago, in the mid-Devonian period, the patch of Earth's crust that is now Ireland was still just south of the Equator. Much of it was a low desert plain of sand and gravel washed down from the raw and fast-eroding Caledonian Mountains to the north. Rain, when it came, was torrential and collected in transient lakes between the rivers. At this shifting interface of sand and water, primitive amphibian creatures derived from lungfish were leaving four-footed tracks in the mud: the first vertebrate animals to walk on land.

Ireland's most significant fossil, found in 1992 by a Swiss geology student, is a trackway of more than 150 footprints, crossing a rippled sheet of hard, purple siltstone on the rocky northern coast of Valencia Island in County Kerry—a slab long known to islanders as *Carraig na gCrúb,* the Rock of the Hooves. The discovery by Iwan Stossel was the first of its kind in Europe and one of only six sites in the world to offer such evidence of early vertebrate evolution. The tetrapod tracks are dated at older than 385 million years, from fossil and radiometric data, and are possibly the second oldest, next to footprints found in west Victoria, Australia. Some of the Valencia prints, on a rock platform above high-water mark, are almost 1 inch (2 cm) deep. Their detail is lost through deformation of the sediment but their spacing suggests a creature like a salamander, about 3 feet (1 m) long, walking slowly on very wide feet.

The flux of Devonian sands was rarely suited to the preservation of fossils, but fine sediment laid down and now preserved in a lake at Kiltorcan in County Kilkenny enfolded vegetation carried in by rivers, together with fragments of strange, armor-plated fish and a modern-looking mussel. The old red sandstone (here actually greenish) preserved more than a dozen different plants discovered in the 1860s, among them fronds of the beautiful *Archaeopteris hibernica*, found elsewhere in the world as a substantial tree fern. Other fossil material is thought to be from the earliest seed plants known from Ireland or Britain.

Amphibians occur again, together with spiders and dragonflies, in fossils in a seam of coal near Castlecomer in County Kilkenny (one of the few Carboniferous coal deposits to survive the massive erosion of the island). Dinosaurs of some sort must have walked the steamy landscape of the Jurassic period, but no trace of them has survived erosion. The life of the sea at that time, however, left fossils preserved in mudstones, among them the nautilus-like ammonites that, at the turn of the eighteenth century, made Portrush (then a tiny fishing village), in County Antrim, known throughout the scientific world.

In this pioneering period of modern geology, the origin of basalts was fiercely contested in Ireland between the Vulcanists, who correctly regarded them as volcanic, and the Neptunists, who saw them as crystalline deposits formed, like limestone, in the sea. In 1798 the claim of a discovery of ammonites in the massive igneous sill of the Portrush Peninsula seemed to vindicate the Neptunists and attracted international attention as samples of "fossiliferous basalt" were circulated among geologists. In fact, as they soon recognized, the ammonites were held in mudstones adjoining the sill and baked by the heat of its intrusion into a dark, fine-grained rock, looking somewhat like basalt.

Ammonites can be examined today on the County Antrim coast (most conveniently in Jurassic mudstones on the shore at Waterloo Cottages, just north of Larne). But the really abundant fossils of Ireland are those of Carboniferous limestone, obvious on every hand in the bare-rock landscapes of the Burren and the Aran Islands and the fossil-packed limestone and mudstone headlands of County Sligo. Even the limestone blocks of Ireland's grander buildings are often flecked with the skeletal remnants of corals, crinoids, and brachiopods.

In University College, Galway, the restoration of the James Mitchell Museum as a "time capsule" of nineteenth-century academic geology matches in atmosphere the Victorian showpiece of the Natural History Museum in Dublin. At the core of its collections, first assembled in the 1850s, are fossils from Permian limestone reefs in England and specimens of Jurassic marine reptiles, *Ichthyosaurus* and *Plesiosaurus,* from Germany and England. Unlike the rock collections, which are enriched by many excellent Irish specimens, the fossils have yet to be augmented by much local material, but the modern displays offer well-organized themes from Earth history.

Among the rock scoured away from Ireland by the last two ice sheets was much of the sediment that might have preserved some record not only of dinosaurs but of Ireland's mammals in the early Pleistocene. While unglaciated areas of neighboring England have yielded fossil mammals dating back over some 1.8 million years (woolly rhinoceros, *Coelodonta antiquitatis,* and bison, *Bison schoetensacki,* among them), few Irish records go further back than 35,000 years. Mammoth and musk-ox *(Ovibos moschatus)* bones discovered in Pleistocene deposits on the shores of Lough Neagh may have pushed the record back to more than 50,000 years ago, but recent radiocarbon dating of museum bones, originally recovered from caves, share the fauna record firmly between 45,000 years ago and subsequent phases.

There was an intriguing diversity among the animals living in late-glacial Ireland, from temperate-zone mammals such as red deer to cold-climate species such as mammoth, arctic lemming *(Dicrostonyx torquatas)*, and arctic fox *(Alopex lagopus).* The oldest radiocarbon date in this series—32,060 ± 630 B.P.—belongs to the bone of a giant deer *(Megaloceros giganteus)* salvaged from Ireland's most important bone cave, Castlepook, in a limestone knoll near Doneraile in County Cork. Tens of thousands of bones were excavated from the sandy floors of the cave's galleries at the start of the 1900s. Among them were those of spotted hyenas *(Crocuta crocuta),* which had left so many tooth marks on bones of woolly mammoth, brown bear *(Ursus arctos),* wolf *(Canis lupus),* and reindeer *(Rangifer tarandus),* and which have themselves been dated to about 34,300 B.P.

Even more exciting dates, from an ecological point of view, were those that show that, at a time when the Midlandian ice extended to the Cork-Limerick border, the arctic fox and lemming still lived in Ireland. Indeed, there are dates to suggest that wolf and bear may have established themselves on the island during late-glacial times, after 12,000 years ago, and survived the "Nahanagan cold snap" to be numbered among the founder members of Ireland's post-glacial fauna.

REVISITING MUSEUM BONES

In the mid-1800s, most of the bones collected as "antiquities" in Ireland had come from the middens of archeological sites or were recovered in dredging the beds of rivers or cutting into bogs. An early marvel had been the finding of woolly mammoth bones in the construction of a water mill in County Monaghan in 1715. It was the discovery of more mammoth remains at Shandon, County Waterford, in 1859 that showed the promise of limestone caves for evidence of a fauna that had long preceded human occupation of the island.

The mammoth bones were found mixed up with those of bear, horse *(Equus caballus)*, and hare *(Lepus* spp.), and one was paraded by a quarry worker through the streets of nearby Dungarvan as that of "an antediluvian giant." To the local naturalist, Edward Brenan, who presented their discovery in a paper read to the Royal Dublin Society, they were evidence of animals sheltering together in the biblical flood, when

> the floodgates of heaven were opened to overflow exceedingly, and all the fountains of the great deep were broken up; all animated nature became terror-stricken; men fled to the tops of the highest mountains, and the wild animals of the forest sought shelter in the caves. There were to be seen, in one confused and panic-stricken crowd, the lion and the tiger, the hyena and the bear, the rhinoceros and the elephant, with the fleet horse and the timid hare; and all things breathing life upon the earth lay down, and there at once together died. . . .

For such catastrophists (this, as it happens, in the year that Darwin's *Origin of Species* finally reached print) the cache of bones at Shandon spoke for, as Brenan put it, "an omnipotent agency, capable of accounting satisfactorily for every physical effect hitherto observed on the face of the globe."

The Shandon discoveries prompted a Dublin surgeon-naturalist, William

Wilde (father of Oscar), to drag out the bones "stowed away among the lumber of the crypts" at the Royal Irish Academy and make an inventory of them. But it was not until the turn of the century that the academy launched a more thorough excavation of the bone deposits in caves in the counties of Sligo, Clare, and Cork. Most of the work fell to Richard Ussher, better remembered today as an ornithologist and primary author of *Birds of Ireland* (1900). He sent some seventy thousand bones to the National Museum from three cave systems in County Clare alone. Among them were reindeer, wolf, and vast quantities of bear, more mammoth, giant deer, arctic lemming, arctic fox, and spotted hyena. The excavation was painstaking, digging and sifting by lantern light through layer after layer of debris, often full of rocks fallen from the roof. "Cave work of this kind is an exact science," wrote Robert Lloyd Praeger, the excavation secretary. But Ussher's great care was often pointless: glacial water, sweeping through the limestone galleries, had jumbled the bones together in piles that defied the best attempts at stratigraphy. Fixing the succession and association of animals through the climate swings of the Ice Age had to wait until the refinement of radiocarbon dating in the 1990s.

Up to that time, few of the bones from excavated caves had actually been dated, so there was little accurate information about the periods of their accumulation. How many were contemporaneous, within short periods: had the woolly mammoth lain down, so to speak, with the hyena? How many had accreted over thousands of years, even bridging several different episodes within the Ice Age? The questions took on even greater weight as the complexity of temperature swings through the later Quaternary became clear, and scientists became dissatisfied with accepted views of mammalian colonization of the island.

The progress of dating had been slowed by the need to sacrifice whole bones in early radiocarbon analysis (and, indeed, its high cost). Although scores of thousands of bones had been sent to the National Museum, many had been discarded for lack of space, and those remaining might never be replaced. But the development of the use of accelerator mass spectrometry for Carbon-14 dating has meant that a small fraction of an ounce of bone is an adequate sample. In the 1990s, a large-scale dating program assessed the chronological range of the mammal fossils from the caves and those found in peat-cutting, engineering works, and so on.

RADIOCARBON SURPRISES

The Irish Quaternary Fauna Project, led by archaeology professor Peter Woodman, turned many assumptions upside down. Almost all the bones that had been found by chance and were dated from their context were now given a quite different radiocarbon age. And there were surprises for zoologists who had assumed that Ireland had a very sparse late-glacial fauna. Some had even thought that, after the final five-century cold snap of 10,000 years ago, the postglacial colonizers were entering a virtually empty island.

The new dates have shown that a wide range of mammals colonized Ireland in the period between at least 45,000 years ago and 20,000 years ago, with the earliest date, from a solitary bone, falling to the spotted hyena, a predator that was certainly not alone. The range of species seems to have built up gradually, using ice or land bridges over the Irish Sea, so that by some 28,000 years ago, one of the warmer periods of the Ice Age, the hyena shared the tundra-like landscape with mammoth, giant deer, wolf, brown bear, reindeer, red deer, horse, and possibly arctic fox, arctic hare *(Lepus timidus),* and the stoat *(Mustela erminea).* After 25,000 years ago the fauna begins to have a more high-arctic character, and at the glacial maximum around 20,000 years ago, when the last ice sheet reached down to Munster, the only animals certainly present in a cave at Castlepook, County Cork, were mammoth, arctic fox, and arctic lemming.

Matched to today's closer understanding of the Ice Age and its often rapid climatic oscillations, the new dates from the Quaternary Fauna Project add many thousands of years to the continuity of mammals in Ireland. So long as zoologists believed that a scanty late-glacial fauna was wiped out by the exceptionally cold period of the Nahanagan Stadial of around 10,500 years ago, a whole group of animal bones had been assigned to a postglacial recolonization of the island by virtually a new range of species. In reality, Ireland was effectively ice-free and accessible by about 13,000 years ago, well before the end of the late-glacial period, and several of a wide range of animals went on to survive the cold snap, some of them continuing even into the fauna of today.

As late as 1985, the only species placed with certainty in the late-glacial period (13,000 to 10,000 years ago) were the reindeer and giant deer. The new dates for the period, mostly from after 12,000 years ago, have added red deer, wolf, brown bear, hare, and stoat, with the arctic lemming entering as the cold returned again. The change seems to have extinguished all three deer, but not wolf, bear, hare, and stoat, which continued into the Early Holocene and were there to greet the first human hunters of the Mesolithic around 9,000 years ago.

The new knowledge has thrown up questions of its own. It had long been assumed, for example, that the red deer, an important native animal in Irish history, must have reached Ireland early in the Holocene. But there are no dates much earlier than 4,000 years ago. No red deer remains have been identified with any certainty from Irish Mesolithic sites, nor are there any antler artifacts to set beside those shaped from flint. Red deer are excellent swimmers, well-able to cross the North Channel from Scotland, and Woodman finds their scarcity in Ireland's Early Holocene "a major enigma."

There are other intriguing footnotes from the dating program. The horse scapula found with mammoth bones in the Shandon cave was indeed of a comparable age (27,000 B.P.), but other likely specimens, from caves, gravels, and peats, belonged well within Ireland's human history, and the earliest postglacial date is 4,000 B.P., in the Neolithic. Remarkably, the lynx *(Felis lynx),* which briefly colonized Britain at the beginning of the Holocene, preying chiefly on the small roe deer *(Capreolus capreolus),* was also in Ireland about 9,000 years ago.

Of all the inhabitants of late-glacial Ireland, the giant deer stood head and antlers above the rest, and the remains found at close to 200 sites across the island testify to its abundance, especially between 12,000 and 10,800 years ago. This remarkable animal, with such significance for studies in evolution and extinction, deserves a story to itself.

EXTINCTION: THE BIGGEST DEER OF ALL

Among the odder events on the rooftops of Dublin city in the spring of 1985 were protracted, slow-motion battles between two extraordinary figures, half-man, half-deer. Confronting each other on the flat roof of the National Museum, Andrew Kitchener, zoologist, and Nigel Monaghan of the museum's staff repeatedly raised and lowered huge antlers, locking the tines at different angles and pushing at each other in a circling struggle on the asphalt. The result, in due course, was a paper in *Modern Geology* that has helped to resolve one of the oldest questions in evolutionary biology: Could the giant Irish elk ever have used its enormous antlers in battle?

The male *Megaloceros giganteus* is the pride of Irish palaeontology, with a skeleton nearly 10 feet (3 m) high from hooves to tine tips. The animal was never exclusively Irish—nor, for that matter, was it an elk—but Ireland was the early treasure-house of its fossil skeletons and antlers, impeccably preserved in lake mud beneath the bogs. It was first recorded as a fossil in 1697 (long before dinosaurs were known), and it featured in fierce debates about extinction and the impact of Noah's flood. In the nineteenth century, it was used to challenge Darwinism and the concept of "fitness"—the idea that through natural selection animals evolve the characteristics most helpful to their survival. The controversy continues to draw scientists to the world's biggest cache of the giant deer's bones. When visitors enter the Victorian splendor of Dublin's Natural History Museum, a pair of magnificently antlered skeletons greet them inside the door, but these are only the mounted representatives of more than 250 animals, their remains stacked up in the museum's stores.

Megaloceros fossils were also found widely in Europe and northern Asia, where the deer wandered after temperate grasslands as these shifted with climate changes in the Pleistocene. The earliest are from British gravels dating from an interglacial period some 400,000 years ago. There are hundreds more fossil sites, ranging from Mediterranean shores in interglacial periods to boreal or subarctic habitats in cold stages of northern Europe, but Ireland offered the exceptional conditions to preserve a rich collection of remains. Here, the most recent radiocarbon date for a bone is a mere 10,600 years ago, the time of the cold snap that interrupted the final thaw of the Midlandian and brought giant deer (as it is now more generally called) to extinction.

Long before scientists considered them, antlers discovered by chance in Irish excavations were prized curiosities, fit for gifts in tribute. A set found in south County Dublin in 1588 was sent to Robert Cecil, Queen Elizabeth's secretary of state: a contemporary drawing is held in the National Museum of Ireland and the antlers themselves are on display in the Provincial Museum of Edmonton in Canada. Thomas Molyneux, the Dublin physician who first described the structure of the deer, records other gifts, to England's Charles II and William III, and antlers mounted like hunters' trophies still decorate the walls of many stately homes in Ireland and Britain.

It was Molyneux who proposed that "the great American deer, call'd a moose, was formerly common in [Ireland]" and thus launched the idea that

Megaloceros was an elk, as the moose is called in Europe. He believed, with many naturalists of the time, "that no real species of living creatures is so utterly extinct as to be lost entirely out of the world" (once created by God, its continuance was beyond doubt) and he found, in travelers' reports of the moose, a likely form of Ireland's giant deer surviving in the wilds of Canada. The question of extinction was, in Stephen Jay Gould's phrase, "the first great battleground of modern palaeontology." In 1812 a minute anatomical description of the Irish elk was used by Georges Cuvier, the great French palaeontologist, to show its uniqueness in the mammal record and thus to demonstrate that extinction actually occurs.

By the late 1800s, after the publication of Darwin's *Origin of Species* in 1859, *Megaloceros* was brought center stage again, as anti-Darwinians searched the fossil record for examples of evolution that could not possibly have served the fitness and survival of the species involved. They believed in a theory they called orthogenesis, a straight-line progress of evolution that natural selection could not regulate, even if its trend should lead on to extinction. The Irish elk was their prime example. Giant deer had clearly evolved from smaller forms, but the usefulness of bigger and bigger antlers evaporated—so it was argued— because they tangled in trees or dragged the stags down into the mud of bogs and lakes.

A Darwinian counterattack was led in the 1930s by Julian Huxley. He observed that, as deer get bigger, their antlers increase in size at a proportionately faster rate, an orderly phenomenon for which he coined the term *allometry*. As natural selection favored the body size of *Megaloceros*, so allometry dictated the enormous size of its antlers, curving out in great spiked shields to a span of almost 13 feet (4 m) and weighing up to 88 pounds (40 kg)—almost a third heavier than those of the modern moose. By Darwinian reasoning, any inherent disadvantage must be outweighed by the benefits of larger size.

Allometry became the standard theory in textbooks of evolution, but, as the young Gould was dismayed to find, it was based on no data whatsoever. Aside from a few desultory attempts to find the largest set of antlers, no one had ever measured an Irish Elk. He spent most of 1971 in Dublin's Natural History Museum, yardstick in hand, measuring and comparing antlers and the length of the attached skulls as a key to original body size.

Most of the museum's great collection of specimens, piled antler on antler in the warehouse, was accumulated during the nineteenth century. Many came from Irish peasant families who, cutting peat slowly by hand for fuel, found tines sticking up from the blue marl at the base of the bog. In the more famous locations, such as Ballybetagh Bog, south of Dublin, commercial dealers probed the bog with iron rods to make bone collections from which to assemble skeletons for museums overseas. (The market for antlers and skeletons continues even today, at auction-room prices high enough to cause concern for Ireland's remaining stock of fossils: the National Museum now sells fiberglass replicas to create an alternative supply.)

The museum's formidable collection, plus a few specimens mounted in big country houses and other museums, enabled Gould to demonstrate the workings of allometry. The antlers, he found, increased in size 2.5 times faster than body size from small to large adult males. But, even as he confirmed this relationship, he came to doubt the merit of supposing that the outsize antlers had no primary value or function in themselves. "The case for inadaptive antlers," he decided, "has never rested on more than subjective wonderment born of their immensity."

He suggested that, rather than being used in actual combat, the broad-palmed antlers were status symbols in ritualized confrontations between males that avoided damaging or deadly battles in competition for a harem of hinds. The idea of such a mechanism, significant for evolutionary biology, had been gaining ground from studies of behavior in red deer, caribou *(Rangifer tarandus)*, and mountain sheep *(Ovis musimon)*. As actual battle weapons, the antlers of the giant deer could seem impossibly unwieldy and ineffectual. A typical opinion came from Ireland's Frank Mitchell in *Reading the Irish Landscape:* the antlers, he wrote, though impressive, "were structurally very weak, with elongated points mounted on the edge of a thin curved plate. They would have been useless in combat and their only function can have been to impress."

This idea seemed to be strengthened by the way the antlers were mounted. In the modern fallow deer, often proposed as the giant deer's nearest living relative) the palms of the antlers are swept back, so that the stag has to sweep its head from side to side to display them. In *Megaloceros*, the palms have a full-frontal span for head-on intimidation—especially impressive, perhaps, in a "parallel walk" between posturing stags of the sort seen today among red deer.

The importance of symbolism in conveying the giant deer's power is supported by the rare and striking Palaeolithic paintings discovered in south central France, in a cave at Cougnac in 1952 and another at Chauvet in 1994.

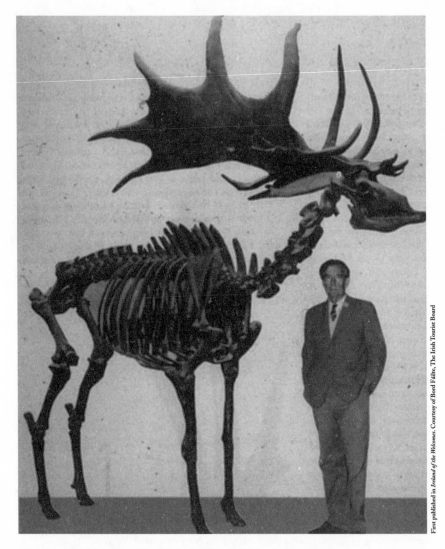

First published in *Ireland of the Welcomes*. Courtesy of Bord Fáilte, The Irish Tourist Board

Professor G. F. Mitchell standing beside the skeleton of the giant Irish Deer.

These show the deer with a large, almost camel-like hump above the shoulders, boldly colored to catch the eye and linked to diagonal black lines across the animals' flanks. On the vertebrae of fossil skeletons there are distinctive dorsal spines, elongated in the males and apparently meant to anchor the muscle that helps to hold up the hugely burdened head. This would have produced an obvious bulge on the back, but, said Gould, hardly the big hump of soft tissue so clearly emphasized in the cave paintings.

Several authors have stressed the value of such a massive hump in display, and for Gould it offered a splendid example of the way in which a physical feature, arising for one function, can be coopted and developed by natural selection to suit another, more specialized, role. Still another possibility, however, is that the shoulder hump was simply used to store fat for energy during the rut: today's red deer stags lose one fifth of their body weight at this time because they are too busy fighting and defending their does to feed properly.

ANTLERS IN COMBAT

Gould came to think later that he was wrong to suggest that the deer did not use their antlers in actual combat. Among work that has helped to change his mind were the practical experiments carried out by Kitchener in Dublin in 1985.

Zoologists agree that, because fighting within a species is such dangerous behavior, elaborate displays have developed to make sure that only animals of well-matched size fight together. The macho posturings of stags and their resonant roaring contests may be more important than the comparative size of antlers. But even on theoretical grounds, it has been argued that no structure so energy-costly as the antlers could be selected for mere display: it had to be backed up by a credible threat of use.

Kitchener accepts that if *Megaloceros* fought like moose—that is, locking the points of the antlers and pushing—then the giant deer's superstructure was impossibly engineered. But there are other ways of locking horns. Up on the roof in Dublin, with the museum's assistant keeper, Monaghan, as opponent, Kitchener raised and lowered his borrowed antlers, testing alternative fits. He found that if the deer tilted their heads almost parallel to the ground the antlers locked together just above the second tine, as happens with fighting red deer stags today. In this position, the force of battle was absorbed by the neck muscles and a thickened skull. The tines on the giant deer's antlers swung into purposeful alignment: some protected the eyes, while others, in a contest of pushing and twisting, would have gored the loser in the flank and neck in the manner seen in red deer fights today. Kitchener's study of the antler bone showed that the main beam of the fossil antlers had crystal alignments similar to those in living deer, adapted to withstand the physical shocks and stresses of fighting. As mere implements of display, the antlers would have been massively overdesigned.

In the course of 400,000 years, and in different parts of the giant deer's ιge, the morphology of *Megaloceros* varied considerably. Early on, the antlers were shorter and more upright than in the Irish, late-glacial sample, suggesting more wooded conditions in the earlier interglacial periods. And deer with longer limbs, such as the Irish animals, all lived in open or relatively treeless habitats where fast, long-distance running might be demanded. One constant was an unusually thick lower jaw, suggesting the huge demands for calcium storage that the annual renewal of antlers made on the body as a whole.

The American mammalogist Ron Moen calculated that the largest antlers would have contained nearly 44 pounds (20 kg) of mineral elements, including not only some 15 pounds (7 k) of calcium but almost 9 pounds (4 k) of phosphorus. These extraordinary mineral demands seem both to have changed the vegetation of late-glacial Ireland and to have doomed the deer to extinction as the climate cooled again.

As modeled by Moen and his colleagues at the University of Minnesota, Duluth, a mature giant Irish stag growing 88-pound (40-k) antlers in 150 days would have needed 138 acres (56 ha) of shrubby tundra just to meet its phosphorus needs. It would also have sought out the shrubs that were richest in the mineral—willow, above all. The Irish pollen record shows that, as the *Megaloceros* population rose and this selective browsing pressure increased, willow was replaced by juniper *(Juniperus communis)* or crowberry *(Empetrum nigrum)* scrub and then by grassland in which willow had virtually vanished. In younger fossil skulls, the teeth show the typical wear patterns of grazing rather than of browsing on leaves and twigs.

The deer's exceptional demand for calcium and phosphorus was to accelerate its problems with the onset of the Nahanagan stadial, the little Ice Age that gripped Ireland around 10,900 years ago. Even in the lushest of mineral-rich vegetation, the animal could not have grown its huge antlers each summer from its current food intake: the phosphorus it needed—more than 1 ounce (30 g) a day—was more than could be absorbed from the gut. As Moen and his colleagues computed it, about 10 percent of the minerals had to be borrowed from the deer's own skeleton and replenished during the winter. When temperatures tumbled, springs grew shorter and vegetation swung back to meager tundra plants, this physiological debt became impossible to sustain.

The intuition that *Megaloceros* died out because its antlers were simply too extravagant has thus been refined to the atoms of body chemistry, and the enduring picture of giant stags becoming mired in icy lake mud and drowning under the weight of their cumbersome antlers has been shown to have a truth of a more metaphorical kind.

The image of drowning stags had been nourished over centuries by the finding of more than one hundred antlered skulls in lake sediments at Bally-betagh Bog, 9 miles (15 km) south of Dublin. They have made up the bulk of the remains collected in the National Museum, and in the 1980s this concentration of males became the focus of taphonomic research by Anthony Barnosky of the Carnegie Museum of Natural History at Pittsburgh.

Taphonomy examines fossil bones in their environmental context—the layers of sedimentary rock, the fossil plants and pollen and fossil insects that allow a detailed reconstruction. Ballybetagh was exceptional in preserving so much of the evidence surrounding an important large-mammal extinction—one in which humans played no part.

The winter segregation of bachelor males to this particular steep-sided valley was quite in keeping with the present-day behavior of red deer as observed, for example, at Killarney. But the Ballybetagh animals, on average, died younger, were smaller, and had abnormally small antlers compared to *Megaloceros* from other localities. All this was consistent with malnutrition and the winter-kill of animals exhausted by the late summer rut.

In their progressively weakened condition, did these stags eventually become mired in the lake mud, weighed down by their preposterous antlers? Barnosky found no evidence for it. First, the Ballybetagh antlers happened to be relatively small; second, the bones were scavenged, trampled, and scattered and lay on the surface for months or years before being washed or kicked into the water. There were no even partially complete skeletons or even the vertical leg bones and feet that would speak for deer mired in mud; nor was there any sign of struggle in the sediment layers.

These almost forensic observations support the conclusion that winter kill alone eliminated *Megaloceros* from Ireland. But as Barnosky pointed out, the giant deer was unable to survive anywhere by the early Holocene, even though their preferred food plants quickly returned with the warmth. He conjectured that the season of the spring green-up, when plants contain their maximum nutrients, may have failed to return to its previous length and intensity. This could have been crucially important to a deer growing such enormous antlers at an almost visible daily rate. But the images of *Megaloceros* among the cave paintings of France suggest that, whatever happened in Ireland, human hunters probably also played some part in extinguishing this great animal across the steppes of Europe.

The giant deer had long disappeared from Ireland by the time the first Mesolithic hunters landed on the island, some 9,000 years ago. A millennium had passed since the last retreat of ice: what wildlife fled from the first human step and how had Ireland been furnished with plants and animals? As David Quammen summed it up in *The Song of the Dodo* (1996): "How they arrive, which kinds of creature turn up earlier and more often than which others, are matters of consequence that shape the biological history of an island." As the next chapter shows, they can also be matters of tantalizing mystery.

SELECTED REFERENCES

Gould, S. J. 1977. The Misnamed, Mistreated, and Misunderstood Irish Elk. In *Ever Since Darwin: Reflections on Natural History.* New York: Norton, 79–90.

Holland, C. H. 2001. *A Geology of Ireland.* Edinburgh: Scottish Academic Press.

Mitchell, F., and M. Ryan. 1997. *Reading the Irish Landscape.* Dublin: Town House and Country House.

Pilcher, J., and Hall, V. 2001. *Flora Hibernica.* Cork: Collins Press.

Ryan, S. 1998. *The Wild Red Deer of Killarney.* Dingle: Mount Eagle.

— *Chapter 4* —

A Bridge Too Far?

The retreat of the last big glaciation left Ireland a largely devastated island, the uplands scoured by glaciers and the lowlands strewn with rocks and silt. But how far it presented a *tabula rasa*, a clean sheet on which nature must inscribe all over again, is still uncertain. In the traditional model of glaciation, unchallenged until the end of the twentieth century, a margin of tundra had been left to the south of a main ice sheet, roughly on a line from the River Shannon to Wicklow, with a separate ice dome on the mountains of Cork and Kerry; there were also ice-free promontories along the northwest coast. At these margins, it was feasible that an arctic flora survived, along with a sparse population of lemmings and reindeer.

In the 1990s, however, new studies of Ireland's eskers and moraines by a Geological Survey scientist, William Warren, proposed a model in which the last glaciation was fed from three domes on Ireland and one in the Irish Sea, leaving no room for life on the island until the major withdrawal of ice began around 13,000 years ago.

A general acceptance of this picture will set a new challenge to some of the conjectures about Ireland's postglacial colonization by plants and animals. The exact sequence and means by which species arrived (or persisted) and the reasons why so many British species failed to do so have long been a frustrating puzzle. The false start of the post-Midlandian warming and the bitter, if geologically brief, return of ice around 11,000 years ago have complicated the story of survival, and the see-saw balance of land and sea levels as the ice melted has made a juggling match of access across the Irish Sea.

Nothing could better demonstrate the broad reconnection of Britain to the Continent during the last cold stage (the Devensian, in British terms) than the trawling of abundant teeth and bones of mammoth, reindeer, and woolly rhinoceros from the bed of what is now the southern North Sea. The British faunas of that period appear to have been similar to those of the adjacent continent, and humans, arriving in England at about 35,000 years ago, seem to have kept a presence there in all but the 5,000 severest years of glaciation. The English Channel was probably flooded by about 9,000 years ago, bringing to an end the natural immigration of land mammals. But the recolonization of Ireland, and the existence and timing of its land links to Britain, pose continuing problems.

The map of the Irish Sea floor between Ireland and Britain shows a trough as much as 328 feet (100 m) deep running up the center. At its northern end, it rises to shallower contours between Islay, the Scottish island, and Malin Head in Donegal. At the middle of the Irish Sea the trough is interrupted by a ridge running east to west between Wales and Wicklow on Ireland's east coast. Even allowing for erosion, the ridge seems to lie too deep—263 feet (80 m) at its shallowest point—ever to have been exposed in the postglacial period.

As biogeography seems so insistent on a land bridge at the right time, a great deal of research has been spent on providing at least the possibility of one. Land bridges could have occurred from causes other than the simple withdrawal of the ocean. In the millennia between the first onset of thawing and the final return to temperate climate—say 13,000 to 10,000 years ago— there were many forces simultaneously at work in the physical chaos of broken rock, silt, and water. Trying to integrate them all in a model of what happened can itself seem a chaotic and precarious enterprise.

As ice sheets melted and retreated northward, the land was relieved of weight and began to rise. As meltwater was released, sea level, too, was rising, but at first more slowly than the land. In the north of Ireland and Scotland the land lifted rapidly. About 12,000 years ago its rise was overtaken by water flooding into the ocean, and at about 6,000 years ago the rise of the land re-

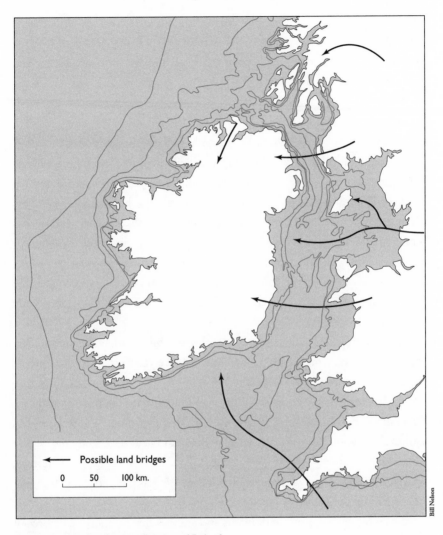

Possible land bridges between Britain and Ireland

asserted itself. It was at intervals of equilibrium in this see-saw of levels that "staircases" of raised beaches, like that at Malin Head in Donegal, were formed.

The earliest pictures of land bridges across the Irish Sea were of moraine ridges left behind by the glaciers and subsequently washed away. But most geologists were adamant that the rapid postglacial rise in sea level would have given no time for temperate plants and animals to migrate across them.

Other transient events, however, could have lifted the seabed for long periods. In one notable theory, the slow northward retreat of the ice was followed at its boundary by a migrating "forebulge," as the Earth's crust rebounded from the weight of the ice cap. As modeled by Robin Wingfield for the British Geological Society, this could have provided a land bridge to southwest Ireland around 11,000 years ago.

MIGRATION OF SPECIES

To Ireland's Frank Mitchell, distinguished Quaternary scientist and naturalist, the rapid postglacial expansion of the island's plants and animals could not have resulted from the mere chance arrival of species, carried by wind or birds or drifting ashore on logs. This, he argued, would not have sustained the orderly ecological development of the Irish flora traceable through pollen diagrams, radiocarbon dates, and dendrochronology. In his book, *Reading the Irish Landscape,* he preferred to picture "organised woods advancing up a dry coastal strip . . . along the shore of the Atlantic, and later a very remarkable type of automatic trackway across the Irish Sea" (that is, Wingfield's mobile forebulge).

In the 1980s, scientific comment on the possibility of land bridges to Ireland was still extremely cautious. Their existence was certainly expedient in explaining plant and animal distributions, but the geology and sea-level data seemed against them. If there had to be a bridge, then the most that might be granted was the "low, soggy, possibly shifting and partially discontinuous linkage" between the Scottish islands and Malin Head reluctantly posited by the Irish geographer Robert Devoy at a Postglacial Colonization Conference in 1983. Even the later work on crustal movement has not persuaded him further than the possibility of land connections.

The rapid early expansion of plants and animals that so impressed Mitchell does not affect the fact that Ireland's list of species was—and remained—impoverished and that a land bridge, if it existed, was a problematic means of immigration. The biogeography of the land bridge issue is often a study of anomalies and paradoxes, some of which seem to point one way, some another.

Ireland has, for example, as its smallest mammal, the pygmy shrew *(Sorex minutus).* But the island lacks the common shrew *(Sorex araneus),* which overlaps with it, and often greatly outnumbers it, in Britain and most of western Europe. The common shrew burrows after earthworms as a major prey. On acid moorland peats, however, there are no earthworms and the pygmy shrew

Dr. George Francis Mitchell (1912–1997), Professor of Quaternary Studies, Trinity College, Dublin, 1965–1979.

predominates, feeding on surface invertebrates. A soggy land bridge would have suited one shrew but not the other.

An example strung about with paradox is the case of the snails of Newlands Cross, near Dublin, reported in the *Journal of Biogeography*. In the early 1980s, excavations uncovered a deposit of tufa, a lime-rich sediment resting on glacial till. The start of its formation had a radiocarbon date of 9720 ± 300 B.P. and it contained a whole sequence of land snails. This began, at the lowest level, with several arctic-alpine species, typical of cold and marshy open habitats. They were gradually replaced in the tufa by a series of shade-demanding snails, coinciding with Ireland's great postglacial spread of hazel trees. At a higher, drier level, where one of these snails made its greatest expansion, was a fresh flake of flint dropped by a Mesolithic human.

The orderly succession of the Newlands Cross snails was exactly comparable to sequences in western Britain, and among them were southern species that never set foot in Scotland (and thus on any northern land bridge). Could random and passive arrival have produced such a sequence of snails and their matching vegetation?

An even more self-effacing group of Irish invertebrates are the moss mites *(Oribatida)* living among lichens on the peaty summit of Mount Leinster in County Carlow. They have been found to mirror the montane species list in neighboring Britain and to show the same mix of arctic-alpine, European-alpine, and lowland elements. Again there seems to be a choice between the mites surviving the glacial periods in refuges, crossing a land bridge, and arriving as passengers. There are difficulties with all three options (the mites' lifestyle, for example, would give them small contact with birds); a southerly land bridge across the Irish Sea would, once again, prove convenient, if it could have lasted long enough for moss mites to cross it.

"LUSITANIAN SPECIES"

The case for passive introduction of nonmammalian species, by birds or humans, has been directed especially at what is called the "Lusitanian" element in the fauna and flora of the mild, moist habitats of southwestern Ireland. Lusitania was the Roman name for an area embracing the Pyrenees and northern Spain and it has come to be used for Irish species with strong biogeographical affinity with southwest Europe and its Atlantic seaboard.

Among sand dunes in Kerry, for example, live colonies of Ireland's only toad, the yellow-striped natterjack *(Bufo calamita)*. This is a scarce and local animal in Britain and much more at home in the coastal dunes and heaths of France and Spain. In his *Natural History of Ireland* (1950), Praeger had no doubt that it was an indigenous Lusitanian species, but later there was skepticism, and consensus grew around the possibility that the toads were dumped out on the Kerry coast with ballast sand from boats that plied between Ireland and Europe. By the 1990s, however, the pendulum had swung again. The British herpetologist Trevor Beebee saw the toad as surviving the last glaciation in a single refuge in Iberia and spreading rapidly north and east from there in the early warming period. In a genetic study of the populations scattered across northern Europe, he has found that the Kerry toads are closely related to those of northwest England and far less so to the natterjacks of Brittany. Once again, his work supports a land bridge from Britain to Ireland about 10,000 years ago.

Nothing quite so precise explains the presence on mossy trees and boulders in Counties Kerry and Cork of a large, spotted slug that malacologists and others find "handsome." *Geomalacus maculosus*, known often in Ireland simply as the Kerry slug, is otherwise found only in parts of northern Spain and Portugal. Its presence in southwest Ireland has figured often in speculation on land bridges to the south. But why, as some have wondered, should such a land link have allowed passage of a slug but not of Spain's rodents and reptiles?

Some of the heathers of Spain exemplify the puzzles among the Mediterranean-Atlantic plants along Ireland's western seaboard. Among them are three plants that do not occur naturally in Britain and have a limited distribution in Lusitania and nowhere else. The tall and shrubby Irish heath *(Erica erigena)* grows head-high in widely scattered thickets in Connemara and County Mayo, but otherwise only in Portugal, a few dry, rocky parts of Spain, and western France. The purple-belled St. Dabeoc's heath *(Daboecia cantabrica)* grows in shady pine woods in Portugal, and high enough in Spain's Cordillera Cantábrica to be covered with snow for several months in winter: in mild, moist Connemara, it twines through the gorse bushes and reaches high on the mountains. The purplish-pink Mackay's heath *(Erica mackaiana)* creeps (by layering) across bogs in Connemara and Donegal and is known elsewhere only from northern Spain.

The fossil pollens of both Mackay's and St. Dabeoc's heaths have shown up in one of Ireland's few surviving interglacial deposits, near Gort, in County

Galway, as part of the evidence for a broad ribbon of heathers stretching along the western shores of Europe from the coasts of Portugal to the Shetland Islands, north of Scotland, in a temperate stage of the Middle Pleistocene. It seems impossible that any of the northern heathers could have survived the Quaternary cold, yet one ingenious possibility was advanced by Mitchell and the palaeobotanist W. A. Watts in a paper they wrote together in 1970.

As the Atlantic fell back from the mountains of Connemara and Kerry, there would have been land at the foot of steep slopes, perhaps 16 miles (25 km) west of, and 328 feet (100 m) lower than, the present coastline. "Here," proposed Mitchell and Watts, "ridges and valleys provided shelter, slopes drained the soils, and oceanic waters provided a moderating influence." Could this have been a refuge for the heathers? If it was, the plants took 3,000 years after the last glaciers melted to reappear in Connemara, and this, as Mitchell and Watts confessed, is "not easy to understand." Even so, botanists of the distinction of David Webb, main author of *An Irish Flora*, continued to keep the faith. "No matter how hard the geologists try to terrify us with descriptions of the ice ages," he wrote in 1983, "[I believe] that with a few exceptions [the disjunct species] survived the last glaciation in Ireland." Disjunct populations are divided by a distance too great to allow a gene flow between them.

The native credentials of *Erica erigena*, the so-called Irish heath (more popularly known in its localities as Mediterranean heather), have recently been thrown into question by the use of powerful scanning electron microscopy. This can identify to species level the heather pollens preserved in cores taken from deep bogs. At the modern western headquarters of Mediterranean heath, on Claggan Mountain on the Mayo coast, the heather's pollen appears for the first time in the peat profile at the radiocarbon-dated level of A.D. 1431, a time of great maritime contact between western Ireland, western France, and northwest Spain. The heather could have arrived packed around wine casks or brought back by Irish pilgrims to the shrine of St. James at Santiago in Galicia.

Conceivably, all the seemingly exotic "Mediterranean" heathers could have been brought in by people. In *The Bogs of Ireland*, John Feehan and Grace O'Donovan speculate on their passage with "the megalith builders of the Neolithic, whose homeland was in these areas." At this point, one begins to feel for Praeger, who once wrote of the Lusitanian riddle (in *A Populous Solitude*,

1941): "I have thought about that and written about it for years without getting any further, and I am almost weary of it."

When it comes to the main thrust of postglacial recolonization, however, the proponents of land bridges have never lost their vigor. Mitchell, in particular, in the third and final revision (before his death in 1998) of his masterwork *Reading the Irish Landscape*, drew what seemed to him the only possible conclusions from the presence of oaks in the south of Ireland 9,000 years ago, of the wild boar *(Sus scrofa)* in the north at about the same time, and of the red deer in the midlands 8,400 years ago. He argued for an orderly migration of oakwoods, complete with forest animals, "rather than to imagine the occasional acorn floating across the sea or being carried by a pigeon across the North Channel, while groups of pigs and deer were swimming across the tidal channels." If his scenario is accepted, land links existed from some time after 13,000 years ago until about 10,000 years ago, when the forebulge bridge from Britain "sinking in level and narrowing as it did so . . . disappeared below the Irish Sea."

The Irish Sea may seem a mere pond compared with the Atlantic, and the regular arrival of West Indian drift fruits and seeds, in viable condition, on Ireland's western shores is a graphic, if somewhat eccentric, demonstration of the colonization of islands (the seeds have washed up, thanks to the Gulf Stream, in a climate too cool for their germination). But in the spread of plants from Britain to Ireland, there are some big constraints on colonization by chance. Surface currents around Ireland, influenced by wind, tend to move west to east, and fewer than 10 percent of Ireland's native species have seeds capable of drifting in saltwater for long enough to cross the Irish Sea. Fewer still are capable of floating for long enough *and* remaining viable.

On most days of most years, the wind in Ireland blows from the Atlantic: the force of the Earth's rotation almost guarantees it. There are enough exceptions—enough strong easterly winds in high summer, enough southerly phases in the storms of autumn—to create the chance of seed dispersal from Britain and continental Europe, but the very stability of Ireland's restricted range of native plants suggests the chance has not been a very strong one. Birds, too, carry seeds to islands, internally or externally, and Ireland attracts many regular migrants from Europe and Africa—mostly passerine species— and waders and wildfowl from the Arctic. Geese flying south from Iceland to winter in Ireland and Scotland appear to have brought seeds of the small, wetland northern yellow cress *(Rorippa islandica)*—but this has been a noteworthy event and thus a comparatively rare one. Charles Nelson, a botanist and taxonomist with special interest in the origins of Ireland's flora, believes that neither ocean currents nor winds nor birds have contributed many species, even over ten millennia: he is another land bridge man, for whom Ireland's native flora became "fixed" when the island lost its corridors to Britain.

At first scrutiny, the list of Irish native plants appears strikingly impoverished when compared with that of Britain. According to Webb, these total only 815 compared to the British figure of 1,172. This is now thought a conservative estimate, and the Irish total may be as high as one thousand. But Webb pointed out that about half the British species absent from Ireland can be ruled out on ecological grounds. Ireland extends through only four degrees of latitude, compared with Britain's ten. Ireland's mountains are lower and smaller, and there are virtually no sedimentary rocks of Mesozoic or Tertiary age—no inland sands and very little chalk. There is a greater ecological homogeneity to Ireland and fewer of the climatic extremes and regional variations found in Britain. In Webb's estimation, perhaps 186 plants are absent from Ireland because the Irish Sea opened before they had completed their northwestward migration through Britain; the rest would not have found the right habitats or conditions. But the poverty of the Irish flora is still very marked when measured against some Continental countries of comparable size and far less ecological diversity: Belgium has 1,140 native species, and Denmark has 1,030.

THE MARCH OF THE TREES

In the accelerated time scales of vegetational history, trees advance in regiments along broad arrows on a map. In the real, postglacial, world, they crept northward over centuries, as the advancing fringe of wildwood struggled to maturity and a first release of seeds. Even with the land bridges now implicit in forebulge theory, the first, unstable, warming did not last long enough for the big canopy trees to make it back to Ireland.

In most of the lake sediments plumbed for fossil pollens, the earliest plant colonists appeared around 13,000 years ago. Among them were several dwarf shrubs—willow, juniper *(Juniperus)*, and birch *(Betula)*—and these supplied essential minerals to the giant Irish deer. Their particular need for phosphorus, for the annual regrowth of their great antlers, is thought to have changed the island's late-glacial vegetation, from a shrub-dominated tundra to grassland (see Chapter 3). This extraordinary impact by a single grazing species was erased (along with *Megaloceros*) in the brief return to ice, and the final, rapid warming to the Holocene of 10,000 years ago launched a more orthodox vegetational succession.

It began, as before, with a surge of rich meadows across the arctic tundra—even, indeed, of meadowsweet *(Filipendula ulmaria)* itself, which foams into blossom today on the moister soils of the west. The first tree was juniper, responding almost instantly to the rapid warming and blanketing the meadows with its prickly thickets. Then came the pioneer birches, blowing in on tiny, winged seeds to create the first forested landscape in 100,000 years. As climate warmed and dried out even more, hazels began to rise in the shade of birch and then to overwhelm it, spreading their golden catkins over almost the whole island within 1,000 years. The pollen of hazel from that period is seventeen times more abundant in Ireland than all other tree species put together (many lake muds are full of the pollen and nut fragments and even whole nuts).

Pollens in the lakes and bogs chart the sequence in which the three main forest canopy trees—pine, elm *(Ulmus)*, and oak— arrived on the island, but it has taken radiocarbon dating to draw the contours of their arrival times. These so-called isochrone maps also strongly suggest the geographical thrust of the trees' migration. While hazel may have crossed the Irish Sea as well as the Celtic Sea, pine, elm, and oak seem to have migrated directly from Spain, up a French Atlantic seaboard, and bypassing Britain altogether. This, of course, needed land bridges, and the migration route and timing coincides very well with the Wingfield forebulge model. Pine had arrived in the southwest of Ireland by 9,500 years ago, and oak and elm had followed from the south within another five centuries. Although the Irish Sea of 10,000 years ago seems not to have been wide enough to be a barrier to hazel, it did seal off the late migrators to Britain—lime *(Tilia)*, which reached the Welsh coast at 7,000 years ago, and beech *(Fagus sylvatica)*, a mere 1,000 years ago. Neither tree got to Ireland naturally, but both have thrived as introductions.

The isochrone maps raise other questions about the way trees spread. In Ireland, Britain, and across Europe, they migrated at a faster rate than would match the simple fall of fruit from the first seed set, or dispersal of seeds on the wind. All the species were spreading at more than 2 miles (3.5 km) per generation, which certainly seems too rapid a progress for trees producing heavy hazelnuts, acorns, or beech-mast. The Irish botanist Fraser Mitchell noted that, in North America, oak and beech crossed Lake Michigan, some 56 miles (90 km) at its narrowest, and passenger pigeons are thought to have "assisted" in the process. In weighing up similar possibilities in Ireland and Europe, as he pointed out, even unlikely events can become near certainties over thousands of years.

In northwest Europe, the repeated glaciations over two million years reduced the number of tree genera from forty-seven to seventeen, selecting out those that were quickest to migrate in response to rapid warmings. All of them were growing in Ireland in the penultimate interglacial of some 240,000 years ago, but about one-third of them failed to make it back after the last major Ice Age. Frost-hardiness, too, may have limited the island's original native species. Noting that introduced and highly successful "southern" trees such as horse-chestnut *(Aesculus hippocastanum)* and sycamore *(Acer pseudoplatanus)* break into leaf far earlier than the native oak and ash, Mitchell suggested that the native tree species descended from those selected out by the frosts of the early postglacial period.

Comparatively few conifer species were among the colonizers, but pine of the kind now known as Scots made substantial woodlands in the rocky, infertile west of Ireland, both on the limestone pavement of the Burren region and in wet, acid soils of the uplands—an opportunism that makes its apparent general extinction as a native tree by about A.D. 900, after more than seven millennia, even more unexpected. Its biggest populations were on uplands such as the Nephin Mountains of Mayo, but the clearances of early farmers and the spread of blanket bog reduced these strongholds by about 4,000 years ago. The primal pines lasted longest among the raised-bog peatlands of the midlands, but dwindled away throughout historic times, their ultimate fate confused by the widespread planting of the species by colonial settlers in the late seventeenth and early eighteenth centuries.

Yew was also an early conifer, but one that seems to have flourished locally only when other tree species—pine or elm—ran into difficulty or decline. Its fossil pollen was recognized only recently, so that its true abundance in the past may be misjudged.

The expansion of common alder, lover of wet soil, at about 7,000 years ago, is significant mainly as a marker for climate change. With the flooding of the North Sea plain the British Isles were finally surrounded by water and enveloped in a warm, moist climate. Called the Climatic Optimum of the Atlantic period, it was the time of highest temperature since the last Ice Age. Initially, the dense cover of wildwood insulated the soil from heavier rainfall, but in Ireland the spread of alder on lakeshores and riverbanks marks the transition from high forest and open lakes to the beginnings of swampy surfaces and formation of bog.

"High forest" may suggest progression to a stable forest climax, but this concept is dismissed by Fraser Mitchell as "untenable even over time-scales of thousands of years." Ireland's postglacial forest history is full of changes driven by climate, competition between species, pathogenic decline, and human impact. Before the spread of alder, a mixed deciduous woodland probably covered four-fifths of Ireland. The midlands were dominated by woods of elm and hazel; oak thrived on the more acid soils of the south and northeast, while pine had colonized the more exposed acid soils of the west. The spread of alder led to a competitive reshuffle of species, especially on wet soils, followed by about 1,000 years of stability—the longest such period in the Holocene. Even the elm decline at 5,200 years ago (probably caused by a pathogen similar to the Dutch elm disease of today) did not dent this inertia unduly, and change did not dramatically increase until the Bronze Age, when the interplay of climate and human activity launched the development of western blanket bog.

ADDITIONS TO THE ARK

The original, rather sparsely furnished ark of Ireland's native postglacial fauna has been given many extra passengers through deliberate and accidental introductions: rabbit, gray squirrel *(Sciurus carolinensis)*, bank vole *(Clethrionomys glareolus)*, and American mink *(Mustela vison)* are certainties; the hedgehog a medieval probable, whether for food or its spiny skin (for carding wool), and the red squirrel *(Sciurus vulgaris)* possibly another, for its skin or as a pet.

But even today, Ireland has only twenty-eight species of wild mammal compared with fifty-five in Britain. The picture is distorted by Britain's own introductions (of, for example, several deer that are absent from Ireland), but there are significant vacancies that may bear on the land bridge question. Moles *(Talpa europaea)*, for example, would not have been tempted to burrow their way to Ireland through soggy glacial silt, or, indeed, to have ventured anywhere near frozen soil in the immediate postglacial period. Tundra-going field voles *(Microtus agrestis)*, on the other hand, could have crossed such a bridge but did not. Certainly the vole should have been ahead of the long-tailed wood-mouse *(Apodemus sylvaticus)*, which never travels north of Europe's tree line. Now common on both sides of the Irish Sea (but known as the "field mouse" in Ireland), the wood mouse was found, as fossil fragments, in the 8,000-year-old tufa deposits at Newlands Cross. Until then, despite a number of records from prehistoric bone caves, *Apodemus* was thought to have been introduced in fodder by the Vikings: the Newlands fossil teeth, from the same deposit as a flint flake from a Mesolithic hunting group, have pushed the same possibility back to an earlier age.

Some evidence for the early presence of Ireland's mammals comes from a small number of well-dated archaeological excavations, beginning with sites that record Mesolithic people in the northeast (Mount Sandel) and the marshy midlands (Lough Boora) just before 9,000 years ago. They had arrived in an island quite limited in animals worth hunting—lacking altogether the large elk *(Alces alces)*, aurochs *(Bos primigenius)*, and roe deer of Britain. From base camps beside rivers and lakes come plentiful bones of mountain hare *(Lepus timidus)* and far fewer of wild boar, along with those of migratory salmon *(Salmo salar)* and other fish and birds, but red deer is surprisingly missing from the large faunal samples at Mount Sandel and Lough Boora. Its bones appear in the cave record at about 12,000 years ago, in the late-glacial, but have not been recovered again until the Neolithic. It may have followed the giant Irish deer into extinction in the cold of the Nahanagan Stadial and have been reintroduced by humans. Finbar McCormick in a recent study, *Early Evidence for Wild Animals in Ireland*, suggested that such deliberate introductions of wild species "seems to have been a characteristic of prehistoric peoples in the British Isles" and points to the probable importation of red deer to Scotland's Outer Hebrides and Orkney Islands.

The brown bear and the wolf were among the island's early predators, and three more—fox, badger *(Meles meles)*, and wild cat *(Felis sylvestris)*—were present by the Neolithic, when people had begun to farm. The otter left no trace of itself until the Bronze Age, although probably well-capable of swimming across the North Channel of the Irish Sea. The absence of any earlier otter record may simply reflect the inadequacy of current archaeological evidence. Yet the Irish otter, like the Irish stoat and mountain hare, is accepted as an indigenous subspecies of the arctic form, noticeably darker than the same species in Britain, and this suggests an arrival much earlier than the Bronze Age.

A carnivore's need for prey raises particular problems for the Irish stoat, especially if it waited out the Midlandian ice in refuges somewhere to the south. What was it eating in the immediate postglacial period? There is no evidence that lemmings survived until then. An adult hare would have been very big as prey, and the Irish stoat is unusually small. There was, according to mammalogist Patrick Sleeman, a "prey gap," crucial for the stoat and to some extent for the later wild cat, pine marten, badger, and fox. Ireland's postglacial complement of carnivores was little short of Britain's: eight species, rather than ten, with only weasel *(Mustela nivalis)* and polecat missing. Yet lacking the vole (a leading prey for both these British mustelids), the only small mammal Ireland had for carnivores to eat was the distasteful pygmy shrew. The Mesolithic wood mouse of Newlands Cross may yet be recruited as a "native," especially if Sleeman's own discovery of a rare rodent flea on a modern wood mouse carries the significance he suggests for it. He finds it improbable that an uncommon nest flea should have entered Ireland on the small numbers of wood mice that would have been introduced by humans. Yet the history of the island's early colonization has obviously had to indulge all manner of improbabilities. A cautionary tale has been the discovery of a Barbary ape *(Maca sylvaus)*, in a late Bronze Age setting in County Armagh—evidence of the constant movement between Ireland and Europe in the trade of later prehistoric times.

Human introduction has been a complicating option for the history of many island mammals. On the large Mediterranean island of Corsica, for example, the animals are entirely of human introduction, most of the native species having been killed off soon after human settlement. Changing tastes in meat, changing needs for fur clothing, desire for new hunting quarry—any or all of these can confuse the origins of island mammals that might well be indigenous. Badger, otter, pine marten, red fox, red squirrel—even the wild cat—are all candidates for early introduction to Ireland for their skins (the first evidence for pine marten is, however, Bronze Age). To establish viable populations, as Sleeman has pointed out, would have needed repeated introductions and many trapping expeditions to Britain in small, primitive boats.

Whatever the Neolithic people's sense of Ireland's size and topography, they knew it was not a small island, and the amount of stubborn optimism neces-sary for repeated trapping, transportation, and release of (presumably rather cross) wild animals into dense forest does seem excessive. Finbar McCormick, however, is persuaded that "early people probably captured and tamed animals on a fairly regular basis" and noted that early Irish legal documents list a sur-prising range of "pet" mammals: fox, wolf, deer, otter, squirrel, stoat, pine marten, badger, and wild pig.

One way of helping to date the arrival of at least some of the controversial Irish mammal fauna will lie in comparing them genetically with populations in Britain. This could confirm, for example, the isolation of the Irish stoat over 40,000 years, or indicate that badger and otter were, indeed, late intro-ductions in the postglacial period. It could strengthen the arguments for a Scottish land bridge or a Welsh one—or for neither. Some initial research seems to show significant differences between the skulls of Irish, Scottish, and English badgers, otters, and stoats, and DNA fingerprinting of fossil material may refine this into something more conclusive.

SELECTED REFERENCES

Curtis, T. G. F., and H. N. McGough. 1988. *The Irish Red Data Book: Vascular Plants.* Dublin: The Stationery Office.

Edwards, K. J., and W. P. Warren. 1985. *The Quaternian History of Ireland.* London: Academic Press.

Fairley, J. 1984. *An Irish Beast Book.* Belfast: Blackstaff.

Feehan, J., and G. O'Donovan. 1996. *The Bogs of Ireland.* Dublin: University College.

Hayden, T., and R. Harrington. 2000. *Exploring Irish Mammals.* Dublin: Town House.

Mitchell, G. F., and M. Ryan. 1997. *Reading the Irish Landscape.* Rev. ed. Dublin: Town House and County House.

Pilcher, J., & Hall, V. 2001. *Flora Hibernica.* Cork: Collins Press.

Yalden, Derek. 1999. *The History of British Mammals.* London: Poyser.

—— *Chapter 5* ——

Climate as Theater

U ntil the dramatic revisions compelled by global warming, "climate" was a very definite term in books of regional natural history. Meteorological statistics collected over thirty or thirty-five years were thought adequate to fix the bounds of "normal" shifts in weather, and there was a strong assumption of constancy in the longer term.

Even within this reassuring matrix, the climate of Ireland has been celebrated for its changeability. Its oceanic location guarantees a generally equable temperature, while the constant grappling of polar and tropical air masses at this latitude has made for a striking alternation in weather regimes. The struggle between the moist, warm air of the Azores high-pressure cell to the south and the spiraling depressions of the Icelandic low to the north is the chief control on the pattern of weather in Ireland prevailing from month to month.

Day to day, however, and hour to hour, the passage of cloud and precipitation is what gives the Irish climate its often delightful sense of theater. Dominated by a west-southwest airflow, moisture gathered from the ocean is released in frequent showers or longer rain spells, and the chase of clouds across the island mediates sunshine into constant and subtle shifts of light. A cool humidity helps to keep the peatlands moist and promotes the abundant growth of ferns, bryophytes (mosses and liverworts), and lichens. Warmth delivered by the North Atlantic Drift lifts winter temperatures some 60°F (16°C) above the latitudinal average, and grass can grow in the southwest almost all year round.

Within the island's small land mass of 32,595 square miles (84,420 km^2), there is room for even more variability. Many of the mountainous areas of the west receive more than 63 inches (1,600 mm) of rain annually, rain falling on more than 250 days a year, whereas Dubliners get less than half that: drier "rain shadow" areas in the lee of mountains are a marked feature of Ireland's lowlands. There is a clear variation in hours of sunshine, with brighter days in the south and east of the island, and another in wind speeds that are often halved in strength by the friction of passage inland. In the markedly colder winter of Northern Ireland's interior, away from the influence of the Gulf Stream, the grass-growing season dwindles to 240 days. All these parameters have helped to shape Ireland's agriculture, restricting grain crops, in particular, to the drier, sunnier, and calmer counties of the southeast. Ireland's balmy summers, when they come, are created by blocking of the westerly flow by a series of anticyclones, or areas of high barometric pressure, stationary over continental Europe. In winter, similarly, the mild southwesterlies may meet a "blocking high" in northwest Europe that holds their warmth outside a cell of extreme cold.

The polar front that dominates Ireland's climate is a turbulent zone in which masses of cold and warm air interleave in opposite directions in the North Atlantic: the cold moving south, the warm moving north. The instability of the front causes depressions to form, tracking eastward across the Atlantic. Most of them pass to the northwest of Ireland, on a path between Iceland and Scotland, while others track to the south of the island. At this stage in their life cycle they are usually in a phase of maturity or decay, but depressions can suddenly deepen explosively in the eastern Atlantic, reinvigorated by a warmer sea-surface temperature—a phenomenon known as the "cyclonic bomb." This can bring particularly severe and damaging storms among the one hundred or so depressions that spin toward Ireland each year.

The most notorious historically was *Oiche na Gaoithe Móire*, The Night of the Big Wind, which struck Ireland on January 6, 1839. More than one hundred people died as hundreds of thousands of trees toppled, thatched roofs collapsed or caught fire, and floods surged over riverbanks to sweep away

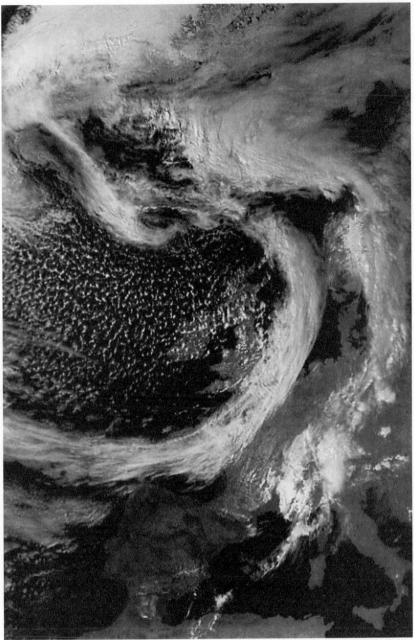

Satellite picture of a cold front bringing a band of cloud and rain, approaching Ireland from the North Atlantic.

homes and haystacks. It was a late catastrophe of the Little Ice Age, which afflicted Europe from about 1450 to 1850 and reached its peak in the frozen winter rivers of the late 1600s. Ireland shared in the shorter summers and extremely cold winters of that period. Intensely wet spells brought bog bursts; increased storminess blew out sand dunes to overwhelm settlements and farmland. Ships of the Spanish Armada, wrecked on west Irish coasts, encountered at least six storms in August and September of 1588, thought to be stronger than any experienced in these waters in recent decades.

GLOBAL WARMING AND CLIMATE CHANGE

Frequent and violent winter storms and flooding from intense rainfall are the chief apprehension in Ireland as global warming takes hold and more heat moves from the Equator to the poles. Between the 1970s and 1990s there was a fall in the number of moderate gales across the northern British Isles, while the number of damaging storms, with winds of between 60 and 80 miles per hour (100 and 130 kph) approximately doubled. On the Irish west coast, gusts of around 100 miles per hour (160 kph) have been experienced in several recent storms. In late December 1999, the worst Atlantic storms in memory swept into France, with gusts reaching 125 miles per hour (200 kph). The winds uprooted or broke in half 270 million trees in one of western Europe's most wooded countries, damaged ancient cathedral spires, and killed almost ninety people in France and neighboring countries.

Such events have brought fears that more hurricanes born in the Caribbean will survive to make landfall in western Europe. At the same time, in a seemingly perverse consequence of global warming, the graver prospect for the region could be the loss of winter warmth from the North Atlantic Drift and a possible plummeting of temperatures by 9°F (5°C) or more. The hypothesis is that melting arctic ice, diluting the saltiness of surface water in the North Atlantic, will stop the water sinking to the ocean floor, a circulation mechanism crucial to the two great "pumps," one in the Greenland Sea and the other in the Sea of Labrador, which draw warm water north from the Gulf of Mexico and send cold water back along the bed of the ocean. The first indications that this may be happening have emerged in ocean research off Scotland, the Faroes, and Norway.

The ecological implications of a relatively more benign scenario—warmer, drier summers; mild, wet, and windy winters—were explored for the Irish government by a scientific team in 1991. A projected doubling of carbon dioxide by 2030, and a rise of some 3.6°F (2°C) was expected to boost grass and tree growth and widen the range of Irish crops. Wild plants would grow more quickly and reach a larger size, and many would continue to grow throughout the winter. A greater biomass, both above and below ground, decomposing more slowly at higher carbon dioxide levels, would enrich the level of humus in the soil. Peatlands, however, could suffer great change if summer drought increases their rate of decay: there could be a gradual shift from a wet, sphagnum-dominated system to a drier heath of dwarf shrubs or grass.

For deciduous trees, the timing of bud burst and leaf fall depends more on length of day than temperature, so that milder and moister winters will have little direct effect on growth. But warm summers would lead to a more regular crop of acorns on Ireland's oaks. This would help to regenerate the island's remaining oakwoods, boost the breeding of small mammals such as the wood mouse, and the numbers of predators such as owls. More abundant insect food in the breeding season could enrich the island's birdsong with more migrant and resident warblers.

Some of the most profound ecological changes would be brought about by rising sea level and a more severe regime of storms. The rising wind speeds of the later twentieth century took a substantial toll of trees, both those blown down directly and the many felled or lopped at roadsides for fear of human injury. The distinctive vigor of Ireland's ivy *(Hedera helix)* often builds up a dense growth of evergreen leaves in the crown of wayside trees such as ash and sycamore, making them unusually vulnerable to storm gusts. In the western half of the island, large, mature broadleaf trees may come to survive only in sheltered areas.

The combination of rising sea level and storm surges seems bound to change both the topography and nature of many "soft" (nonrocky) Irish shores and estuaries. The first effects are likely to appear in the southern half of the island, which has been tilting for many millennia (see Chapter 2). The area of land at risk from flooding and erosion is believed to be relatively undramatic (247,100 a, or 100,000 ha, by 2030 in the 1991 scenario) and there is broad agreement among coastal engineers that accepting a natural, self-adjusting coastline, rather than building big seawalls and other defenses, makes ecological, as well as economic, sense.

On the high-energy coasts of Ireland, erosion of beaches and islands could be balanced by a piling-up of sediment in the inner reaches of estuaries and bays. Many existing wetlands and coastal lagoons will probably be lost and others created. New salt marshes, of the kind reclaimed for farmland in the nineteenth century, could be fed with estuarine silt by heavier winter rainfall.

In the longer term, the balance of habitats lost and recreated should leave the diversity of wildlife at least no worse off. However, the loss of long-established intertidal habitats and mudflats, especially on the east and south coasts, would be seriously disruptive for winter wildfowl and wader populations. In Dublin Bay, beside the capital city, the loss of North Bull Island would rob the birds of an exceptional refuge and staging post.

Given the significance of wind speed in the future climate of the island, it seems proper to note the Irish origin of two notable figures in the meteorology of wind. Sir Francis Beaufort, creator of the Beaufort Scale of Wind Force, was born in Navan, County Meath, in 1774 to a family of French Huguenot origin. His great achievement, in 1803, was to offer a set of criteria of wind speed that all sailors could recognize and agree on. The "Force 10" of a modern storm may no longer be framed in terms of its effect on a British naval "man-o'-war," but the description of the behavior of wind and sea is still essentially Beaufort's.

The means of precise measurement of wind in miles per hour was the inspiration of Thomas Romney Robinson, born in Dublin in 1792. He was a clergyman-scientist who became director of the Armagh Observatory. In 1846, he completed the first working model of the "rotating cup anemometer," with two whirling cups. Today's instrument has four: like Beaufort's wind scale, a great idea could only be perfected.

VOLCANIC ASH: THE LONG SHADOW

For all its temperate norms, Ireland's postglacial climate has provided scientists with strikingly precise clues to catastrophic events, preserved within the island's peatlands. They hold layers of tephra—ash blasted into the air in volcanic eruptions and eventually falling to Earth as glassy shards less than 100 microns in diameter. Valerie Hall and Jonathan Pilcher, palaeoecologists at Queen's University, Belfast, have found that much of the tephra has come from volcanoes in Iceland (notably Hekla in the south of the island) some 621 miles (1,000 km) to the northwest of Ireland. But uncertainty remains about the origin of tephra from a violent eruption at about 1159 B.C., which is thought to have created "nuclear winter" conditions across much of the northern hemisphere, making the uplands of Ireland and Britain too cold and wet for farming and forcing the abandonment of some late Bronze Age villages.

Nonetheless, precise dating of Irish tephra, by geochemical analysis that ties the ash layers to particular Icelandic volcanoes and their known eruption dates (for example, that of Oraefajokull in 1362), also pinpoints the age of fossil pollen layers directly below and above the tephra in cores taken from the bogs. Thus, the "historic tephrachronology" pioneered by Hall and Pilcher is another new tool in reconstructing changes in the Irish landscape. As an ecological dating system, it meshes with such indicators as peaks of acidity in the Greenland ice cores and extremely narrow growth rings in Irish subfossil (bog) oaks.

TREE RINGS AS HISTORY

The precise dating of tree rings as a corrective calibration table for radiocarbon dates had its origins in America, where researchers in the mid-twentieth century overlapped the patterns of rings from living bristlecone pines with those of surviving ancient stumps to produce a dendrochronology stretching back to beyond 6000 B.C. Archaeologists in Europe, however, were unhappy with a calibration using American trees. The Queen's University of Belfast acquired a radiocarbon laboratory in the mid-1960s, a time when several thousand subfossil bog oaks were being dug up in drainage schemes and highway construction. With this potential for extending ring samples back into prehistory, the original project was for a 6,000-year chronology that would recalibrate the radiocarbon timescale for the Old World. By linking modern trees to the rings in historic building oak samples, the record soon stretched back to A.D. 1380. Some gaps were difficult—the period of the Black Death (1348–1370), for example, brought a halt to new construction with oak. Some Early Christian timbers were recovered by archaeologists from the artificial lake islands (called crannogs) in Fermanagh. The prehistoric samples were from oaks that had colonized the surfaces of peat bogs for many centuries at a time. These ancient trees grew slowly, with hundreds of tightly packed rings (in a few trees, up to almost 500 years), so that long, well-replicated site chronologies were possible. The final 7,272-year record was completed by 1982, and Belfast's high-precision radiocarbon calibration curve was published in 1986. As Queen's University's Mike Baillie could boast in the journal *Technology Ire-*

land, it offers Ireland "an absolute timescale against which historians, archaeologists and environmentalists can place their deductions."

A horizontal slice of oak-tree trunk, polished to a shine, can be a thing of beauty—also of some awe when the annual growth rings spreading out from the core count the passing of many human lifetimes. Over the centuries, a few years stand out sharply in every Irish oak tree growing at the time, creating especially wide or narrow rings in response to climatic extremes, but long periods pass in which the variation lacks such landmarks as years "without a summer." Nonetheless, a sequence of rings is as reliably unique to its period as the bar-code of a supermarket purchase.

Their evidence for dust veils and their impact on climate and vegetation have also produced some dramatic conjectures. In the period A.D. 536 to 545, Irish oaks, Sierra Nevada foxtail pines *(Pinus balfouriana),* and Fitzroya conifers *(Fitzroya cupressoides)* from Chile show years of extreme cold—the only years in which rings narrowed from severe weather match up across the three. There was no volcanic eruption to account for such global chilling, and Baillie together with some astrophysicists have proposed a catastrophic encounter with cometary debris, an event recorded not only in tree rings but in Britain's Arthurian myths and Celtic folktales.

It is the striking, hour-to-hour variety of Ireland's weather, its miniature dramas of sun and shadow or sudden wind, that gives one an appetite for such longer views of climate. Despite the mild and temperate tenor of most of its days, almost any variation becomes at least imaginable. Thus Irish weather insists on a place in any reflection on the human shaping of the island.

SELECTED REFERENCES

Baillie, M. 1998. *Exodus to Arthur.* London: Batsford.

Hulme, M., and E. Barrow, eds. 1997. *Climates of the British Isles: Present, Past, and Future.* London: Routledge.

McWilliams, B. E., ed. 1991. *Climate Change: Studies on the Implications for Ireland.* Dublin: Department of the Environment.

Wheeler, D., and J. Mayes, eds. 1997. *Regional Climates of the British Isles.* London: Routledge.

— *Chapter 6* —

The Well-Worn Island

Most of Ireland's most beautiful landscapes reflect, in their ecological history, a combination of human wear and tear and a relentlessly rainy climate. The bare gleam of limestone in the Burren or a thrilling sweep of moorland in Mayo can call up, along with a fine aesthetic reverie, thoughts of deforestation, erosion, overgrazing, the impoverishment of soil. Even at its wildest corners, the island's surface is a palimpsest, worked over and over by farmers and their stock. Much of the small-farm west and south still carries the mark of hands and spades: of digging, of piling one rock on another. In this often untidy regulation of nature there is a saving awareness of human scale, of people intimately at work in a small island.

Among the many postglacial enigmas is the true date of human arrival. In the first late-glacial spell of warmth, when deer in several sizes were roaming the island, the grassy landscape was entirely empty of hunters—or was it? Perhaps human traces from that time have been lost beneath rising sea levels, the growth of peat, and silting up of river valleys. No Pleistocene human remains have been found in the "bone caves" and no implements anywhere that might be diagnostic. Despite all the caveats, Peter Woodman, in his report on the Irish Quaternary Fauna Project, holds that "the discovery of even a very limited Paleolithic occupation in Ireland would be more in keeping with the rest of the mammalian presence in Ireland during the Pleistocene."

The earliest radiocarbon dates for Mesolithic people are taken from the finds of charcoal in the hearths of a hunting base camp at Mount Sandel, beside the River Bann in Antrim, and on a storm beach at an ancient lake, Lough Boora, in the very heart of the midlands. They belong to the period between 8,400 and 9,000 years ago, some five millennia after the onset of postglacial warming.

Few animal bones were recovered at either camp. It was clear at Mount Sandel that Ireland's first settlers were fisher-gatherers rather than hunter-gatherers, and relied greatly on salmon and eels *(Anguilla anguilla)* to supplement small game and the occasional wild pig. The many Mesolithic kitchen-middens discovered on the coasts confirm a heavy dependence on fish and shellfish: cockles *(Ceratoderma edule)*, limpets *(Patella vulgata)*, and periwinkles *(Littorina littorea)*. The hazelnut harvest was important, too, as a portable and concentrated food that would keep indefinitely. Hazel shells littered the Mesolithic firesides, together with the fruits of white and yellow waterlilies *(Nuphar alba* and *Nuphar lutea)*, roasted in the embers like sunflower seeds.

It is appealing to picture the first Stone Age adventurers crossing in small, hide-covered boats to Ireland at the narrowest neck of water: the deep, dark channel between Scotland and the distant headlands of Antrim. Some of them may have landed literally beside the greatest treasure they could imagine: whole beaches of flints, eroded from the chalk cliffs, from which to chip razor-edged flakes for arrowheads, knives, and scrapers—the "microliths" that mark their culture. Archaeologist Michael Ryan in *Reading the Irish Landscape* has pointed to comparable microliths and axes in Mesolithic sites in Denmark and at sites stretching westward to the Pennines in Britain and the Isle of Man: "We cannot . . . overlook Denmark in our search for the original home of the first men to reach Ireland."

Historian Liam De Paor in *The Peoples of Ireland* also pointed to the affinities, in their surviving equipment and traces, with Mesolithic material from Atlantic Europe. He sees a Mesolithic "stock pot" of a population, topped up to some tens of thousands by arrivals from diverse origins, and forming the basic genetic stock of the Irish people.

The first settlers, arriving near the temperature peak of the so-called Climatic Optimum (4,000 to 8,000 years ago), found Ireland's woodlands at the peak of their development, with oaks towering up, unbranched, to 98 feet (30 m). In Britain there was widespread burning of forests, both to drive game and to open upland clearings to attract herds of red deer, but in Ireland the Mesolithic picture is of a largely coastal life, with little activity in the uplands, and forest clearance only immediately around settlements. The charcoal found so widely on soil beneath the upland bogs belongs mostly to later phases of woodland clearance and burning to grow fresh grass and heather for grazing.

The traditional view that Mesolithic hunter-gatherers were swamped by the culture of immigrant Neolithic farmers has been losing much of its certainty. Radiocarbon dates have pushed back the tomb-building Neolithic tradition: the passage-grave cemetery at Carrowmore, County Sligo, for example, is now dated to the turn of the fourth millennium B.C., which makes it the oldest in Europe. A domesticated cattle bone from a Late Mesolithic site at Ferriter's Cove in the Dingle Peninsula has been dated to before 4300 B.C. So it seems increasingly likely that organized food production evolved directly, if gradually, from the island's native hunter-gatherer communities and not from any new wave of settlers. It is even possible that sea-faring contacts brought food production directly to Ireland from continental Europe and then across the Irish Sea to western Britain.

Contact between the two islands would have presented no problem; mountain peaks are visible from both shores. The Antrim coast is only 15 miles (24 km) from the Mull of Kintyre; from Down to Galloway is only 25 miles (40 km); and the Isle of Man is roughly 37 miles (60 km) from either coast.

A coastal people would have soon learned the sea lore, tides, and currents of the Irish Sea: that the tides from the Atlantic enter it at both ends and retreat the same way. They would have known that the two tides meet and disarm each other peacefully just south of the Isle of Man.

THE EARLY FARMERS

The story of Ireland's early forest clearance has also been changing. In traditional models of prehistoric society, farming began with a shifting, slash-and-burn agriculture—the schoolbook image of the tribal group making its small clearing in the woods. But mapping of prehistoric landscapes preserved under blanket bog has given a startling new picture of the scale and cohesion of some Neolithic farming. At Céide (kay-jeh), on the cliffs of north Mayo, probing of the hillside peat, often 13 feet (4 m) deep, has revealed an extensive network of fields with long, parallel stone walls — an ancient, settled cattle-farming landscape strikingly similar to the surrounding small-farm countryside of today.

The Céide Fields have been mapped across more than 2,500 acres (1,000 ha) of windswept bog as the most extensive Stone Age monument in Europe, unique in the record of Neolithic farming. In other forest clearances of the time, boundaries are irregular and follow the contours of the land. Only at Céide are the walls so long and straight — laid out, very probably, to a coordinated plan within a few decades. The pollen record shows a sharp decline in tree pollen at about 4,900 years ago and the rise in the pollens of grasses and the herbs of open land: plantain *(Plantago)*, buttercup *(Ranunculus)*, and clover *(Trifolium);* there is also a level of charcoal that may mark forest clearance by fire. The site's chief archaeologist, Seamas Caulfield, believes that some quarter of a million pine trees may have been destroyed by burning, ring-barking (cutting a ring around the bark of the trunk), or felling with stone axes.

The axhead of polished stone was a prized implement in the tool kit of the first Irish farmers, often chipped into shape from a bar of rock quarried at Tievebulliagh, a hillside above the coast of Antrim, or at Brockley, further north on the island of Rathlin. At both outcrops, Jurassic clay was baked by a subsequent lava flow to form porcellanite, a rock as dense as china yet cleavable to a fine, tough edge. The quarries supported a trade in roughly hewn blanks and finished axes that extended through Ireland and Scotland and reached as far as southeast England. The finished axhead was often D-shaped in cross-section and mounted in a curving haft, an ideal design for tree-felling.

Only a single round house has been excavated at Céide, so little is known of the dwelling settlements of the people who laid out the walled fields. But the regular discoveries of substantial Neolithic houses elsewhere in Ireland and their relationship to a cultural landscape of tombs and ritual sites suggest a far more settled and local pattern of living than existed in neighboring Britain.

Céide's highly organized landscape may have been repeated elsewhere in Neolithic Ireland. On the east coast, in the valley of the River Boyne, the massive circular tombs of Newgrange, Knowth, and Dowth are thought to have been the focus of a cleared and enclosed landscape of quite dense population, farming grains, and livestock. In northern Ireland, the forest clearances at several Neolithic sites lasted long enough — 700 years or more — to suggest yet more walled farm landscapes at a remarkably early time. But it is in those areas of good land and the longest farming record that clear evidence has been lost.

In other parts of Ireland, the pattern of Neolithic farming seems to have been very different. In Connemara, for example, where much of the landscape is, like that of Céide, sealed under bog, there is no sign of similar field systems or of fire being used in forest clearances at that time. The pollen record in lake sediments suggests small clearances in dense, tall forest, maintained for five or six generations. The first grain pollen, at 3850 B.C., is as early as any found in Ireland and it precedes, by some decades, the local decline of elm trees.

The apparently simultaneous destruction of elm across much of northwest Europe was first credited to climate change, then to the advent of Neolithic farming, either through stripping of foliage to feed domestic herds or because elms, growing on good land, were felled selectively. But radiocarbon dates for the elm pollen decline now range over nearly 1,000 years between Ireland and Scotland alone and are found even in areas innocent of human settlement. The Neolithic clearances and fodder practices may have encouraged the spread of *Scolytus scolytus*, the beetle that, in recent times, transmitted the fungus of Dutch elm disease throughout Britain and Ireland.

Dramatic and mysterious declines also hit the human populations of prehistory. The people of the Céide fields, for example, seem to have flourished for 500 years, then dwindled away within fifty. The Connemara farmers endured for some 150 years and then vanished from the pollen record for seven centuries while yew and ash crowded into their clearings. Indeed, the late Neolithic saw a remarkable regeneration of forest cover, including the elms, which lasted in the southwest of the island for almost a millennium. In the late Bronze Age and the Iron Age, human efforts to penetrate and colonize the central lowlands seem to have ended in retreat, and there was another resurgence of forest, this time a secondary growth with smaller trees and shrubs.

As a cooler, wetter climate promoted the spread of bog, cultivation gave way to a pastoralism in which cattle ate the leaves of trees as readily as grass. It was often supposed that the spread of blanket bog must have forced the abandonment of some of the earliest western farming settlements, but cause and effect were not so neatly meshed. The Neolithic people of Céide, for example, departed some two centuries before the bog engulfed their land: in between, a warm, dry episode even let pine trees grow on the surface of existing peatland.

But the later spread of blanket bog across vast areas of the west and the up-lands was certainly assisted and accelerated by the impact of human activity. The fact that bog began to form in different areas at widely different times strongly suggests that climate change alone was not the cause. Forest clear-ances exposed soil to the rain in quite a new way, leaching out its mineral nu-trients. Heather took over the new, open habitats and acids released in its decay helped to create an impermeable, iron-rich layer underground. This hardpan often played a big part in later waterlogging of the soil. But quite as significant was the regular burning of vegetation to stimulate regrowth of heather and grass. Fine particles of soot, washed down into the earth, clogged its drainage pores and waterlogged the surface, inviting in the first clumps of rushes and sphagnum moss (*Sphagnum* spp.). From Antrim to Connemara, from the Sperrins in Tyrone to Kerry, the endless fires of the early farmers have left a brown-black, greasy layer, the *"mor,"* trapped between the mineral soil and the base of the bog. It is rich in seeds of rushes and particles of char-coal—a tell-tale stratum, like the "iron pan" that so often runs through the impoverished soil beneath.

In Connemara the main phase of blanket bog development was from about 2500 B.C. to 1000 B.C., and local investigations suggest the events that helped it along. On slopes that now fall within the Connemara National Park, oakwoods rich in mosses and liverworts were cleared in the Bronze Age, at 1300 B.C., and the heathy vegetation that followed was regularly fired to stimulate the growth of grass. Soot washed down into the soil and stopped the growth of woody plants that would have absorbed water. Within a century or two, sphagnum mosses invaded the waterlogged ground and bog began to build.

The early degradation of the Burren's limestone hills was an almost per-versely opposite process, in which deforestation and overgrazing literally washed the soil away into crevices in the bedrock. Here, too, much of the first wave of erosion and loss seems to have occurred over quite a short period in the late Bronze Age (see Chapter 8).

The population of Ireland at the close of the Neolithic period has been put at between 100,000 and 200,000 people, widely spread across the island but most of them living in small communities on high ground above the wooded valleys. These may have gathered up a few extended families, with the women beginning to cultivate in small gardenlike fields near their settlements. Some of their houses were rectangular and of wooden-post construction, supporting

The setting sun appears to roll down the flank of
Croagh Patrick on August 24, as seen from a
prehistorically inscribed stone near Westport,
County Mayo.

rafters and thatch. The excavated outline of one can be seen at Ballyglass near
Ballycastle in County Mayo, and a reproduction, walled in hazel wattle, stands
in the Ulster History Park at Omagh in County Tyrone. It is a house design
that, in basic form and living space, lingered into modern times in rural Ireland. De Paor commented that up to 100 to 200 years ago, the working lives
of large numbers of people were not very different from those of their distant
Neolithic ancestors, and many were to live at an even lower level of comfort.
In early Victorian Ireland, he suggested, hundreds of thousands of people
were actually worse off than their forebears of 5,000 years before.

If the remains of Neolithic dwellings are now traced only in relief on hillsides and grassy fields, the mysterious standing stones, singly and in various
formations, are a fairly common punctuation of the Irish landscape. Some
offer topographical alignments or keep equinoctial appointments with sun or
moon. At Boheh near Westport in County Mayo, for example, on August 24,
a huge cup-marked rock offers a view of the setting sun seeming to "roll"
down the western scarp of Croagh Patrick, a holy mountain since pagan
times. But most of these monuments, at this remove, are quite inscrutable.
There are at least 240 stone circles in Ireland, with particular concentrations
in west Cork and Kerry and around the Sperrin Mountains of Ulster. These,
with the great Neolithic tombs and ritual centers, command continuing reflection on the Ireland of prehistory.

TOMBS AND RITUALS

Megalithic tombs number about 1,500 and come in two distinct forms: large, round, passage tombs that are chambered mounds, most famously found in the River Boyne Valley of County Meath and in County Sligo, and long, rectangular court and portal tombs, some up to 197 feet (60 m) long, numerous in a wide belt of countryside from north Connacht to east Ulster. They are often clustered into "cemeteries" on hilltops, but the best known complex of passage tombs sits in a valley in the lusher landscape of Leinster.

Brú na Bóinne (Bend of the Boyne) in County Meath is perhaps the most dramatic prehistoric landscape in Ireland. On a ridge within a loop in the river is a collection of some forty passage tombs, among which are the three great megalithic mounds: Knowth, Newgrange, and Dowth. All stand directly on knolls of rock, and the land around them is deeply mantled by glacial gravels rich in clay and lime. Fertile as the soil is today, it may have been even richer in Neolithic times. This verdant "island" within the loop of the river drew generations of farmers to the site and provided them with the wealth to start building their great monuments at around 3300 B.C.—a communal effort, said de Paor, "comparable to that involved in the erection of Chartres and Rheims in the extraordinary age of cathedral-building."

Dúchas, The Heritage Service

Decorated kerbstone at the reconstructed entrance to the passage tomb at Newgrange in the Boyne Valley.

The mound at Newgrange is 43 feet (13 m) high and covers a little more than an acre (0.5 ha). Like Knowth and Dowth it is intensely decorated, the great stone slabs of its kerbs and passages swirling with engraved designs that some have seen as images induced by drugs or by a mental state heightened by shamanistic ritual. But Newgrange has another astonishing feature that marries the magical with a sophisticated understanding of the natural world. When the sun rises in the winter solstice on December 21, it shines through a stone aperture built in the roof and reaches down the passage to strike the end of the tomb—a golden signal of the arrival of winter's shortest day. The mound at Knowth, on the other hand, with its exceptionally long passages, suggests in some interpretations a significance bound up with calendrical observations of the moon. Indeed, the linear "decorations" of the slabs of all three mounds may be linked to their use for astronomical alignment.

THE ADVENT OF METALS

In the later Neolithic a new type of tomb was adopted in Ireland: the wedge-shaped structure of slabs given dramatic profile in photographs of dolmens on the Burren, such as that at Poulnabrone. Wedge-shaped tombs are widespread along the western seaboard, and in Cork and Kerry they are found near the region's distinctive stone circles and also near the copper deposits extensively mined in Munster in the Earlier Bronze Age. A score of drinking cups of "Beaker" pottery (a distinctive style that had spread across Europe), used at the campsite of copper miners at Ross Island, at Lough Leane near Killarney, are striking evidence of the transition from the age of stone to metallurgy. The Ross Island mines, in a stratum of Lower Carboniferous limestone, have been dated to between 2400 and 2000 B.C., which sets them among the oldest in Europe. Others are well-preserved on Mount Gabriel, above Schull in County Cork, where more than thirty shallow mines tunnel into the sandstone. The ore was fractured by setting fires to break up the rock and smelted in pit furnaces, so that the need for fuel was substantial. Over the two centuries of the Mount Gabriel mining, some 15,000 tons of timber may have been hauled up the hillside.

The miners were supported by a settled community, and farming continues to be the principal human activity in the Bronze Age. We can mark such things as a rise in pig breeding, the establishment of oats as a grain crop suited to the Irish rain, and the introduction of the horse as an agricultural and tribal status symbol. But the story of human impact on nature and landscape through the Bronze and Iron Ages in Ireland is as erratic as the peaks and troughs of the pollen diagrams that recorded it. The transient clearances of the typical Neolithic farmers, regularly reclaimed by trees and bushes, are echoed in later centuries by cyclical ebbs and flows of farming that speak of repeated, often lengthy, setbacks. The close of the Bronze Age and the start of the Iron Age that followed were marked by regenerating woodland, the spread of bogs, and a swing from cultivation to pastoralism. This was the period in which a change of climate to cooler, wetter conditions found many tracts of soil degraded and misused. The farmers' changing fortunes, as historical geographer Fred Aalen concluded in *Man and the Landscape in Ireland* (1978), reflected their difficulties "in achieving a stable adjustment to the unstable Irish environment." Sometime between 500 B.C. and A.D. 300 the peoples known as the Celts arrived in Ireland. It is not known if an invasion took place, although this could be inferred from the number of hoards of gold and bronze artifacts deposited around 600 to 500 B.C., but these could be explained just as easily by local turmoils. At that time, however, people called Celts, who spoke Indo European languages, had settled widely in central and western Europe and could well have made their way to Ireland.

The legends of the time describe them as heroic, quarrelsome, boastful barbarians who were formidable in war, drank copiously, and feared nothing. They were described as tall, blond, and blue-eyed—but this also described the barbarians from northern Europe who were feared by the Mediterranean peoples. De Paor has maintained that there was no Celtic race and that "Celtic" societies in Europe were probably formed by the intrusion of Celtic-speaking war bands, or perhaps migrating tribes, into the more numerous indigenous or longer settled populations of Europe. However, by A.D. 600 Irish, a Celtic language, was dominant throughout the country. In Europe the Celts were overtaken by the Romans, but although the latter invaded the island of Britain, they did not annex Ireland; Celtic society continued there for the first millennium A.D.

The Celts are credited by some with launching the Iron Age in Ireland at around 200 B.C.; others say that the technology of iron evolved from that of bronze. About that time Ireland entered a kind of dark age that lasted until around A.D. 250. Paleobotanical evidence suggests that there was a decline in farming and climate and probably also in population. It was a tribal and rural society, and people probably lived partly in independent, single farmsteads and partly in nomadic herding of their cattle, pigs, sheep, and goats. But the Iron Age deepened the human imprint on the landscape in its fortifications, ritual sites, and linear earthworks.

Hillforts—imposing circular enclosures, smoothly ramparted in stone— were already being built in the Late Bronze Age as relative power and status began to structure society. More than sixty forts are known to archaeology, many of them grouped in a pattern of strategic locations spanning the island from Counties Wicklow to Clare. Aerial photography is still adding to the palimpsest of the fainter hilltop contours, but some forts are dramatically intact, notably Dun Aengus, perched at the cliff edge on Inishmore in the Aran Islands, and Grianán Ailleach on a hilltop at the Inishowen Peninsula in County Donegal.

At the apex of the settlement hierarchy were sites of great ritual and political or "royal" significance, such as Raith na Righ at Tara, Eamhain Macha (Navan Fort) near Armagh city, and Knockaulin near Kilcullen in County Kildare. The role of Eamhain Macha as a powerbase is reinforced by linear earthworks that form a loose defensive boundary to Ulster, from Monaghan to Donegal. The most famous is the Black Pig's Dyke in County Monaghan,

Dúchas, The Heritage Service

Dún Aengus, a promontory fort on the western shore of Inishmore, Aran Islands.

radiocarbon-dated to between 500 and 100 B.C. In Armagh, a stretch called the Dorsey earthwork is dated by an oak post that was cut down around 95 B.C., essentially the same date as the massive ceremonial timber structure built at Eamhain Macha.

The Romans invaded Britain in the first century A.D. and brought a new civilization and political organization to the southern part of that island; they also developed cities, towns, roads, and industries. They did not annex the neighboring island of Hibernia, so that Ireland was largely bypassed by these enormous changes, but increasing contact with Britain and its technologies prompted a dramatic upsurge in Irish farming from A.D. 250. The new coulter plow was shod with iron and drew an iron knife to cut through the matted roots of old sward; grains soared in the pollen profile. Forest and hazel scrub that had crept in over the Iron Age pastures were hacked back again. Indeed, Frank Mitchell in *Reading the Irish Landscape* saw this as the point at which destruction of major woodlands became "a slow and unreversed process."

EARLY CHRISTIAN AND MEDIEVAL FARMSTEADS

In early Christian Ireland there were hundreds of small chiefdoms or *tuath*, each with a chief called a *rí* (which translates as "king") and with varying degrees of independence. They made changing alliances, forming generally into four main regional groups or provinces (knowledge of which owes as much to mythology as to any ascertainable history). Conflicts between chiefdoms, cattle raiding, and competition for well-drained land saw development of farmsteads that had at least the potential for defense. These were circular raths, or ring forts, with earthen banks, about 66 feet (20 m) in diameter, and crannogs in the lakes. The sheer number of raths—more than 45,000 (perhaps many more, as aerial surveys add to the total)—has made them the most widespread and familiar of ancient remains in the Irish countryside. That so many remained intact into modern times was largely the result of a rural culture of the supernatural that placed them in a fairy underworld (a belief encouraged by the frequent underground passages or souterrains). In a less imaginative era, thousands of unexcavated raths have been destroyed in field enlargement and building development in defiance of conservation.

The enclosures were elaborated according to status: a plain (but substantial) farmer had one circular bank with its outer ditch and entrance causeway; a nobleman had two outer rings, built of stone; and a king had three (the stone-walled structures are often known as cashels). Many of those excavated have held remains of timber buildings on the flat ground inside, and Ireland's exceptional development of dendrochronology has allowed most oak samples to be dated to within a year. This evidence, together with radiocarbon dating, shows that most of the enclosures were occupied and probably constructed during the 300 years of the seventh to the ninth centuries.

The crannog farmsteads, quite clearly defensive, seem to have been constructed in two intense phases within the same period, notably in small, shallow lakes among the drumlins north of the central plain. Their typical foundation was a platform of rocks and brushwood, held within a circular palisade of wooden pilings driven vertically into the lake bed (there are plausible reconstructions in theme parks at Craggaunowen in County Clare and near Omagh in County Tyrone). Some had causeways to the shore, often concealed underwater, and most had wooden dug-out boats. The crannogs were substantial undertakings—they were occupied, sometimes, for centuries—and protected the homes of a wealthy elite whose cattle were penned and guarded ashore.

The picture of the Irish landscape in early historic and medieval times is being brought more sharply into focus through pollen studies given firm dates by the chronology of layers of volcanic ash (tephra) discovered in the Irish bogs (see Chapter 5). They conjure up an early Christian countryside rich in variety, with cattle and sheep grazing through an intimate mix of rough grasses and scrub that resulted from a burst of woodland clearance about the fourth century A.D. It was this herb-rich grassland, with ample browsing on alder, willow, and hazel, that supported a sophisticated dairying economy from the late fifth century.

Dairy cattle and their well-being dominated the economics and culture of early Ireland in a manner unimaginable today. "Virtually everyone in that society was preoccupied with cows," wrote the National Museum folklorist A. T. Lucas, in *Cattle in Ancient Ireland.* "They were in the mental foreground of king and peasant, cleric and layman, warrior and poet, young and old, men and women, and they touched the lives of everyone from sunrise to sunset and from birth to death." Cattle-raiding inspires the epic tales of Middle Irish (roughly two thousand years ago), such as the mythological epic, *Táin Bó Cúailnge* (The Cattle Raid of Cooley), and cows inhabit fairy tales and mythology for a thousand years and more. A respectable calculation of the early "national herd," quoted by Lucas, is that in 1272 there were 1,650,000 cattle in Ireland (about one fifth of today's total), each needing 4.5 acres for maintenance. Just as today, intensive concentrations produced erosion around ring forts, and overgrazing of the thinly turfed Burren saw a whole landscape worn to the rock (see Chapter 8).

As with other mammals, domestication of the aurochs (probably on continental Europe around 6000 B.C.) produced a marked reduction in size, and the cattle bones recovered from Ireland's Neolithic sites suggest small, short-horned animals, similar in size and build to today's black Kerry cattle, hardy and thrifty animals maintained by conservationists as an old Irish breed. Even in pre-Norman Ireland, however, the aristocracy were breeding competitively, and cattle bones at royal sites tended to be bigger than average.

Most of the early cattle, however, trod more lightly on the earth. When the main spring growth of grass had been eaten, they were taken to the hills or other rough land in the care of young people or professional herders, who lived with them in makeshift shelters and watched keenly for wolves. This practice, called in Ireland "booleying," continued in the west of Ireland until

By kind permission of An Post, the Irish Postal Service

An Irish postage stamp celebrating the native Kerry Cow.

the early nineteenth century and is still the tradition in some upland areas of countries such as Norway, Switzerland, and Spain. As Fergus Kelly noted in his *Early Irish Farming*, it is "in harmony with the animal's natural tendency to migrate seasonally in search of fresh pasture." In autumn the cattle returned to eat the tall grass left unmown all summer and the stubbles of harvested grains. In medieval Europe, hay-making was standard practice, but the mildness of Irish winters allowed sufficient growth in grass to feed cows kept at low density, together with woodland browsing on evergreen holly and fallen leaves. Archaeology finds no evidence of scythe, hayrake, pitchfork, or mower's anvil before the coming of the Normans.

If all 45,000 ring forts were occupied at the one time during the seventh and eighth centuries, a secular population of around 450,000 would seem a conservative guess and one counting only the wealthier people. Add in the inhabitants of church settlements and monasteries and the core figure rises to 500,000. The big ecclesiastical enclosures—some six hundred of them, spread most densely in the central plain—grew up after the conversion to Christianity that began in the fifth century. They were the proto-urban centers of Early Christian Ireland, and the eighth-century population at the monastic settlement of Armagh was big enough to be described as a town. The early bishops attempted to apply Roman organization to the Irish Church, and in the absence of cities they adopted the *tuath*, or chiefdom, as the basis of bishoprics— mostly of no more than 300 or 400 square miles (800 to 1,000 km²).

Many churches and monasteries were founded on waste and marginal land, and while ring forts kept to the high ground, monasteries commonly sprang up on lowland sites near rivers and routeways: Clonmacnoise, built on bog beside the River Shannon, is a prime example. Others, with Derry *(doire)* in their names, took over forest clearances or woodland islands in the middle of the bogs. Remote settlements such as the sixth-century monastery on the Great Skellig off the Kerry coast were marginal to the powerhouse of early Irish Christianity and represented a minor strain of religious asceticism. Great areas of land were ultimately owned by the Church, and the most impressive and monumental of monastic and church buildings, mostly surviving as ruins, were built in the twelfth and thirteenth centuries. The abbeys of Holy Cross, Graiguenamanagh, and Ballintubber are now restored to use, and Mellifont, Jerpoint Abbey, and Clonfert remain mute witnesses of conflicts and interactions between religion and politics during the last millennium.

The arrival of the Vikings, momentous for Ireland's culture and social geography, had little ecological impact. The Scandinavians first came as raiders to Ireland's northern and western coasts at the end of the eighth century and continued their attacks for several decades. These were the true Vikings; the word comes from the old Norse *víkingr*, which means sea robber and should not properly apply to the traders and colonizers who came later. Larger Scandinavian expeditions arrived from A.D. 830 onward and established defended coastal sites along the east coast from which they made forays inland. Before the end of the century these centers were destroyed and the settlers driven out; the one at Dublin was abandoned at the beginning of the tenth century.

A more serious Scandinavian settlement took place in the first quarter of the tenth century when the northern settlers, now called Ostman, established Ireland's first urban centers: Dublin, Wexford, Waterford, Cork, and Limerick. They had their most extensive settlement around Dublin from where they spread out to cover the modern county of Dublin and part of Wicklow. The legacy of the occupation remains in the many place names of Scandinavian origin: Skerries, Lambay, Howth, Dalkey, Wicklow, and Arklow, to name a few. From Dublin the Norsemen controlled the Irish Sea and from there and the other port settlements they established international trading links.

Excavation of Viking Dublin and study of the seeds in its refuse pits and privies has added a few strokes to the rough woodcut of medieval life. The grain harvest the town consumed, for example (wheat, barley, and oats—all cut with sickles) took up at least 10,000 acres (4,000 ha) in the hinterland. But its people lived closer to nature in some unexpected ways. Great quantities of sphagnum moss were brought in from the bogs to serve as toilet paper. And the floors and foundations of the timber and wattle-walled houses were a deep mass of wood chips that, slowly fermenting and decaying, provided a gentle heating. A steady scatter of hazelnut shells helped to pave the floors in autumn and winter. Indeed, archaeologist Siobhan Geraghty described the Viking settlement in *Viking Dublin: Botanical Evidence from Fishamble Street* as "a very large and rich compost heap, over 18 hectares in area and more than three meters deep. The town which flourished on top of this bacterial and fungal power station . . . was quite unlike any habitat which now exists."

By the start of the eleventh century A.D. the Ostman settlements had been brought under the control of the Irish kings *(rí)*, and their political power waned after the defeat of the Dublin settlement in 1014. But the Scandinavians remained to develop Irish industry and expand international trade from the towns and ports they had founded.

THE COMING OF COLONIZATION

Changes from subsistence farming were already taking place in the twelfth century, notably at the abbeys of the Cistercians, a continental monastic order that introduced new methods of agriculture and animal husbandry and farmed industriously for profit. The monks' enterprise was endorsed in the grant of large estates both by Irish chieftains and by the lords of the Anglo Norman colonization that began in 1169.

Aalen described the Anglo Norman invasion in *Man and the Landscape in Ireland,* with its annexation of prime land in the south and east of Ireland, as the first important immigrant movement to Ireland for more than a millennium. It flowed from a great surge in population and migration in continental Europe and the massive reclamation there of wooded and boggy land. The Anglo Norman barons and the families that followed them came mainly from Wales and soon conquered some two thirds of the island, leaving the native lords to the bleaker regions of the north and west and upland enclaves within the "occupied" zone. The invasion was led by entrepreneurial barons, but they brought large numbers of Welsh and English farmers and rural artisans and Flemings from their settlements in Wales.

With them came Giraldus Cambrensis, the clerical nephew of one of the Norman leaders, whose treatise on the topography and natural history of Ireland, given its inaugural reading at Oxford in England around 1188, has been a colorful (and sometimes wildly fanciful) historical source. *Topographia Hiberniae* seems to confirm at least part of a medieval warm period and the lack of much need, or perhaps opportunity, to make hay:

> The soil is soft and watery, and even at the tops of high and steep mountains there are pools and swamps. The land is sandy rather than rocky. There are many woods and marshes; here and there are some fine plains but in comparison with the woods they are indeed small. The country enjoys the freshness and mildness of spring almost all year round. The grass is green in the fields

in winter just the same as in summer. Consequently the meadows are not cut for fodder, and stalls are never built for the beasts. The land is fruitful and rich in its fertile soil and plentiful harvests. Crops abound in the fields, flocks on the mountains, wild animals in the woods, it is rich in honey and milk.

·He found the island richer in pastures than in crops, and in grass rather than grain, as wheat so often failed to ripen. "What is born in the spring and is nourished in the summer . . . can scarcely be reaped in the harvest because of the unceasing rain. For this country more than any other suffers from storms of wind and rain."

Although the Anglo Norman lords left the less fertile parts of the uplands and the west and northwest to the native Gaelic community, in many places the Irish workforce was retained, living in groups of wood, clay, or wattle houses on the estates. After the monastic settlements, the Anglo Normans were the first commercial farmers, bringing agribusiness to a mainly subsistence economy. They introduced, in the drier east, the saving of hay for winter fodder; their oxen plowed bigger fields, and grains grown on rotation yielded large surpluses of wheat for export. Larger, white-fleeced sheep were imported to improve the Irish stock (which Giraldus Cambrensis called, disdainfully, "black").

The new settlers cleared a lot of new land, felling virgin forest in the southern river valleys of the Barrow and Nore. Along the Slaney they chopped away the dense forest of Duffry, which lay between the river and Mount Leinster Ridge and gave shelter to bands of defiant natives. Elsewhere, their forest clearances provided safe passage between settled areas and brought the word "pass" into place names such as Tyrrellspass and Milltownpass in County Westmeath and Poyntzpass in County Armagh.

The barons introduced a feudal system of administration that divided the country up into manors, baronies, and shires. They built some major towns inland, such as Kilkenny, and set up villages to attract settlers in the more prosperous eastern counties. This confident development was protected initially by a large network of earthwork-and-timber castles built to overawe the native population. About 350 of them were built on mounds (called mottes) with a surrounding ditch. Some were built on the sites of existing raths and others on eskers or rocky outcrops. After 1200 many of these were replaced by the superior castles of stone, and within a century these stood at most large centers of population, such as Dublin, Kilkenny, and Limerick, and strategic sites, such as Athlone on the River Shannon and Trim on the Boyne.

The 1300s, however, saw a weakening in the central authority of Dublin and a growing ethnic fusion of the Anglo Norman settlers, in their scattered enclaves, with the Gaelic Irish around them. Local lords and landowners began to build their own small, defensive "castles"—the single stone tower-house, three or four stories high, sometimes with a large, walled courtyard, or bawn, into which cattle could be herded. Perhaps seven thousand of these towers were built between the fourteenth and seventeenth centuries, and several thousands still survive in picturesque decay (more than 400 in County Limerick alone).

The Gaelic resurgence of the fourteenth century was helped by a brief but ferociously destructive Scottish incursion into Ireland, led by Edward Bruce, and the subsequent exodus of many of the demoralized Anglo Norman colonists. Later in the century, the weather deteriorated into the "Little Ice Age," and Black Death, a bubonic and pneumonic plague, swept across Europe and reached Ireland. There are no Irish death-rate figures, but the urban centers and grain stores of the Anglo Normans, with their heavy complement of rats, may well have attracted a heavier mortality than the pastoral Irish suffered. Many manors and small hamlets were completely abandoned, and land left unattended returned to scrub and brambles.

De Paor asserted that the Anglo Norman conquest brought about the first great dilution of the Celtic races in Ireland.

> Irish and Normans intermarried freely. . . . The custom on both sides of
> tracing descent primarily through the male line and indicating patrilineage
> through surnames, conceals the fact that in the mid-thirteenth century
> many of the principal Norman magnates were the grandsons of Irish twelfth
> century kings, and many of the Irish kings were the grandsons of twelfth
> century Normans.

This ethnic mix was diversified further by the Tudor "plantations" of the sixteenth and seventeenth centuries, intended as an instrument of Britain's colonial control. As discovery of the New World turned Europe's eyes across the Atlantic, Ireland became crucial to provisioning the seaborne trade with beef and butter and to supplying fat cattle to England's expanding economy.

From the 1550s onward, colonists from England were assigned land by the Crown, much of it confiscated from the Church in the dissolution of the monasteries or from rebellious "Old English," descendants of the assimilated—and Catholic—Anglo Normans. In the post-Reformation struggles of faith, religious affiliation was now to mark the boundaries of ethnic apartheid and class. At its Anglicized, Protestant heart was the Pale, the stretch of rich Leinster farmland, north and west of Dublin, obedient to English laws and ringed by fortified towns against all threat from the rest of the country.

The last wave of Gaelic resistance to the Tudor reconquest, led by the Earl of Tyrone, Hugh O'Neill, and Earl of Tirconnell, Hugh O'Donnell, was played out in a landscape of woodland, bog, and mountain still monstrously wild to English military eyes. "Elizabethan soldiers," wrote the modern historian Roy Foster in *Modern Ireland 1600–1972,* "hated the terrain with vehemence." Ulster was the wildest province in every sense and left to last in the military assault on Gaelic Ireland. The first Tudor plantations, in Leix and Offaly, were not successful. Plantations in Munster brought in some twelve thousand new English settlers to colonized lands. But the Ulster plantation, carried out in 1609 after years of war that may have halved the province's population, left the most lasting mark. With the "flight of the earls" to Spain, the six counties of Ulster were opened to a colonizing plan on the scale of the English migrations to the New World. Along with the grants of confiscated lands to mainly Scottish, with some English, Protestants, and segregation of the Irish to the poorest corners of the new estates, came imposition of English law, which turned land into a marketable commodity.

Anglo Norman Ireland had left the southern half of the island with a framework of walled provincial towns, mostly on estuaries and involved in foreign trade. The new assertion of British domination came equipped with a program for plantation towns, twenty-five of them in Ulster alone. Most of these ended up as pocket-sized rural boroughs with tiny houses, but Londonderry and Coleraine emerged as important fortified towns. In the rural landscape, new plantation villages clustered beside the landlords' castles, and the horizons of their hinterlands were changed by the felling of oaks for construction and the clearance of woodlands and scrublands seen by the military as refuges for the "woodkerne"—outlawed landless men and former soldiers of O'Neill. In the dwindling stretches of secondary woodland, wolf and goshawk *(Accipiter gentilis)* drew back a little further among the leaves.

Oliver Cromwell's savage conquest of Irish insurrection in the mid-seventeenth century saw a ruthless displacement of people now remembered in the phrase "to hell or to Connacht" (hell being transportation to Van Diemen's Land penal colony in Australia). The lands of Catholic owners were confiscated and the families forcibly removed to smaller holdings in the counties west of the River Shannon. This was for strategic, not economic reasons: Ulster, not Connacht, was still considered the poorest province. Protestant entrepreneurs and the officers of Cromwell's army were given the lands in return for war loans and in lieu of pay, and at the end of the plantation three quarters of the agricultural land of Ireland had moved into Protestant ownership. Most landowners believed that Protestantism and improvement went hand in hand, and they also encouraged Protestant farmers from continental Europe to settle on their estates. In 1709 the Palatines, who suffered religious persecution in Germany and were noted for their progressive farming, were invited to settle in Tipperary, Limerick, and Kerry.

The plantations made further inroads on the woodlands, as settlers found a valuable and "neglected" resource on their newly acquired lands. This is discussed more fully in Chapter 13 in the context of the Killarney oakwoods, but it is worth repeating the conclusion of the historical ecologist Oliver Rackham in *The History of the Countryside* that the severe deforestation of Ireland has a much longer history than is popularly supposed and that clearance for farming was the primary cause, not rural ironworks, British shipbuilding, or even the vast demand for barrel staves. Within five years of Cromwell's arrival, the Civil Survey of Ireland (1654–56) became the island's first equivalent of the "Domesday Book" (William the Conqueror's famous inventory of England), and from its meticulous details of many thousands of woods, Rackham totaled 420,000 acres (170,000 ha) actually under trees, or little more than 2 percent of the island.

Nonetheless, the pressure on woodland that flowed from the Tudor plantations was intense: the building of Londonderry alone took 150,000 oaks, 100,000 ash, and 10,000 elm. As timber grew scarce its use was restricted, and the first of many Acts of Parliament was introduced to compel or induce the planting of more. As castles and defensive tower houses gave way to the grand and confidently windowed mansions of the eighteenth century, their estates became islands of trees in a largely unenclosed and often treeless landscape.

Many of these "demesnes" (the word comes from medieval manorlands

farmed directly by the lord) were surrounded by massive stone walls enclos-
ing lawns and formal gardens, parkland, and ornamental woods. At Carton,
County Kildare, around 1,200 acres (500 ha) were enclosed by a great wall
about 5 miles (8 km) long. The landlords' importance and power was expressed in grand avenues of imported lime trees or Dutch elms; fields mar-
shaled in checkerboard grids; and formal, geometric pools and canals. But by
the mid-1700s a new, Romantic view of the human relationship with nature,
expressed in parklands of Arcadian imagery, arrived in Ireland from England.
As Terence Reeves-Smyth wrote in *Nature in Ireland: A Scientific and Cultural
History,* "The ideal now was smooth, open meadows, dotted with clumps of
oak or beech, sweeping lakes in which the house and park were flatteringly
mirrored, and tree-lined glades with animals grazing peacefully in the shadow
of romantic ruins, temples and pavilions." Within a century this reverie dominated the design of some 800,000 acres (320,000 ha) of parkland, with more
than seven thousand houses featuring large ornamental landscapes. They were
artfully planted with great numbers of trees, to frame views and heighten their
aesthetic appeal through manipulation of perspective and the flow of light
and shade. This reinforced the presence in Ireland of many alien tree species
from England and the continent of Europe. In the wake of Norman colonization, beech, sycamore, and sweet chestnut *(Castanea sativa)* had been introduced perhaps as early as the thirteenth century, and these were now widely
planted, along with horse chestnut, the English elm *(Ulmus procera),* and a variety of conifers. The ingeniously "natural" appearance of the new parkland
helped to absorb these species into the landscape, whereas the later collections
of trees from far-flung temperate zones, so fashionable in the Victorian era,
remain impressive but unashamedly exotic.

The Royal Dublin Society, founded in 1731, stimulated landlords to take
an interest in the management of their estates. Irish farming systems lagged
far behind English methods in every respect, and the landlords set in motion
an agricultural revolution in the northeast and south of the country that
would change the appearance of the Irish countryside. The land was surveyed, the open-field system was consolidated and subdivided, and new
planned field systems were laid out; ditches were dug to improve drainage,
banks of stones and sods were thrown up and secured by hedges of hawthorn
(Crataegus monogyna) to provide shelter, to keep stock from wandering, and to

control grazing and manuring. These boundary fences were substantial structures that have largely survived to the present and are the dominant feature of
the lowland landscape (see also Chapter 12).

EIGHTEENTH- AND NINETEENTH-CENTURY ECONOMIC ACTIVITY

Outside the demesne walls, however, the developed landscape of Ireland was
shaped by the economic forces of the eighteenth and early nineteenth centuries, an often ruthless dynamic explored with particular vigor by the historical geographer Kevin Whelan. The newly established West Indian sugar
plantations, for example, were provisioned by trade from Munster, and the
hinterlands of the ports of Cork and Waterford, with their navigable rivers,
became enduringly prosperous. Cattle were bred on the poor, boggy uplands
and fattened in the limestone valleys, where they often displaced the current
inhabitants. When, in boom times, cattle ranches began to encroach on the
hills, resentment spawned secret agrarian societies—the Whiteboys and
Rightboys—and "houghing," or cattle-maiming, became a violent form of
protest.

Demand for beef and butter for the London and New World markets
produced a similar pattern of land use and social stratification over much of
Ireland. The limestone lowlands of north Leinster and inland Connacht became the great core areas of cattle-fattening, where villages were dissolved
and people had little place. In the west, tenants were turned off the good
grassland and pushed into mountain farms. Whelan described in *A History of
Settlement in Ireland* a ranching economy, in Counties Mayo and Sligo especially, that created a landscape desolate to the human view, "where the lonely
box-Georgian grazier houses shadowed the crude cabins of the herds." For
wildlife, however, these ill-kempt tracts of grassland, only moderately grazed
by modern standards, must have provided a rare interlude of stability.

Along with beef and butter, the third great Irish export of the eighteenth
century was linen, woven from the long, supple fibers of the cultivated flax
(Linum usitatissimum). More than half a century before new spinning ma-
chinery began to concentrate the industry in Belfast, many thousands of small
farmer–weavers in Ulster and elsewhere gave their best land to the pretty,
blue-flowering plant that had probably come originally to Ireland from Roman Britain. It was a greedy crop, needing a long rotation, and was pulled la-

boriously by hand. To rot the stalks, the sheaves were left in long ponds filled with stagnant water, thus creating a farm effluent that must have been quite as damaging to streamlife as many of today's oxygen-hungry pollutants.

The demands of flax cultivation particularly suited the small farmers in the swathe of hilly drumlin country, with its many streams and ponds, stretching across Ireland from Ulster to north Connacht. At the Atlantic fringe of Ireland, from Donegal to west Cork, the pre-Famine small-farm landscape was strikingly different and created by forces as dramatic as any in European history.

Today enough of its physical fabric remains, in the web of stone walls and "deserted" villages, the grassy corrugations of old potato ridges, to speak of exceptional and desperate human effort. Modern farming reclamations often cloak in benevolent pasture the original soils of leached and stony glacial drift. But they still have not wholly erased the pattern of village settlements clustered tightly on pockets of arable land, surrounded by bog and mountain and within reach of vital resources on the shore.

This evocative weave of the wild and human has had an iconic significance in Irish culture and nationalism and in efforts to preserve the Irish language. At the mid-twentieth century it was to mesh with academic visions of a timeless, aboriginal settlement of the west, with continuity from remote prehistory: a surviving peasant refuge on the furthest fringe of Europe. In popular books such as *Irish Folkways* (1957), the geographer E. Estyn Evans saw the clustered pre-Famine settlements of the west—"clachans," as he called them—as a residual, living museum of what the human habitat had once been like all over the island.

Today most historical geographers would agree with Whelan that, far from an enduring refuge, most pre-Famine landscapes of the west were "newly settled, an adventitious and desperate veneer born out of grotesque, unbelievable, bizarre and unprecedented demographic circumstances." The population had doubled during the seventeenth century to around two million and had doubled again by the end of the eighteenth century. It increased at an even faster rate in the first half of the nineteenth century as eight million people were recorded in the 1841 census, and reached 8.5 million by the time of the next great social upheaval—the Great Famine. This exponential surge was supported by the easy cultivation, reliability, and high food value of the potato.

Most of it took place in the west, pushing settlement out inexorably into increasingly marginal land. The custom of subdividing holdings by settling plots on married children meant that more and more people were moving on to the margins of bogs and rocky offshore islands. Only the resilience of the potato, the virtual creation of soil, and an unlimited supply of labor made it possible. All came together in the brilliantly adaptive method of the "lazy-bed"—a term perhaps invented by a disdainful Englishman looking down in bemusement from his horse. In the Englishman's country, potatoes were sown in rows in well-cultivated, weed-free soil.

The lazybed, still in use in Ireland's traditional cultivation, is a long, deep ridge of soil about a spade's length wide. It is made by slicing and turning over rough sods, then spreading seaweed or farm manure on top, into which the seed potatoes are set. They are covered with soil dug from between the ridges, thus draining the ground through a network of furrows and raising the crop into the sunlight. The method buried weeds, reclaimed buried minerals, and paved the way for further crops, such as oats. In the coastal strip, the ridges were renewed annually by the layering of farmyard manure, seaweed, and

Grassed-over remains of the long, deep ridges traditionally used for potato growing in Ireland.

shell sand from the shore and peat from the bogs, so that a cultivated, or "plaggen," soil built up over large areas. Indeed, the general use of shell sand and maerl (sand of calcareous algae) for improving acid soil reached a remarkable pitch of activity: in Munster alone, in the 1840s, an estimated one million tons a year were dredged from the sea or dug from beaches and often carried far inland by horse and cart.

In the pioneering of the clachan settlements, by a system known as rundale, the arable land around the thatched cabins was shared out between families in unfenced strips to give a fair distribution of good and bad soil. Outside this "infield" was an "outfield" of rough grazing held in common, where cattle were tethered or watchfully herded by children. The landlords in their remote mansions welcomed this labor-intensive assault on their marginal "wastelands" that actually increased their rent rolls and found a home for surplus or "inconvenient" people.

FAMINES AND DISASTER

At the end of the eighteenth century, the rundale villages proliferating in Connacht were sharing in an unprecedented commercial boom in which young cattle, flax, herrings, and *poitin* (a raw spirit distilled from barley or potatoes) were all in strong demand. To many at this time, according to Whelan, the region must have seemed "the best poor man's country." The decades that followed, however, brought a cascade of disasters. With the end of the Napoleonic wars, farm prices collapsed. New technology moved the spinning and weaving of flax into factories. Shoals of herring *(Clupea harenqus)*, abundant for thirty years, deserted the west coast. Even the weather turned around in a succession of wet summers and bad harvests. By 1822 increasing poverty and food shortages were threatening the equilibrium of the overcrowded rundale villages, and mass emigration was already under way. At the same time, in the southeast of Ireland, the post-Napoleonic slump in farming brought misery to the cabins of agricultural laborers, the "cottiers," huddled along the lanes around the tillage farms. Their displacement swelled a great shifting underclass on the very eve of the Famine, many squatting in what amounted to shanty settlements in the most inhospitable corners of the landscape and totally dependent on the potato.

The blight fungus *(Phytophthora infestans)*, which took Europe by surprise in the cold, wet summer of 1845, had been ravaging potato crops in North America in the previous two seasons. It was noticed first in Belgium in June and by mid-October had followed humid, thundery depressions through much of northwest Europe. In Ireland it reduced perhaps 40 percent of an otherwise abundant crop to a blotched and pulpy disaster. The blight of 1846, which struck the west of Ireland much earlier in the season, devastated the sole food of several million people. Wandering and ragged bands existed on nettles, blackberries, roots, and shellfish as the island headed into a winter of bitter, snowy gales from the northeast. The "great hunger" wiped out the squatters in their sod huts on bogs and seashores and a large proportion of the smallholders, those with tiny plots and insecure tenure. From 1846 to 1851 around one million people died, mostly of disease, and 1,500,000 emigrated. By 1848, huge tracts of the countryside were empty of inhabitants and cattle, and even substantial farmers, demoralized by the disaster and its aftermath, were selling up their farms and leaving for North America.

Abandonment of myriad small fields and cabins and the long destruction of bushes and trees for fuel created a bleak landscape ruined as if by war. Many estates collapsed financially, more trees were felled to raise money, and schemes for new plantations were forgotten. As the population fell steadily, grass and scrub crept back to heal the wounds. In post-Famine reconstruction, most clachans were demolished and their clustered cabins replaced with dispersed, thatched houses that became the iconic whitewashed "cottages" of Irish landscape paintings. Fragmented rundale holdings were consolidated into fields enclosed with banks and boundary hedges.

The landlords were widely blamed for their apathy, if not reckless indifference, during the Famine, and the agrarian disturbances of the Land War in the 1880s helped to speed a series of reforming land acts. In what Whelan called "the legislative euthanasia of the Irish gentry," these promoted the transfer of land to the tenants by a generous system of long-term state loans, or annuities. Almost 11,000,000 out of 17,000,000 acres (4,500,000 out of 7,000,000 ha) were purchased in this way, and by the beginning of World War I, two thirds of former tenants owned their land. Over much of the country, the process had little impact on the landscape, but it did spell destruction for much surviving woodland, as landlords selling their land called in traveling sawmillers to realize the timber value of their trees.

CHANGE IN LAND OWNERSHIP

The Land Acts of the late nineteenth and early twentieth centuries eventually created a peasant proprietorship throughout the country. In the west, where the number of people on the land was still at unsustainable densities, the Congested Districts Board set out to reorganize farm structure. Rundale holdings were rearranged, using side by side parallel stripes of perhaps 25 acres (10 ha) to create fair, if still very small, parcels of varied soils. Many of these "ladder farms" run from high on the hillside to the valley bottom or to the sea, with the houses spread out along an intersecting road. This history of land division explains the lack of nucleated, English-style villages, with their central pond, church, and village store, over much of the countryside of the coastal west, in particular. It also set a precedent for the "ribbon" development of roadside housing that has suburbanized the look of so much of today's rural Ireland.

The massive reorganization of land, especially in the west, continued in the independent Irish Free State that took control of twenty-six of the island's thirty-two counties in 1921. An additional 1,235,500 acres (500,000 ha) were redistributed in the ensuing decades, overlaying old field patterns with new geometries and dividing the open parkland of estates with barbed-wire fences. The felling of trees also continued, an opportunism sometimes colored by hostility to every symbol of the Anglo Irish Big House.

In the new regime, however, a state forestry policy begun in 1903 was accelerated as both economic and social instrument. The woodland areas of estates taken over for division were planted thickly with conifers, often swamping the original trees (fine oaks among them) and leaving the house a gaunt and roofless ruin at the heart of the plantation. "Good land" could not be sacrificed, but rough grazing in uplands all over the country was planted with stands of Sitka spruce *(Picea sitchensis)* and lodgepole pine *(Pinus contorta)*. In the west this was extended to remote areas of mountain peatland to provide work and industry for depressed small-farm communities. Much of this forestry imposed a dark, hard-edged monoculture on the landscape, to be regretted and somewhat amended in today's more ecologically sensitive times, and its economic value was often undermined by extensive windthrow and poor growth rates.

The development of the peatlands for fuel and the generation of power is often identified with the modernizing aims of the Irish Free State. In the mid-1930s Russian and German methods of peatland mining were brought to bear, leaving an almost Soviet-style image of rows of great harvesting machines shaving the peat from the surface of the midland raised bogs. Their impact, while locally spectacular in the creation of brown swathes across the landscape on the scale of Ohio wheatfields, merely continued a centuries-old assault on "turf" as a resource. By the time Bord na Móna, the state turf board, took over the great Bog of Allen in the midlands, almost half of it had already been cut away by hand, part of an island-wide excavation of peat over the centuries conservatively estimated at some 741,000 acres (300,000 ha).

The early decades of the Free State (declared a republic in 1948) were a struggle toward self-sufficiency in which mixed farming was the rule. A guiding political ideal was to settle as many families on the land as possible (in conditions of "frugal comfort," to use a much-quoted phrase from Eamon de Valera, then Taoiseach, or prime minister). But a 25-acre (10-ha) farm could not provide a future for its large families, so that massive emigration became the norm, and in the long run the holding could not compete in the mechanized world of large-scale farming. The land-owning structure had been rigidly fixed by the transfer to owner-occupancy, and the physical fabric of small fields and scattered farms remained unchanged, even in Northern Ireland where the British wartime market brought rural prosperity. In the South the newly won security of small farmers engendered a fierce possessiveness of land, so that little passed on to the market and sons inherited late.

In the 1940s about half the land on a typical 25-acre (10-ha) farm was pasture and rough grazing, about a quarter grew crops, and the remainder was hay meadows. The farm held cattle, sheep, horses, pigs, and several kinds of poultry, and animal manure was essential to grow half a dozen different crops. The regime was "organic," without herbicides, pesticides, or artificial fertilizer (then unnecessary and not readily available). The farm's stone walls and hedgerows were natural habitats, and the rougher corners of surrounding landscape (scrub, stream, turlough) were left largely intact. For natural diversity, such a traditional farming system was something of a golden age, studied now for guidance in the management of Irish nature conservation.

The modernization of the 1960s and 1970s, and access to the European Economic Community, saw Ireland drawn increasingly into intensive livestock farming, with pasture and silage (fermented, storable grass) the dominant land use. Tillage was concentrated into larger and larger fields in the warmer, drier southeast (in the 1970s a 297-a, or 120-ha, farm in County Carlow, already big by Irish standards, was cleared completely of field boundaries, leaving one large field for continuous cultivation of grains). By 2000, Irish grain yields, in a national crop of more than 2.2 million U.S. tons (2 million metric tonnes), were said to be the largest in the world: about 3.3 U.S. tons (3 metric tonnes) per acre of spring-sown barley and almost 4.4 U.S. tons (4 metric tonnes) per acre of winter-sown wheat. New fungicides were protecting the exceptional yields made possible by Ireland's moist climate.

Funding from Europe supported the final phases of drainage schemes, started in the mid-1800s, which had changed the ecological regime of about five million acres (two million ha), or almost one-quarter of the island. The Republic's arterial drainage program of the 1960s and 1970s was particularly primitive ecologically, canalizing smaller rivers without thought of impact on fish stocks or other wildlife and leaving the banks carelessly heaped with spoil. In Northern Ireland, too, major drainage of the River Blackwater, which flows into Lough Neagh, left banks scraped back to a stable 45 degrees and drained more than thirty wetlands among the silty drumlins of Tyrone and Armagh. In the west of Ireland, drainage has destroyed many turloughs, with their distinctive flora and special value to migratory wildfowl.

Thus, agriculture and forestry have steadily eroded the island's biodiversity, through the spread of grassland and conifer monocultures, degradation of peatlands, eutrophication and pollution of waterways, and overgrazing of the hills. As economic success reverses emigration and the population grows, the ecological pressure comes directly from numbers of people and the impact on nature of their increasingly mobile and elaborate lives.

For more than a century, the frozen structure of land ownership, minimal off-farm employment, and an acceptance of high emigration all kept the Irish population to a notably sparse level by European standards (from a low point of 2.8 million in 1961, the Republic's population rose to 4 million in 2002; and Northern Ireland's from 1.4 million to 1.6 million).

At the start of the twenty-first century, however, the future of the Republic was suddenly transformed, as economic initiatives primed by European Union structural funding and the attraction of foreign (largely American) investment in the computer and pharmaceutical industries reached a critical mass for growth. For the first time, the Irish Exchequer was filled to overflowing, and crucial labor shortages arose at every level. Much of the massive investment in infrastructure that was launched around 2000 was destined for cities, towns, and new highways, but commuter house-building sprawls out into a countryside liberally webbed with accommodating, leafy back roads. Already in Ireland, almost half of country dwellers have no direct contact with the land.

As off-farm jobs multiply, even smaller farmers feel secure in selling land at high construction prices or planting it with private forestry for long-term financial return. The number of farmers has been falling steeply (dropping by half in three decades, to little more than 100,000) amid constant market crises, the disruptive menace of diseases such as bovine spongiform encephalitis (BSE) and foot-and-mouth disease in Britain, and a widely accepted pessimism about the future of traditional small-scale farming. More general college education and rising social expectations mean that few young people choose farming as a way of life. In this first phase of retrenchment, land has been retained within extended families as elderly farmers have died or retired. But the dramatic changes in the Irish economy have given land a new, non-farming value and diminished the emotional symbolism of possession.

Land as a tradable commodity, in an island of rising population, will shatter old farm structures whose boundaries are now preserved in hedgerows, stone walls, and field banks. Fewer, larger farms with bigger fields on moderately good grassland and a checkerboard of mixed forestry and low-density housing at the margins seems the probable rural trend in the new century. Comparatively little of Ireland is beyond use for construction: even much of the remaining peatland is shallow and easily cut away. The room left for nature will increasingly be a matter for deliberate conservation policies and an ecological sensitivity in all forms of planning: a prospect reserved for discussion in Chapter 17.

SELECTED REFERENCES

Aalen, F. H. A. 1978. *Man and the Landscape in Ireland.* London: Academic Press.

Aalen, F. H. A., K. Whelan, and M. Stout, eds. 1997. *Atlas of the Irish Rural Landscape.* Cork: Cork University Press.

Barry, T., ed. 2000. *A History of Settlement in Ireland.* London: Routledge.

De Paor, L. 1986. *The Peoples of Ireland.* London: Hutchinson.

Foster, J. W., and H. C. G. Chesney, eds. 1997. *Nature in Ireland: A Scientific and Cultural History.* Dublin: Lilliput, 549–572.

Kelly, F. 1997. *Early Irish Farming*. Dublin: Institute of Advanced Studies.

McCracken, E. 1971. *The Irish Woods since Tudor Times*. London: David & Charles.

Mitchell, F., and M. Ryan. 1997. *Reading the Irish Landscape*. Dublin: Town & Country House.

Praeger, R. L. (1997 reprint). *The Way That I Went*. Cork: Collins.

Rackham, O. 1986. *The History of the Countryside*. London: Dent.

— *Chapter 7* —

The Brown Mantle

P eatland in one form or another still covers a substantial portion of the Earth's land surface—some 1.9 million square miles (5 million km^2). From the tundras of the permafrost regions to the wetlands of rainy coasts from Chile to Kamchatka, peat stores more of the planet's carbon than any other terrestrial vegetation. Even though local extinction of peatland is regretted in the crowded and highly developed countries of western Europe, there are still great swathes of raised and often wooded bog across the north of Europe and Canada. Yet tiny Ireland is thought of as an especially boggy place—as, of course, it is. This curious, microcosmic association has to do with essentially literary impressions, derived from Ireland's unhappy peasant history. The drama of that intimate human involvement with the bog, so often somber and wretched, gives way now to fascination with peatland itself and the plants and wildlife adapted to its harsh conditions.

Ireland has, indeed, something special to offer in the variety of the peat-lands that cover some 3.2 million acres (1.3 million ha), or 16 percent of the island's surface, and in the immediate sense of their links to the Ice Age and an oceanic climate. In a lowland landscape rolling above the debris of eskers, drumlins, and moraines, the strange domed topography of heathery raised bogs is a companionable land form. A Connacht mountainside veiled in rain dramatizes the role of soft but steady moisture in the rise of blanket bog as a soggy brown cloak to the hills. At a windy shore of western oceanic bog, with the taste of salt on one's lips, the flow of mineral nutrients in Atlantic showers is that much simpler to accept. It is the extreme temperate oceanicity of Ireland that makes its peatland ecosystems different from those in the harsher climates of, say, Canada or Eastern Europe. And it is the chemistry of the waters feeding the systems that shapes the vegetation of different kinds of bog on the island.

Peat consists of the dead remains of plants that have built up over thousands of years, their decomposition inhibited by lack of oxygen. Rain brings minerals to plants growing at the surface of the peat, and fills its pores with water to stifle the decay of plants when they die. In the droughts of a warmer global climate, bogs would shrink, crack, and decompose, to grow ordinary shrubs and grasses—"the most deleterious outcome of climatic change in Ireland," as one government study *(Climate Change: Studies on the Implications for Ireland)* put it. What was once considered bleak wilderness, fit only for mining as fuel or "reclamation" for forestry or grassland, is now a valued ecosystem.

RAISED BOGS

The raised bogs of the central lowlands—now greatly reduced and degraded—had their beginnings in the lakes and basins and hollowed floodplains of the big rivers, after the ice had gone. In this wetland maze the underlying rock was limestone, like much of the glacial debris and silt, so that the water was rich in dissolved calcium. At the lake margins, massed plants of submerged and primitive stoneworts *(Charophyta)* drew on the calcium for carbon dioxide, leaving their branching stems crusted with lime. Water snails thrived, and the sift of their coiled remains added to the shell-mud mounting in the shallows. Higher plants could root in this creamy colored marl: floating water lilies and pond weeds, then bulrushes *(Typha latifolia, Typha angustifolia)*, and reeds *(Phragmites australis)*. As their root systems spread, the lakes began to close in from the sides, becoming rustling reed swamps, and the fall of leaves and stems built up dense layers of debris in the water.

In this natural succession of lake vegetation, peat bog is the inevitable long-term outcome. The process accelerates in a climate in which rainfall exceeds evaporation and transpiration, and the onset of a warm, moist "Atlantic" climate around 7,000 years ago, when the sea resumed its full circuit around the British Isles, triggered the development of rain-fed bog right across northwest Europe.

In Ireland there was another significant factor: a sudden drop in the water level of the midland lakes, which drained the wide floodplains of most of their standing water. This was caused by the rebound of the land after the ice retreated, a recovery effected in a series of movements along old geological faults. It is thought to account for the widespread and simultaneous acceleration of the normal in-filling of lakes, which created waterlogged fen (a kind of lowland marsh) and promoted the first formation of peat. Evidence for the fall in water level can be seen in groups of limestone boulders, deeply undercut by wave erosion, which can be found, high and dry, around the midland rivers, lakes, and bogs.

Some of the early stages of in-filling can be seen at lakes on the modern periphery of the limestone plain. Lough Carra, for example, in County Mayo, is a shallow lake of 4,200 acres (1,700 ha), only lightly fringed with reeds, and the white marl in which they grow gives a special color to the water: a "wonderful pale pellucid green" in Robert Lloyd Praeger's description in *The Botanist in Ireland.* The limestone pavement around Carra has little glacial drift cover, but in drumlin regions of the north and west the hollows abound in small, impounded lakes, filling with plant communities at various stages of succession. The three large lakes of the River Shannon, often bordered with dense reed swamp, once covered a far greater area, and the myriad small lakes that surrounded them have long vanished under the peat they helped to generate. Actual calcareous fens are now rare in Ireland, as their rich soil tempts reclamation for farming. A few survive in the midlands: notably at Pollardstown, County Kildare, which is the largest intact fen, and at Scragh, in County Westmeath, a small, very wet quaking fen (one that shakes underfoot). A range of good examples of surviving raised bogs and fens can be found among the nature reserves listed in the appendix, under the counties of Kildare, Offaly, and Westmeath.

The plant communities of fens vary quite widely, according to habitat, but dominant herbs typical of a rich fen might include black bog-rush *(Schoenus nigricans),* grass of Parnassus *(Parnassia palustris),* and the insectivorous common butterwort *(Pinguicula vulgaris).* As a fen firms up above the water table, plants such as purple moor-grass *(Molinia caerulea)* and tormentil *(Potentilla erecta)* may blur the boundary with adjoining wet grassland. At ground level, the fen bryophytes (mosses and liverworts) foreshadow those plants that can, in time and in the right climatic conditions, begin to lift the fen into a raised bog.

While the fen-peat remains shallow, capillary action draws water up through it, replenishing the nutrients for surface plants. As calcareous minerals in the soil become exhausted, conditions begin to turn acid, and the vegetation is taken over by plants adapted to a leaner regime. Some of them are heathers— ling *(Calluna vulgaris)*, cross-leaved heath *(Erica tetralix)*, and also the cranberry *(Vaccinium oxycoccus)*. New and stringier sedges appear, and insectivorous sundews (*Drosera* spp.) offer their sticky palms to passing flies. Bog-cotton (*Eriophorum* spp.) joins the purple moor-grass, the scattered white tufts of their flower heads suddenly defining the airy distances of the fen. As the peat builds above water level, birch, alder, willow, and pine often spread out across it, creating a transitional carr, or tangled, wet woodland, with a waterlogged floor on which sphagnum mosses develop.

It was these specialized bog mosses that secured the final transformation of the postglacial fens. Their rapid growth and build-up of dead tissue lifted the whole ecosystem up from soggy platforms of sedge-peat into domes of raised bog often many feet deep and hundreds of yards across. In the process, the surface vegetation lost its connection with the already meager nutrient supply from groundwater and came to depend, instead, on minerals dissolved in rain.

Of the world's several hundred species of sphagnum, Ireland has about twenty-five. A dozen or more of these may be woven together on the surface of a raised bog, each species a different, often vivid, color and adapted to a particular niche in the mosaic of hummocks, hollows, and pools. The bright-red *Sphagnum capillifolium*, for example, has small, close-packed leaves and joins species with similarly tiled or interlocking foliage to weave a tight surface for the hummocks, shielding them against desiccation. The bright-green and common *Sphagnum cuspidatum* (drowned kitten moss) forms floating mats in the open pools that serve as rafts for the support of less aquatic species. In the level lawns of the bog, red, copper, and yellow species are normally growing some 6 inches (15 cm) above the water table: a soft, moist carpet of mosses in which seedlings of bigger plants can germinate and grow.

Whatever their color or precise leaf pattern, the feathery tendrils of sphagnum have crucial characters in common that help the mosses to dominate the raised bog and control its acidity and hydrology. To feed on the nutrients in falling rain, the plant draws in water through pores in large, nonliving, tubular cells in its leaves. These allow sphagnum to hold water that is twenty times the plant's dry weight—an extraordinary absorbency. Indeed, in World War I, the Society of United Irishwomen formed units to collect sphagnum for wound dressings, for which it was far more absorbent and effective than cotton wool. Almost a million tons of sphagnum were collected all over Ireland, much of it the vividly pale-green *Sphagnum cymbifolium*, which grows in cushions on both raised and blanket bogs.

As it passes water through its system, sphagnum treats it in a way that maintains the bog's acidity. It used to be thought that sphagnum released its own organic acids, as, indeed, it does when decomposing without oxygen further down in the peat. As a living plant on the bog's surface, however, it uses chemicals in its cell walls to exchange ions of hydrogen for the scarce mineral ions potassium, magnesium, and calcium in the surrounding water, and the concentration of hydrogen ions produces acid conditions. The initial acidification of the fen peat began at the point where rainfall became predominant in the upper layer. A high level of carbon dioxide in peat and water seems, indeed, to be crucial to invasion by sphagnum and to its vigorous growth.

The specialized "dead" cells of the mosses hold only a fraction of the water retained in a raised bog. Although 90 percent of the volume of the bog is water (there is more solid content in an equivalent measure of milk), most of this is trapped, as in a sponge, between the stems and leaves of the mosses and in pools in the hollows. The bog allows rainwater to run away only when totally saturated. It may also be fed by a spring that, welling up full of nutrients, feeds a "soak," or pocket of richer, fen-like vegetation. A bog may also trap a large lens of water between peat layers of differing densities. This "shaking" bog, trembling underfoot rather like a water bed, sometimes bursts asunder after very heavy rainfall, a phenomenon occasionally found at the edge of blanket-bog on a hillside. In *The Way That I Went*, Praeger gave a vivid description of a bog burst in the hills near Killarney in 1896 that swept away a farm and killed a family of eight. The displaced mass of peat "with the abundant stumps of pine which it had contained, was spread for miles over the lands below. . . ."

A single raised bog, swelling up from its original lake, may be isolated as a

The stumps and roots of trees that died in the growth of blanket bog thousands of years ago.

roughly circular dome of peatland surrounded by a fringe of wetland, the "lagg" often a tangled mass of willows and birches. But where the edge of a raised bog meets a slope, this wet "gutter" of fen wood would continue to accumulate peat. As this builds up, the bog may expand and slowly climb the gradient. At the top of the rise it could meet the dome of a neighboring raised bog and merge with it, so that this form of peatland may then extend, in gentle undulation, over large areas of the lowlands, varying in depth from 10 to 40 feet (3 m to 12 m). Frank Mitchell, studying the history of the Great Bog of Ardee near his home in County Louth, found a chain of bog domes, large and small, stretching out across 4.3 miles (7 km). But such an amalgamation was dwarfed by major complexes such as the Bog of Allen, which covered most of County Offaly and part of County Kildare, and by the vast expanse—more than 2,300 square miles (6,000 km²) of the central limestone peatlands as a whole.

The great advances of the raised bogs, upward and outward, followed climate swings creating cooler and moister summers, a development frequently marked, in a section through a bog, by a change from older, well-humified sphagnum peat to a younger, less oxidized layer. This often followed a climatic change around 2,500 years ago, the start of the "Sub-Atlantic" period, but it also began at different times in different bogs, prompted by local conditions. The fresh growth could be extremely rapid. Hummock-forming sphagnum can grow by 1.5 inches (3.5 cm) a year, and mosses in wet hollows can billow up by as much as 16 inches (40 cm) in a year (a more general average for the growth of peat bog is, however, about 24 in., or 60 cm, in a century).

As the bogs invaded and engulfed wide areas of low-lying forest and wetland scrub, human movement in many parts of the midlands became progressively more difficult. The early farmers laid down trackways of brushwood or heavier timbers to link green islands of drier mineral soil. At Corlea, near Kenagh in County Longford, a complex of some sixty trackways has been investigated, some dating to the Neolithic but others to much later periods. In one of these, crisp cuts from a bronze ax were found in alder stems dated to within a decade of 2249 B.C., a remarkably early date for metal tools in Ireland. In another, the most substantial, a road of some 1.2 miles (2 km), was made from heavy oak support timbers cut from trees felled in 148 B.C. Some of the lesser trackways may have led to seepage areas where deposits of bog iron, a rich and easily processed form of the ore, welled up as a rust-colored mud.

BLANKET BOG

The raised bogs began as lakes and fens, but they have needed a generous humidity to sustain them. At the few bogs that survive undrained and intact a mean annual rainfall of 30 to 39 inches (750 to 1,000 mm) keeps the sphagnum growing by half an inch or more per year, and moss already chokes the trenches at bogs where turf-cutting was abandoned a mere generation ago. The blanket bogs of the mountains and the western seaboard, however, have a far wetter regime: a mean annual rainfall of 49 inches (1,250 mm) or more, spread over some 250 days in the year.

Peat has been forming in badly drained places, high and low, almost since the Ice Age cold disappeared. At some bogs in the west, the basal layer of peat, at perhaps up to 16 feet deep, was formed from birch debris some 9,000 years ago, and hazel pollen and nut fragments from that time are also quite common in cores taken from older peat basins.

These existing bog hollows and other reed swamps served as nuclei for much of the great expansion of blanket bog that accompanied the shift to a wetter climate (from boreal or subarctic to Atlantic) about 7,500 years ago.

But the blanket also formed independently, where ground was sufficiently waterlogged by high, continuous rainfall. This process started at different times, even in quite adjacent areas, and often seems related to the degree of human impact on the land, through tree clearance, burning, overgrazing, and cultivation (see Chapter 6). This is indicated particularly in the pollen records of northern Ireland, where blanket peat began to expand about 4,500 years ago, and in those of Connemara and Kerry at somewhat later dates. In western Ireland, peat spread across the lowlands, often right to the edge of the sea, and covered the mountains to the highest ledge and summit. Further east, blanket bog is confined to the uplands above 500 feet (150 m), a zone of higher rainfall and reduced evaporation.

FLORA OF THE PEATLANDS

Lowland (or oceanic) blanket bog looks very different from raised bog. Treeless as tundra, it robes the land in an undulating, seemingly uniform shimmer of grasses and sedges, dull green in summer but turning in autumn to a tawny gold shot through with pink and purple. Two deep-rooting species dominate its character: purple moor-grass (only the flowers are purple—this is the dying, lion's-mane grass of the fall) and the black bog rush *(Schoenus nigricans)*. Elsewhere this is a rush of richly nourished fens, but in the west it is fed by the extra minerals in the rain and wind-blown sea spray.

The blanket bog sedges that catch the eye in summer are the bog-cottons flowering in drifts of silky white tufts that have the consistency, but not the strength, of cotton wool. Bog-cotton grows also in raised bog, but it can dominate whole stretches of blanket bog, crowding out even the sphagnum, so that the peat is made up almost entirely of its remains. Sphagnum does relatively poorly on the lowland blanket bog: outside of the very wet pool areas it builds its water-conserving hummocks at intervals among the moor-grass.

The gradients of wetness that govern the lives of bog plants are full of subtlety and seeming paradox. Lowland blanket bog depends on rain to keep it waterlogged, yet its abundance of sphagnum is limited by the frequent drying out of the top few inches of its surface, and the peat of these bogs is actually composed of grasses and sedges, not moss. Heather *(Calluna vulgaris)* is in Ireland generally thought of as a bog plant, but it flourishes best in the drier and better drained banks and roots only shallowly into wet and poorly aerated peat.

With their wiry, woody stems and narrow, folded-down leaves, the ericaceous shrubs of the bogs are equipped to survive drought, wind, and fire as much as any excess of moisture. The cross-leaved heath prefers a wetter root run than heather and grows abundantly in blanket bogs alongside black bog-rush. Here and on the raised bogs its shoots can be strikingly green and vigorous against the sooty debris of recent burnings. Fires are a natural hazard of peatlands, but many are begun deliberately to promote the regrowth of young heather shoots for sheep and cattle, and strip burning is traditional in the management of moorland for red grouse *(Lagopus lagopus scoticus)*. With care, the heather survives. But where the plants are old they may shelter as many as twenty-five species of epiphytic lichens and liverworts. These perish in the wave of fire and may take decades to return.

Rooted in moisture, the bog's ericaceous shrubs are strikingly designed to survive drought and water loss—they are xerophilous, to use the botanical term. Drought is not synonymous with heat: cold, windy weather can both slow down the absorption of water by the roots and speed up its loss by evapotranspiration through the leaves; and water held in evergreen leaves can freeze and kill the cells. The design of heather and its relatives deals with these problems so efficiently that they can survive the winter on tundra peatland far to the north of Ireland.

Their wiry, woody stems transmit little water at all, and their leaves are rolled or folded to shield the stomata on the underside. An early stage in this strategy is shown in the in-rolled margins of the leaves of bog-rosemary *(Andromeda polifolia)* and cranberry, both characteristic of the wettest parts of raised bogs and both with beautiful pink flowers.

Heather, like most bog plants, finds help in making a living from the peat by forming mutually beneficial alliances with fungi coiled in its roots. These mycorrhizal associations, in which fungi help plants to obtain essential nutrients (nitrogen and phosphorus, in particular) in exchange for carbohydrates they cannot make for themselves, are common to most higher plants, but they take on a special importance where food is scarce. Most, perhaps all, orchids are mycorrhizal plants. Their tiny seeds have no stored food, so that the bogland orchids, twayblades *(Listera ovata, Listera cordata)*, and marsh helleborine *(Epipactis palustris)* must find their fungal partners in the peat even to begin life as seedlings. Another plant that uses mycorrhizal help is the bog asphodel *(Narthecium ossifragum)*, which in July and August can carpet the peat in spikes of starry golden-yellow flowers. The *"ossifragum"* in the name means "bone-breaking" in Latin and recalls a belief that eating it gives brittle bones to cattle. Livestock forced to graze the bog fringes where bog asphodel is at its most abundant would, indeed, find themselves short of calcium and of many other nutrients.

Isolated from minerals in the soil and denied the nutrients locked away in a mass of inert, dead fibers, some peatland plants are ingeniously adapted to trap small invertebrates and digest their remains. The most abundant carnivorous plants are the sundews of which three species may share a bog. They are variations on one form: a low rosette of leaves flushed scarlet, each with a halo of hairlike tentacles ("all whiskery, like a bee's leg," as a Connemara man described it to Praeger). The tentacles are tipped with glands that exude a viscid fluid, glistening like beads of dew. This and the color, perhaps also a scent, tempt small flies, beetles, or ants—sometimes even a damselfly or wasp—to alight on the leaf, whereupon it folds and the tentacles begin to bend toward the center. Smaller insects are quickly dead, their tracheae choked by the fluid. The glands that tip the tentacles also secrete digestive enzymes that break down the protein in the soft tissue of the victim and leave its chitinous parts to blow away. Five small insects a month appears to be an average intake.

The most abundant and adaptable of the species is the common round-leaved sundew *(Drosera rotundifolia)*, which grows as a low and solitary plant, carrying its round spoons of leaves almost flat to the ground. The "great" sundew *(Drosera anglica)* has tall, upright leaves and often grows in massed ranks near the surface of water in bog holes and old drains. *Drosera intermedia* forms dense, crimson mats at the edge of bog pools or on the wet peat of blanket bogs. All depend so completely on their carnivorous diet that their roots are reduced to a few short branches to anchor the plant and draw up water.

The flowers of sundew are small, unremarkable, and white, but those of the equally carnivorous butterworts (*Pinguicula* spp.) include blossoms ardently praised for their elegance and intensity of color. In his *Flora of County Kerry* (1916), Reginald William Scully wrote of the large-flowered butterwort *(Pinguicula grandiflora)*: "No one who has seen its groups of deep violet flowers— sometimes over an inch in diameter—on the black dripping rocks of Conor Hill, or on the boggy roadsides between Killarney and Kenmare, will deny its claim to be considered the most beautiful member of the Irish flora." As fly traps, the rosettes of the butterworts work more like old-fashioned flypapers.

Through glands in the surface of the yellowy-green, tongue-shaped leaves they secrete a remarkably viscid glue that can be drawn out in threads nearly 18 inches (45 cm) long. Other glands exude the digestive fluid, and the margins of the leaves curl in slowly to wrap more glands around small insects stuck near the edge.

Again, there are three butterworts. The common bog violet *(Pinguicula vulgaris)* is found widely on wet soils, both acid and alkaline, in Europe and Siberia and much of North America. Far smaller, and often rather rare away from the extreme west of Ireland, *Pinguicula lusitanica* is more exclusively a plant of bogs, its tiny (.75 in.; 2 cm), almost transparent rosettes tucked down deeply between heather and moss. It shares much the same native range as its much larger and more showy relative, the large-flowered butterwort: through western Spain, Portugal, and France to the western shores of Ireland and Britain. This exquisite Kerry butterwort, with flowers often more than an inch (3 cm) across, actually has its most northerly stations in the limestone of the Burren, County Clare. A solitary plant with white flowers has managed to survive in a muddy, hillside flush above Ballyvaughan, among roaming cattle drinking from the pools.

The third group of carnivorous bog plants, the bladderworts *(Utricularia)*, float just below the water surface in pools and drains and gather their nutrients by an intricate insect trap. Among their feathery fronds are leaves modified into tiny bladders. Each is tipped with a cluster of stiff bristles formed into a cone and glands exuding mucilage and sugar that attract small aquatic animals (minute crustacea, insect larvae, freshwater worms). The bladder creates a partial vacuum internally, and when the animals touch the sensitive bristles, a hinged valve opens like a trap door and a rush of water sweeps them inside. The valve snaps shut again and the prisoners die of suffocation or starvation, their nutrients absorbed by star-shaped glands in the wall of the bladder.

A plant introduced from the New World completes the suite of Ireland's carnivorous species. The huntsman's cap or purple pitcherplant *(Sarracenia purpurea)* is distributed widely in North America, from New Jersey to the Northwest Territory of Canada, and survives not only arctic temperatures but the wettest conditions of all the native Sarracenia species. It was introduced to Ireland in 1906, initially to raised bogs in Roscommon and Westmeath and later to Woodfield Bog in County Offaly, where a large colony became established and is still expanding. More large colonies of up to one thousand plants were discovered on one of the few surviving raised bogs in County Kerry, near Listowel, in 1988.

Now obviously well-naturalized in Ireland, *Sarracenia purpurea* is a pitcher-plant with hollow, urn-shaped leaves, carried upright like a cluster of purple-streaked drinking horns. A lid-like flap at the mouth does not close but serves as an umbrella to protect the nectar secreted inside the lip. It admits enough rain, however, to drown insects when they lose their footing on the slippery interior and fall into the water (whereupon a wetting agent, secreted into the water, makes it impossible for them to lift off again).

Unlike the straightforward origin of the pitcher plants, the American credentials of another aquatic Irish bog plant now begin to seem extremely doubtful. Pipewort *(Eriocaulon aquaticum)* grows in the shallows of bog pools and peaty lakes near the west coast, raising white buttons of flowers on simple stems above the water. Apart from a similar, scanty showing on the west coast of Scotland, its main center is in eastern North America. Evidence from pollen studies near Roundstone shows a presence in Ireland nearly 6,000 years ago, so human introduction is out of the question: this seems to be one of the few plants with a very limited base in Europe and an extensive range in the New World. In the western Irish lakes it often keeps company with the water lobelia *(Lobelia dortmanna),* whose delicate, pale violet flowers also spring up from a submerged rosette of leaves. This is also a plant shared between Europe and America but one much more widespread to the east of the Atlantic.

Among the very local, sometimes rare, plants of Irish peatland are heaths of Pyrenean-Mediterranean—or "Lusitanian"—origin whose eccentric distribution has added to the challenges of Ireland's postglacial biogeography. Ecologically, much of their interest lies in their survival in conditions apparently far different from their hot, dry, and often stony origins. The Mediterranean or "Irish" heath *(Erica erigena)* may well have been introduced through medieval trade or carried home by religious pilgrims to Spain, impressed with its robust size and spring flowering at holy places in Galicia. At Claggan Mountain on the Mayo coast, one of its prime modern locations, pollen analysis in peatland finds the heather appearing for the first time in the mid-fifteenth century, when the mountain had been burned for farming and the bog surface was drying out. Today it flourishes in well-drained peat at streamsides and lakeshores or in cut-over moorland.

The *"erigena"* in its name derives from Duns Scotus Eriugena, a ninth-century Irish philosopher. St. Dabeoc's heath is sanctified by reference to a medieval Donegal saint of shadowy origin and variable spelling but is common only on heaths and rocky ground in Connemara in County Galway, where its big terminal bells of reddish-purple weave through the golden carpet of low, dwarf gorse *(Ulex minor).* This heather, as one of the troublesome Lusitanians, drew an unusual hint of exasperation from Praeger in an essay on the flora of Roundstone Bog in Connemara in *A Populous Solitude:* "The main problem with it—the puzzle of the occurrence in Ireland of a whole group of Pyrenean and Mediterranean plants and animals, absent or nearly so, from all the intervening lands—I pass by, for I have thought about that and written about it for years without getting any further, and I am almost weary of it." Roundstone Bog, near Clifden, a unique tract of oceanic blanket bog strewn with about a hundred lakes, is also one of the few locations of Mackay's heath only known elsewhere from northern Spain. It chooses the drier parts, where it spreads by layering, mysteriously refusing to set seed at any of its sites in Ireland.

The new work on the history of *Erica erigena* has encouraged speculation that the other Mediterranean heaths returned to Ireland postglacially with Neolithic settlers—either accidentally or as a calculated source of springy cushions and mattresses—and that they were given their start in the warmer climate of the time. They have since found their own well-drained niches for survival in the bogs, but one could be forgiven for thinking that they continue to grow largely despite the ecosystem that dominates Irish peatland, not because of it.

The sense of peaceful, watery wilderness that impresses visitors to Roundstone Bog is evoked even more intensely in the somber, windswept expanse of the low bogs of northwest Mayo, where some 296,520 acres (120,000 ha) of peat cover glacial till deposits up to 33 feet (10 m) deep. Along with the tough covering of purple moor-grass and black bog-rush, so typical of oceanic blanket peat, these Mayo bogs often carry a crust of mucilaginous algae, 4 inches (10 cm) deep in places, which protects them against desiccation in summer.

The tundralike atmosphere of the region is accentuated by the frieze of dark mountains that fringe the bogs to the west and south, but the bogs are not nearly so featureless and uniform as they can seem from a distance. Botanist G. J. Doyle, in a paper on their plant communities, found them "complex mosaics of deep peat, dissected by a variety of drainage channels, with surface features including pools and lakes of varying sizes, in-filling depressions, lake island and swallow holes—the entrance to subterranean drainage systems that traverse the bogs in places." Recent botanical discoveries suggest, indeed, that within this great sweep of peatland may be the last vestiges of Ireland's ancient, postglacial fens.

Several rich fen mosses that flourish today were formerly known in Ireland only as subfossil fronds recovered from the lowest organic deposits at the base of raised bogs. They belong to the northern moss flora abundant in the mires of the postglacial boreal period in Ireland and Britain but largely lost from both islands with the change to an oceanic climate about 7,500 years ago.

The Irish discoveries began in the late 1950s, when the relic moss *Meesia triquetra* was found growing at a rich, iron-stained flush in north Mayo. It was the only living plant of the species ever seen in either island and has never been refound. But in the 1980s the lure of its rarity led bryologist Neil Lockhart to search other fens in the vicinity. He found two other relic mosses, the rare *Homalothecium nitens* growing in abundance and creeping stems of *Leicolea rutheana* (new to Ireland), and began to recognize, in the iron-stained, calcium-rich pools and their quaking lawns of moss, "truly the last remnants" of a once-widespread postglacial landscape.

As if in confirmation he discovered, in 1998, the robust and beautiful northern moss *(Paludella squarrosa)*, never found in Ireland as a living plant and extinct at its few sites in England since about 1916. "A most curious moss, of unbelievable appearance" is how it is described in the current *Mosses of Eastern North America*, a response to the intricate folding-down of leaf-tips that make each stem resemble a baroque bracelet or necklace of woven hearts. But in Ireland its interest is also strongly palaeoecological. This was one of the sturdy species that, growing in abundance in the mossy lawn of a postglacial fen, played a big part in the transition to peat bog. By insulating the top of the bryophyte cover from the mineral-rich groundwater, plants like *Paludella* helped to lay the foundation for the growth of sphagnum mosses, nourished entirely from rain.

The boreal associations of these secluded Mayo fens do not stop at mosses but include other beautiful higher plants that today more naturally belong to Scandinavia or Iceland. The search that discovered *Paludella* went on to find the narrow-leaved marsh orchid *(Dactylorhiza traunsteineri)* and the yellow-flowered marsh saxifrage *(Saxifraga hirculus)*, a rare and threatened species both in Ireland and Britain. Despite its northern range, *Saxifraga hirculus* is not an alpine but a plant that needs constant saturation with moving water in flushes or bogs, and its decline in Ireland is directly a result of the drainage and exploitation of its habitat.

ANIMALS OF PEATLAND

To trudge the high blanket bog of an overgrazed Irish mountain may indeed be to cross a "wet desert," apparently lifeless and silent but for the wind and the wing beats of passing ravens. This is peatland at a hard-pressed extreme. But the complex variety and condition of the bogs should by now be clear: they range from the bleakest of anaerobic brown muds to flowery lakelands brimming with dragonflies and skylarks. The range of animals and birds that use the bogs for feeding, breeding, and passing refuge is considerable. Do we include among peatland fauna the badgers digging burrows in dried-out peat banks at the fringe of farmland or the herons *(Ardea cinerea)* scooping up frogs from their peatland mating pools in spring? Animals of peatland have to belong in some sharper sense, be part of the ecosystem in some more essential way.

Frogs

The common frog *(Rana temporaria)* would, indeed, suggest itself as a good focus, as it eats a wide variety of the invertebrates that thrive in bogs, and is itself, as tadpole and adult, a regular food of other peatland predators. Yet a study of frogs in bogland habitats of the west of Ireland found them much less inclined to get wet than one might imagine. They do, of course, spawn in water, early in the year, when the surface of bog holes and temporary pools glistens with vast outpourings of eggs. But they do not feed in water, and thus deny themselves the teeming, aquatic larvae of aerial insects. Indeed, they are highly terrestrial as adults, moving away from the pools to hunt for ground beetles and the big black slug of the bogs, *Arion ater.*

Beetles, Mites, and Springtails

The beetles, often large *Carabus* species with a beautiful metallic sheen, are themselves among the bog's more conspicuous predators. They feed mostly on tiny detritovores, minute animals that in turn feed on dead organic matter both plant and animal and that live in the top few inches of peat and the hummocks of sphagnum. A study of mites on blanket bog in Mayo found 106 species, with densities of nearly 15,000 per square yard (18,000 per m²) in some of the hummocks. Springtails *(Collembola)* can also reach extraordinary densities: up to 54,000 per square yard (64,000 per m²) around the roots of purple moor-grass, which are also the focus of bacterial activity. The bog's springtail species show an almost exclusive association with the root area of particular plants; distinct communities are also found with the roots of black bog-rush and with Cladonia lichens on heather stems. This shows the specialized nature of many of the microhabitats of peatland, functioning almost independently of any overall animal community. Plant detritovores and decomposers find themselves in an extreme environment in bog. Its acidity rules out lumbricid earthworms and small crustaceans, its waterlogged and anaerobic depths are inaccessible to mites and springtails, and its low temperatures further inhibit the activity of microfungi and bacteria on which detritovores feed.

Biting Midges

The surface of wet peat is nursery to the larvae of midges and other small flies at many millions to the acre. They are minute, wormlike creatures, some of them dyed vividly red by hemoglobin, needed to bind the oxygen so scarce within the bog. The first midges to emerge as winged adults, at the end of May, are mostly male, but their swarms are soon joined for mating by newly emerged females, some species of which need a blood meal to promote development of their eggs. Of Ireland's thirty or so biting midges, only four are a torment to people, and the commonest of these on the bogs is *Culicoides impunctatus*. This is also the notorious Highland Midge of Scotland, where its impact on tourism, farming, and outdoor industry has prompted many decades of research. It includes a finding that the most important factor prompting biting in *Culicoides impunctatus* is a decline in the sun's radiation to below 260 watts per square meter, either toward twilight or in cloudy weather.

Peatland pools and bog holes are a ferocious subaquatic world in which the larval stages of many aerial insects prey on each other. They are joined by beetles and spiders that are specially adapted for hunting on or under the water. Especially impressive is the great diving beetle *(Dytiscus marginalis)*, more than 1 inch (3 cm) long, a polished, streamlined oval gliding through the water with a wave of legs fashioned into paddles. It must come to the surface to breathe from time to time, but, with its equally murderous larva, it consumes great quantities of tadpoles: perhaps twenty large ones in a day.

Aquatic Spiders

Among the forty or so species of spider that find niches in the bogs, two of the largest are also aquatic. The unique water spider *(Argyroneta aquatica)* is widely distributed in the ponds and ditches of Europe, and the old bog holes, densely fringed with sphagnum, make a perennially undisturbed retreat. This spider contrives an almost totally submerged life by creating a series of "diving bells" beneath domed webs anchored among the water plants. It fills them with air brought from the surface, trapped as a silvery skin among the fine hairs on its abdomen and then stroked off in bubbles beneath the webs. The bells are used as separate quarters for feeding, courtship, and overwintering. In August the females spin their white egg cocoons just below the surface, which is all that betrays the spider's presence from above.

The bog or marsh spider *(Dolomedes fimbriatus)* is one of Europe's largest, with a body up to almost an inch (2 cm), dark brown, with a broad, yellowish stripe along each side, and long, stout legs. It lurks in the mosses at the edge of the pool, ready to run out across its surface as if the water were its web. It is covered in fine, water-repelling hairs, which both give it buoyancy and provide a long-lasting film of air when it dives for safety or to hunt. This plastron holds enough oxygen for hours, as it constantly absorbs the gas dissolved in the water.

Dragonflies

In the dark, copper-colored water of the bog, the ferocity of great diving beetle larvae in pursuit of tadpoles is entirely equaled by that of the nymphs of dragonflies and damselflies, nourishing a growth that swells through a dozen moults of skin. Hoisted finally into the air for a brief month of flight the dragonflies are still formidably carnivorous, but their hunt is now buoyant and beautiful to watch. Areas of old bogs near village settlements, cut-over for fuel by the community, have a wide assortment of acid pools and bog holes and are, as John Feehan pointed out in *The Bogs of Ireland,* among the most important dragonfly habitats in Ireland: "When the time of year and the weather is right, as many as six or seven different species of these magnificent creatures can be seen on the wing at the same time and in the same place." One of the commonest of Ireland's twenty-nine species is the large four-spotted chaser *(Libellula quadrimaculata),* a strongly territorial dragonfly often seen hovering and circling over boggy pools. It is a darter, making repeated sorties from a perch. The common hawker *(Aeshna juncea)*, with a wingspan of almost 4 inches (10 cm), devours great quantities of midges and mosquitoes as it patrols its beat above ponds or streams in blanket bog.

The rarest Irish dragonflies are both emeralds—actually a striking bronze green—and both occur along the shores of the Killarney lakes, an overlap in habitat unique in Ireland or Britain. But the downy emerald *(Cordulia aenea)* is on the wing from mid-May to the end of July, and the northern emerald *(Somatochlora arctica)*, a glacial relict species, flies later, from mid-June to mid-August.

Damselflies

Among the damselflies, the slender, less assertive, relatives of dragonflies, the Irish damselfly *(Coenagrion lunulatum)*, unique in its pattern of bright blue and black, went undiscovered until 1981. This species is rare in western Europe but still common north of the arctic circle in Finland. In Ireland, its colonies are widely scattered through the northern half of the island, at small lakes and bog pools with plenty of floating leaves and fringing reeds.

An island-wide survey of Ireland's dragonflies began in the summer of 2000, and almost immediately birders watching coastal lakes and fen-fringed pools in County Wexford, at the island's southeast corner, were making the first recorded sightings of the big and beautiful emperor dragonfly *(Anax imperator)*, a migrant species from southern Europe. This and other sightings suggest that global warming is already bringing new migrants northward to Ireland and that colonization may have already begun.

Marsh Grasshopper

The association of many dragonflies with peatland habitats does not make them tyrphobionts, species that are obliged to live in bogs—other kinds of aquatic habitat can meet their needs just as well. One fascinating species that shows its affinity for peatland both in Ireland and Britain is the large, but rare and secretive, marsh grasshopper *(Stenophyma grossum)*. In Britain it is known from a few sites on quaking bog, having disappeared from former haunts in the fens of East Anglia. In Ireland it belongs to very sheltered and wet lowland bogs in the west, where it feeds on purple moor-grass and bog-cotton, and there have been recent finds in raised bogs in the midlands. The grasshopper's striking colors (orange, green, and black) and locustlike silhouette in flight make it an exotic insect to meet in an Irish setting. The female, often more than 1 inch (3 cm) long, lays her eggs at the base of grassy tufts and covers them with foam. Its call is scarcely audible and monosyllabic rather than the typical grasshopper's chirp.

Viviparous Lizard

The marsh grasshopper is really a southern insect, at the northern edge of its range. But the unconsidered attraction of bogs for sun-loving insects is also implicit in the widespread presence of Ireland's only native lizard *(Lacerta vivipara)*, the viviparous lizard. It is another extremely wary creature, but is almost as likely to be glimpsed basking among heather on a dry hummock in the bog as in a hollow of sand dunes or on a limestone wall in the Burren or on Aran. The species' main adaptation to northern climates is in nurturing its eggs internally and giving birth to its young alive (usually in July), but like all lizards it needs warming by the sun to generate the energy for hunting invertebrates; otherwise it remains motionless. While it certainly thrives on bogs and can tolerate wet conditions, it is far from tied to a bog habitat: given a sunny, undisturbed, and well-drained habitat, it is equally at home in the rough grass of a railway embankment or in open woodland.

Smooth Newt

The smooth newt *(Triturus vulgaris)* is often confused with the common lizard—when, that is, the ordinary Irish country dweller has noticed the existence of either. The newt, terrestrial outside the breeding season, has also often been presumed to share bogland pools with the frog, but a study of the breeding sites chosen by each gives a more complex picture. The newt does, indeed, go out of its way to breed in ponds that also attract frogs, probably because frog tadpoles are important in its diet. It is Ireland's only newt, and is thus spared the competition it would have in Britain from, for example, the great crested *(Triturus cristatus)* and palmate newts *(Triturus helveticus);* it has not expanded its range into the more acidic ponds that the palmate newt, in particular, finds acceptable. Indeed, it seems to avoid bogland, while using the calcareous waters of cutover fens that fringe many midland bogs. It especially likes ponds with a fringe of scrub and rotting logs and tree stumps for cover.

Moths and Butterflies

Dependence on particular bog plants for food is a definite kind of belonging. Among the Irish Lepidoptera, some 150 kinds of moths and butterflies live on peatland, each in a different niche, but few of them could not find their food plants elsewhere. The large heath butterfly *(Coenonympha tullia)* is considered the one true bog species, laying its eggs on bog-cotton and the white-beaked sedge *(Rhynchospora alba)*. Rather drab and peaty even in its coloring, and always sitting with its wings closed, it flutters abroad for a few weeks of mid-summer, zigzagging low over the surface of the bog.

The beautifully patterned marsh fritillary *(Euphydryas aurinia)* is endangered throughout Europe because of the loss of so many wetlands; Ireland holds some of its strongest populations. Its colonies need an abundance of the nectar-rich devil's bit scabious *(Succisa pratensis),* a much smaller and more delicate, purple-flowered relative of the teazel *(Dipsacus fullonum).* This plant grows in wet meadows and hedgebanks as well as on bogs, so that the butterflies are not confined to peatland, though much identified with it. The butterfly flies only in sunny weather, chiefly in June, and the black caterpillars hibernate for the winter under a web spun among the leaves of the food plant.

At the midland raised bogs, in particular, dense thickets of gorse—in Ireland known variously as furze or whin—set a spiny margin to the peatland proper, and in spring erupt in a froth of golden, coconut-scented flowers. Among them flies the small but lustrous green hairstreak butterfly *(Callophrys rubi)* for which gorse is a primary food plant.

True tyrphobionts among the peatland Lepidoptera are found among the many moths with heather as the food plant. The caterpillar of the much-admired emperor moth *(Saturnia pavonia)* is itself quite princely, donning a brightly banded doublet just at the right time to blend with the green, pink, and black of the flowering heather. Cryptic coloration is well-developed among all the heather moths and their larvae, exposed in such an open and well-lit habitat. The lovely yellow underwing *(Anarta myrtilli)* keeps the color of its hind wings hidden until the moment of flight, when the sudden flash of yellow disconcerts a predator. In the rusty-colored peatland willow *(Salix cinerea)* the substantial caterpillar of the puss moth *(Cerula vinula)* relies not only on camouflage but on a range of defensive behavior that has fascinated naturalists for centuries: first, a humped and menacing "threat" posture, with black spots mimicking eyes; then a brandishing of red, whip-like filaments from prongs at the end of its body; finally (this against Ichneumon flies that are its special, parasitic enemies) an acid spray squirted from its thorax.

Red Deer

Among the vertebrates of peatlands the red deer gets to grips with bog in the most physical, even passionate, way. In his graphic field study *The Wild Red Deer of Killarney*, Séan Ryan describes the agitated wallowing by the stags in wet peat that forms part of their ritual behavior in the rutting season. This frenetic activity of digging and rolling in a peat slurry, which leaves the antlers darkly stained, is quite separate from the regular summer wallowing of the hinds and calves, most probably to get rid of parasites. But wallowing may be the limit of the deer's actual use of peatland, and the behavior could be duplicated in other marshy habitats. Like the Irish mountain hare, the deer certainly thrives on open peatland and finds security there, but both animals prefer to eat soft grasses more characteristic of mineral soils. Research on the Irish mountain hare has made a point of the animal's widespread distribution, from sea level to high mountain heather. But where the heather has disappeared, as on so many of the overgrazed western uplands, this winter browse is lost to the hare, along with its shelter and warmth. On many hills in Galway and Mayo, there is not enough vegetation to retain the impress of the animal's "form" or lair, and hares have joined the sheep in eating silage doled out into troughs on the bare mountain.

The persecuted red fox frequents remote and open bogs as a predator on hares and nesting birds and a consumer of sheep carrion. It sometimes makes burrows in dry peat mounds, but more usually commutes from dens in conifer plantations. Able to range freely, it would probably opt for more comfortable and productive habitats in farmland or even suburbia.

Birds of the Bogs

Among birds that use the bogs, the Greenland white-fronted goose *(Anser albifrons flavirostris)* has a specific nutritional link, arriving in winter to probe for the bulblike storage organs of the white beak sedge *(Rhynchospora alba)* and common bog-cotton *(Eriophorum angustifolium)*. But disturbance and destruction of bogs have pressed the geese increasingly into protected wintering grounds and a diet of cultivated crops and grass.

The red grouse is the one bird that remains on peatland all year, feeding almost exclusively on the shoots and flowers of heather and depending on it also for shelter and concealment. A race of the circumpolar willow grouse *(Lagopus lagopus)*, it is unique to Ireland and Britain and keeps its brown plumage even in winter. As a gamebird, its numbers can respond dramatically to the large-scale management of moorland to create a patchwork of heather of different ages. By comparison with the heather on the commercial shooting moors of Scotland, that of Ireland is generally poor in production and food value, even on the well-drained hillsides of the east. There have been major long-term declines in red grouse over most of the bird's range, but in Ireland overgrazing, peat mining, and the loss of moorland to conifer forestry have helped to reduce the grouse population dramatically to perhaps fewer than five thousand pairs. The excessively high sheep numbers of the late 1900s also sustained more upland predators: foxes, hooded crows, and ravens *(Corvus corax)*.

Grouse hide their nests in old, tall heather, but share the need for camouflage with other ground-nesting birds of the bogs. Snipe *(Gallinago gallinago)*, curlew *(Numenius arquata)*, skylark *(Alauda arvensis)*, and meadow pipit *(Anthus pratensis)* all have closely streaked plumage that continues the weave of wind-bowed moor-grass and sedge. On the lonelier bogs of the west in spring, the skylark may be the only nesting bird in evidence, soaring up in a pinnacle of song and then falling back to silence and invisibility. The black and yellow brocade of the breeding golden plover *(Pluvialis apricaria)* is both beautiful and cryptic, but few in Ireland have seen it. Habitat loss, overgrazing, and predation have reduced the species, as an Irish breeding bird, to a thin sprinkle of nests among eroded peat on western mountain tops or spaced out between the sphagnum hummocks of the wettest lowland bog.

Blanket bog is accorded more species of birds than raised bogs, embracing so many more habitats at its various levels between mountains and sea: ring ouzel *(Turdus torquatus)*, for example, breeding in high rocky gullies (in Kerry and Donegal), or—more generally—the dipper *(Cinclus cinclus)* hunting underwater for insects in rapid moorland streams. But there is still no great richness of numbers or diversity, and birds of prey are correspondingly sparse.

Peregrine Falcon

On the uplands of the west, pairs of peregrine falcon *(Falco peregrinus)* are widely scattered. In the whole of Connemara there are perhaps ten pairs, three or four of them in the mountains and the rest on coastal and island cliffs.

County Kerry probably has fewer than twenty pairs, and many apparently suitable crags are left without tenants. The severe exposure to wind and rain on these mountain cliffs may help to explain the lower survival of the peregrine chicks, compared to those in nests elsewhere.

In Ireland as a whole, however, the story of the peregrine is one of triumphant recovery from the international population crisis caused by organochlorine insecticides that weakened the falcon's eggshells. At their lowest point, at around 1970, the Irish birds were down to about seventy breeding pairs, By the end of the century, they had passed 350 pairs and their population was still growing steadily. Away from the mountains, there was a striking expansion into quarry sites for nests, even where noisy operations continued. In northern Ireland, no fewer than thirty-five quarries had been colonized by 1997, and it seemed that the pairs using them were actually more productive than those nesting on natural cliffs.

The trend to quarry nesting on low cliffs reflects a general public sympathy with protective wildlife laws. Another sort of bird enthusiasm, however, has brought mixed consequences for the peregrines. A spectacular increase in their numbers in the southeast is unmistakably linked to the routes taken by racing pigeons. On the cliffs of the Waterford coast, peregrines actually doubled their numbers through the 1980s, eventually nesting at an average of a mere 1.2 miles (2 km) apart. Pigeon owners, angry at the disappearance of their homing favorites, have retaliated by leaving poison-laced pigeon corpses on the headlands and by destroying some of the more accessible nests.

Moorland Merlin

Thus, despite the popular image of peregrines perched watchfully on crags above mountain moorland, their expansion in Ireland has brought them into many different settings. The island's smallest falcon, however, the merlin *(Falco columbarius)* still spends its summers mainly in moorland and blanket bog. Not only is it scarce, but its hunting flight, tucked well below the bog's brown horizons, is notably discreet. It perches on rocks to watch pipits and skylarks and flies low and fast in long pursuits of them, or dashes after dragonflies and large, day-flying moths. In his *Natural History of Connemara*, Tony Whilde noted how vulnerable the merlins are to collisions with the many new sheep-wire fences erected across the western moors. The disappearance of heather from overgrazed hillsides has reduced the density of little birds available as prey for the merlin and has influenced its own choice of nesting sites. In Connemara, one of the merlin's strongholds, most birds resort to small islands in the bog lakes, where tall, dense heather gives them cover. Elsewhere, the loss of moorland to forestry has taken merlins off the ground altogether and into old nests built by crows in conifer plantations.

Kestrels and Hen Harriers

The commonest bird of prey on the bogs (and in Ireland in general) is the kestrel *(Falco tinnunculus)*, typically seen hovering as a still, dark speck perhaps 130 feet (40 m) above the ground. Spotting its prey in the vegetation, it glides steeply and drops the last few feet with outstretched talons, seizing small rodents, young birds, frogs, or even beetles. The luxuriant ground vegetation in the early stages of conifer afforestation can produce a sudden upsurge in mice and shrews and small birds and has provided temporary bonanzas both for the kestrel and the hen harrier *(Circus cyaneus)*.

This large and graceful upland bird of prey quarters the moors intently at a low level, pouncing on grouse, pipits and snipe, or young hares. It is easily shot and a particular target of gamekeepers, which helped bring it close to disappearance from Ireland in the early 1900s. Its recovery began at mid-century, when the state's forestry program planted large tracts of peatland with spruce and pine. The young plantations provided the hen harriers not only with extra prey but secure nest sites. Numbers of breeding pairs rose in the southern counties, only to fall again twenty years later as the forests matured and closed in and more heather moorland was reclaimed for farming. Thus, the harrier's fortunes seem destined to swing to and fro with vegetation changes. Although the bird was a widespread species of bogland before the drainage and reclamation of the early nineteenth century, it has never reestablished itself in the west, despite intensive conifer forestry. A typical breeding stronghold is the Slieve Bloom Mountains of the southern midlands, where the cycles of forestry and supply of prey seem to provide a stable habitat. But the harriers roam widely in winter and may turn up in most parts of the island.

THE DARK FORESTS

The challenge of the bogs as wastelands, inviting drainage and reclamation, runs deep into Irish history, and attempts to grow trees on peatland reach back for more than two centuries. In the west of Ireland, in particular, improving landlords experimented with several kinds of pine and other conifers,

Bob Quinn

The bare landscape of a Connemara shore today.

and in 1891 nearly 2.5 million trees of assorted species were planted, at state expense, on a windy bog in Connemara. A few stunted and crooked Scots pines are all that survive there today.

Modern programs of state forestry on blanket peat gathered pace in the mid-twentieth century with the aid of new technology: low-ground-pressure vehicles and a novel plow that could make a deep drain and spew out a mounded ribbon of peat on which to plant the trees. Within a few decades, more than 494,200 acres (200,000 ha) of largely undamaged peatland, north and south, were shaded by a second dark blanket: close ranks of conifers grown from seed from western North America, including the Sitka spruce and the lodgepole pine, a pioneer species for the poorest of acid peats.

In the Republic, the planting of bogs had political and social advantages. It fulfilled a highly symbolic goal of national endeavor without annexing agricultural land, and it created jobs in regions of high emigration. The penalty of this "social" forestry was a widely fragmented pattern of plantations, many on hillsides where growth was stunted and storms toppled the unthinned, shallow-rooting trees. Purchased parcels of bog were planted to the boundaries, so that straight-edged blocks of spruce came to disfigure many of Ireland's most admired mountain valleys.

By the 1990s, ecological and scenic concerns were helping to produce a more rational policy, directing forestry to the rushy grass of marginal mineral land where Sitka spruce, in particular, grows phenomenally well in the moist Irish climate. There was little new planting on the virgin blanket bog that still made up one-third of the state's forestry land bank. As peatland forests were clear-felled and replanted, they were redesigned to contours more in sympathy with the landscape and dressed with a greater diversity of trees. Coillte, the state's forestry company, was declaring a new appreciation of the multiple roles of forestry in a countryside as intimate as Ireland's. But for the 370,650 acres (150,000 ha) of blanket peat that had already disappeared under Coillte's conifers there was no going back, and grants to private forestry continued to encourage peatland planting outside established conservation zones.

The ecological changes that afforestation brings, and their consequences for peatland wildlife, have been well-studied, notably in Britain, where coniferization of the Scottish moorlands has been intensely controversial for its impact on breeding birds. In the first phase of a plantation—drainage, planting, fencing, and fertilizing with rock phosphate—the luxuriant growth of sedges, moor-grass, mosses, and heather is often in dramatic contrast to an overgrazed hillside outside the fence. This rank flush of vegetation provides food and cover for small mammals and birds, and the young, bushy trees become song posts or even nesting places. The calls of stonechat (*Saxicola torquata*), willow warbler (*Phylloscopus trochillis*), meadow pipits, and other passerines provide a dawn chorus on hillsides that had been largely silent. For perhaps a decade, this prethicket stage is strikingly productive and diverse. Then, as the trees reach out to shade the ground, the herbage is suppressed and the birds of open moorland have entirely disappeared. Their place is taken by woodland birds, twittering higher and higher in the canopy. Siskin (*Carduelis spinus*), goldcrest (*Regulus regulus*), and coal-tit (*Parus ater*) are typical of little birds more easily heard than seen in the tall aisles of trees.

In this hushed phase of mature growth, with the peat long buried under a dense mattress of pine needles, the biodiversity of a conifer plantation might seem at its lowest ebb. But the model of avian change outlined already for Scotland needs to be considered for Ireland with two important differences in mind. First, the northern and western Scottish moorland is an important breeding ground for waders (such as greenshank, *Tringa nebularia*) and divers that are virtually absent from Ireland, which is mostly south of their breeding range. Second, the island's conifer forestry is a much more broken mosaic: the average forest plot is much smaller than those in Britain and has a relatively greater edge length. The ecotones, the narrow strip where two or more different comunities of plants meet, are generally more diverse in their flora and fauna than either habitat considered separately.

Ecologists from the National University of Ireland, Cork, have been assessing bird diversity in the mature conifer forests of southwestern Ireland. In the average acre of trees in the breeding season, about half the birds are goldcrests, but the rest are drawn from thirty species, most of them common birds of the farming countryside. They include the valued songbirds of hedgerows along with specialized seed-eaters such as the crossbill *(Loxia curvirostra)*, currently colonizing the forests in periodic eruptions from Europe. Over the course of a year, thirty-eight species were drawn to the conifers, with the strongest diversity in Norway spruce *(Picea abies)* and Douglas fir *(Pseudotsuga menziesii)* rather than the Sitka spruce and lodgepole pine that form the great bulk of the forests. A biodiversity component should figure in selection of tree species, the Cork team suggested, together with more broadleaves, more light for the shrub layer, and changes in the size and shape of forest plots to give a longer edge.

None of these suggestions seemed to fit particularly well with the traditional disciplines of commercial conifer forestry. But in the last decade of the twentieth century, international concern for the protection of forests resulted, in Europe, in a ministerial consensus known as the Helsinki Process. This put maintenance of biodiversity high in the definition of "sustainable forest management." Biodiversity also emerged as the first concern about Ireland's forests in a wide-ranging consultation process conducted by Coillte. The outcome, in 2000, was a management program with a range of environmental guidelines relating to water quality, archaeology, and landscape, as well as biodiversity and the environmental impact of harvesting.

Coillte has also acknowledged the damage done to hill streams and fisheries by the establishment and growth of upland forests. When the trees are first planted, new drains through bog can produce surging spates of water that scour out natural watercourses and smother downstream riverbeds with peat silt. Leaching out of phosphate fertilizer and pesticides can both enrich and poison the water. As conifers grow, their scavenging of chemical ions from the atmosphere into soil water can, on some bedrocks, increase acidity, which in turn raises levels of aluminum, toxic to stream invertebrates.

All these effects have been demonstrated in Ireland. Dramatic falls in pH have been registered after heavy rain in areas such as Connemara, where granite bedrock gives the streams a low buffering capacity. In west Cork, forest pesticides were found in the eggs of the dipper, and acidification has upset its underwater insect supply.

In older hill forests, trees were planted right to the edge of streams, robbing them of sunlight and of wind-blown organic and mineral debris that supplied much of their nutrients. Coillte has now cleared the banks, opening streams to the light and a more normal fringe of vegetation. It has stopped felling trees by machine in very wet weather and puts sediment traps in the streams to catch peat silt that once smothered trout-spawning grounds. And it has changed pesticides for control of the pine weevil *(Hylobius abietis)*, finally relinquishing its more persistent poisons.

The upland conifer forests have not been a totally negative change ecologically. They have attracted birds such as long-eared owl *(Asio otis)*, crossbill, siskin, and jay *(Garrulus glandarius)* to areas of the west that did not have them, and lush vegetation at ragged, windblown forest edges inside the fences has given shelter to otters, pine martens, hares, and other mammals. It has also served to dramatize the bleak and lifeless expanses of overgrazed bog on the other side of the wire.

OVERGRAZING: MISADVENTURE IN THE MOUNTAINS

The western uplands are a largely human-made landscape, their contours bared to centuries of rough grazing and the stealthy advance of blanket peat. The long poverty of the inhabitants, the calamity of the Famine, and the steady depopulation thereafter created an aura of bleakness that ran in tandem with the modern cultural relish of "wild" landscape viewed from afar. Thus, the steady degradation of the western hills through the closing decades of the 1900s was only slowly admitted and confronted, and not before grave ecological damage had been done.

The causes lay first in a decent concern for people. A drift from small farms on poor land has been a general European problem, and the schemes of the European Community (now the European Union) were drawn up specifically to encourage families to remain in "less favoured areas." Among these, in 1980, were payments or subsidies on sheep per head that encouraged Irish farmers in the western counties to build up their flocks of ewes. But the EC planners seem not to have realized—or been told—that most Irish hill farmers were wintering their sheep on upland commonage on fragile, peaty soils with easily degraded vegetation.

By the 1990s more than two million animals—one-quarter of all Ireland's sheep—were located in Galway and Mayo. "During the wet winter months," recorded botanist Andrew Bleasdale in an unpublished doctoral thesis, "the appearance of the Connemara uplands is at its most depressing, with starving sheep and vast expanses of bare peat." Rain swept the peat silt into rivers, and some hillsides were eroded into moonscapes of bare rock and gravel. From a distance, the blackened, bruised appearance of the uplands was obvious to tourists and locals alike.

Even where actual erosion was absent, overgrazing degraded the vegetation in ways charted by Bleasdale and others. First to disappear was the heather. Then the moor-grass was depleted, leaving islands of bare peat between the tussocks until the coarse and unpalatable mat grass *(Nardus stricta)* spread slowly in its place.

The impact on upland plant communities and the birds, mammals, fish, and insects of the peatland ecosystem has been severe. Most of the overgrazed peatlands are in Special Areas of Conservation chosen for the European Natura 2000 network (see Chapter 17), and the European Court of Justice has castigated Ireland for failing to protect peatland habitats that the European Union's own schemes had helped to despoil. In the final decade of the twentieth century new subsidies were introduced—this time to reduce the flocks on the hills—and the western small farmers were being persuaded into new regimes of husbandry under the Rural Environment Protection Scheme. It may now take decades to restore some measure of biodiversity in the uplands and a pastoral balance in which sheep cease to be "woolly maggots," in the angry epithet of some naturalists.

THE CUTAWAY: SPACE FOR A NEW WILDERNESS

To travel by helicopter from Dublin to Galway above the Irish midlands is to fly past hard-edged tracts of chocolate-brown plain, criss-crossed with drains, and sprawling machines that crawl across them. Glistening at intervals are long ridges of milled peat, shaved from the surface of the bog and piled up under plastic sheeting. This is the industry of peat extraction, mechanized for more than half a century as an Irish state enterprise. Peat pulverized for power station fuel, for home-stove briquettes, for gardener's composts and mulches, and for poultry litter has steadily erased the greater part of more than 741,300 acres (300,000 ha) of raised bogs.

The national awakening to their scientific and wildlife interest and the long rearguard battle for peatland conservation are for discussion in Chapter 17. In this chapter my concern is with the future of the industrial cutaway—the great, created surface of hollows and hillocks in which postglacial gravels and fen peats are exposed for the first time in 10,000 years. As more and more bogs reach a stony bottom, after as much as fifty years of milling, some 185,000 acres (75,000 ha) of prehistoric landscape are up for grabs. Pilot projects already suggest a vigorous cooperation from nature in creating new habitats for wildlife.

Bord na Mona, the state peat company, began research on the possible uses of cutaway as long ago as the 1950s. The potential for growing biomass fuel, vegetables, and other tillage crops proved disappointing. Good grassland was possible on the higher, free-draining areas, and commercial forestry seemed an obvious option, but one seriously frustrated, as it proved, by the midlands' late spring frosts. Thousands of acres of cutaway are now under squares of grass and conifers, but by the 1990s it was obvious that most of the total area was suitable only for "nonproductive" uses. The virtues of a parkland mosaic, in which recreation and tourism was integrated with farming and wildlife areas, found strong appeal, both with Bord na Mona and with local communities concerned for their future livelihoods. A pilot management project— the Lough Boora Parklands—was launched in County Offaly, where the oldest area of cutaway was already well-advanced in a natural recolonization by plants and birds.

Turraun, now a nature reserve, took twenty years to create its own mosaic of habitats. On limestone boulder till, with a skin of woody fen peat, grew self-seeded woodlands of birch, willow, and Scots pine, open moor-grass grasslands, and areas of heather and moss. On lower ground, where reed peat carpeted a deep layer of shell marl, enclosure quickly created a shallow lake, reed beds, and winter wetlands. Marsh arrowgrass *(Triglochin palustris)* was the first colonist of wet, bare peat, and its tubers helped to feed the arriving winter flocks of whooper swans *(Cygnus cygnus)* and groups of Greenland white-fronted geese. Thousands of lapwing *(Vanellus vanellus)* and golden plover have come, too, foraging on the new grassland fields and roosting on islands in the newly created lakes of the parklands.

By 1998 some 130 species of birds had been recorded on the cutaway, among them the last colony of Ireland's wild gray partridge *(Perdix perdix)*. Songbirds breed there in high densities, especially in heathery areas at the fringe, and this provides the merlin with good hunting. The young forestry plantations are now one of the few places in Ireland where the nightjar *(Caprimulgus europaeus)* can be heard churring on summer nights, encouraged by the abundance of moths and other insects.

Creation of artificial lakes and parklands is costly, and ecologist John Feehan has argued that the great bulk of the cutaway should be left to develop spontaneously as "a new midland wilderness, a mosaic of fen and reed-marsh and water, pine and birchwood and heath." The cutaway's diversity is already often richer than the bog it replaces, prompting Feehan to quote the Rockefeller scholar René Dubos who wrote in *The Wooing of the Earth* on the potentiality of damaged ecosystems to "undergo adaptive changes of a creative nature that transcend the mere correction of damage." Left to their own dynamic processes, Feehan insisted, the "ravaged" bogs will regenerate wild landscape with an ecological vitality and diversity as great as any in the past 10,000 years.

SELECTED REFERENCES

Carruthers, T. 1998. *Kerry: A Natural History.* Cork: Collins.

Feehan, J., and G. O'Donovan. 1996. *The Bogs of Ireland.* Dublin: University College Dublin.

Gibbons, D. W., J. B. Reid, and R. A. Chapman. 1993. *The New Atlas of Breeding Birds in Britain and Ireland: 1988–1991.* London: Poyser.

O'Connell, M. 1994. *Connemara: Vegetation and Land Use since the Last Ice Age.* Dublin: Office of Public Works.

Pilcher, J., and V. Hall. 2001. *Flora Hibernica.* Cork: Collins.

Whilde, A. 1994. *The Natural History of Connemara.* London: Immel.

Chapter 8

The Magic of Limestone Country

S een across Galway Bay from the Connemara shore, and often underlined in indigo by a cloud shadow on the sea, the limestone hills known as the Burren raise a profile to the south that is modestly high (985 ft; 300 m, at most) and smoothly rounded, yet promising, even from this distance, some sort of topographical drama. For painters of the Burren, an imperative is to set a ridge or abutment against a shadow or dark horizon, to bring out the luminous brilliance in the stone.

Pinned down like this between sky and sea at the northern corner of County Clare, the Burren also gains a definition it can lack in some other, inland, approaches. Like Connemara, it has no border posts: *An boireann* is "a rocky place," and while there is absolutely nowhere rockier than the cliffs and terraces at the bare heart of the Burren, its southern perimeter is vague and untidy: slow hints of change that condense at last, among thinning meadows and thorn-set walls, into stone and stone upon stone.

Geology draws its own boundary to the Burren: the line at which adjoining strata of flagstones and shales were eroded, exposing softer limestone to the gentle corrosion of water. These tougher, younger layers of Carboniferous sediment form the upper battlements of the dramatic Cliffs of Moher nearby and thrust northward into the Burren to cap the high tableland of Slieve Elva. At the edges of the shale and its overlay of peat bog, the runoff of acid streams joins Atlantic showers in the slow work of dissolving the limestone and returning it to the ocean.

This deep weathering of the Burren's limestone was begun in the subtropical climate of the Tertiary, when water began to sculpt a karst landscape of flat pavements and deep fissures, of hollows the size of valleys and of egg cups.

Just as words derived from Irish (like "drumlin" and "esker") have provided science with terms for Ice Age landforms, so "karst" has borrowed from the languages of central Europe to make a name for a limestone wilderness: *krs* is Serbo-Croat for a rock or stone, and *kras* is Slovenian for a bleak and waterless place. The karstic uplands of Europe were once forested, however thinly, and their modern bleakness is the product of erosion, brought about by early human settlement. Much of the Burren, too, was laid bare by prehistoric farming of a dry but deceptively fragile terrain.

Ancient pollens drilled from lake muds at the margins of the Burren conjure the original pine woods that grew on its limestone pavements and the hazel that flourished in their shade. Soils retrieved from under ancient walls, stretching today across bare limestone, show the cover that was washed underground through cracks and fissures after the trees were cleared and burned by early farmers. It is all circumstantial evidence, but convincingly endorsed by the density of ancient structures on the most barren reaches of the limestone. Nowhere else in Ireland can be found such an extraordinary overlay of field systems, tombs, forts, and stone enclosures, beginning in the Neolithic and the Bronze Age and continuing, often inscrutably, into the early Christian period and beyond.

The Burren has more than seventy megalithic or great stone tombs, originally covered with earthen mounds, the earliest built about 3800 B.C. Some of them survive as irreducible cairns—dolmens—of massive slabs, held in place by the capstone like a weighty house of cards. Perhaps the most striking (certainly the noblest in tourist-poster silhouette) is the Poulnabrone portal tomb, excavated by Ann Lynch in 1986. A jumble of bones from a score of individuals were, she concluded, those of "special dead," buried in the course of six centuries. The skeletal fragments bore the signs of a lifetime's hard physical labor, and the concentration of wedge tombs and hut circles on the higher

Erratic boulder on a pedestal left by the solution of the underlying limestone.

ground of the Burren conjures a dense population of farmers in pastoral settlements, herding cattle, sheep, and goats.

The ruins of hundreds of ring forts or "cashels" on the limestone bluffs and terraces (circular, stone-built enclosures that housed farm families) take the Burren's pastoral story on into the Iron Age and the Early Christian period. One of the most elaborate of the cashels, Cathair Chomáin, is perched dramatically at the edge of a 100-foot (30-m) cliff, overlooking a ravine densely wooded with hazel. It was thoroughly excavated in 1934 as part of the Third Harvard Archaeological Expedition in Ireland. The cashel's concentric walls and massive inner rampart have a military appearance, but the robustness of the masonry probably had more to do with status than defense, and the artifacts that helped to date it (to about A.D. 600) were exclusively domestic. With its crowded hearths and magnificent view, Cathair Chomáin was home to perhaps forty or fifty, probably an extended family group, and the cattle bones they left behind were eloquent of their pastoral way of life.

The Burren soil research suggests that much of the first wave of erosion and loss occurred over quite a short period in the late Bronze Age (coinciding, perhaps, with the "nuclear winter" conditions thought to have followed a distant volcanic eruption in 1159 B.C.; see Chapter 5). There seems to have been another big increase in forest clearance after about A.D. 300, a time of marked expansion in Ireland's farming, and this, perhaps, was to compensate for the grazing already degraded or destroyed. But even with a much more sparse vegetation, the value of the Burren's stony uplands for nurturing and feeding livestock were to shape an exceptional pattern of seasonal husbandry.

In much of Europe, the peasant farming tradition included the practice of transhumance—the moving of cattle in summer to graze the upland pastures. In the west of Ireland generally this booleying (from *buaile*, a milking pasture) persisted until late in the last century as part of an open-field system of tillage. It exploited the rough grazing of the hillsides while these were moderately dry and accessible and kept the cattle out of unfenced plots of grain and (later) potatoes. In the Burren, however, from perhaps the Middle Ages onward, the custom was remarkably reversed. Along with flocks of sheep, the cattle were—and still are—driven up stony roads to winter in the winds of the high plateau.

The winter grazing may be sparse in general, but plants grow freely in the sheltered crevices and crannies of the rock. Their season of growth is long—

ten months or more in the year—because the bare limestone acts like a great storage heater, gradually releasing the warmth it has absorbed all summer. The plants are rich in calcium, which builds strong bones, and in the trace elements that go with a varied herbage. The stone remains dry underfoot, even in the wettest of winters, so there is little chance of picking up the parasites, such as fluke, that are common in the wet lowland pastures. Sturdy, healthy, and fluke-free, Burren-wintered cattle have always fetched top dollar, whether in the economies of Celtic clans or the echoing auction marts of the modern County Clare. The winter grazing, shared with the Burren's flocks of feral goats, now helps to preserve an ecosystem originating in catastrophe, a giant rock garden, created by chance and full of fascination and beauty.

THE BURREN FLORA

The floral excitement of the Burren is first of all, and irresistibly, aesthetic. Just as the simple thrill of its strange moonscape sculpture, its sensuous texture and light, comes well ahead of thoughts about geology, first encounters with its plants can bring a candid rush of pleasure and surprise, quite postponing many proper ecological questions. Some of these encounters are with a whole hillside, as in the first May blossoming, cream and gold, of mountain avens. "He who has viewed the thousands of acres of the Arctic-alpine plant in full flower . . . from hill-top down to sea-level," promised Robert Lloyd Praeger in *The Botanist in Ireland*, "has seen one of the loveliest sights that Ireland has to offer to the botanist." But more of these spectacles are miniature and contemplative. Ferns and wind-dwarfed shrubs in the rain-beveled fissures and hollows of limestone seem composed for revelation of their texture and form; individual orchids, fit for the cushions of a jeweler's window, work even better in close, eye-level focus against gray stone. It is this soft setting, more light than color, that makes such exotic accents of the flowers: intense magenta splashes of bloody cranesbill *(Geranium sanguineum)*, bright medieval gems of spring gentian *(Gentiana verna)*—exactly the color, as the taxonomist Charles Nelson distills it in *The Burren*, "of the darker part of a clear summer sky at sunset."

In the wake of such old-fashioned exaltations one can get down to the Burren's ecological conundrums, of which a single image gives the flavor. On a slope above the sea near the cape of Black Head, a spring carpet of mountain avens catches the sun on its gold stamens, just as it would on the summer tundra in the valleys of northern Greenland. And pushing between the leaves of this arctic-alpine plant is the creamy-green flower spike of the dense-flowered orchid *(Neotinia maculata)*. It is not the most charming of the Burren's orchid species but remarkable as a southern plant, far more at home beside the Mediterranean Sea or in the Canary Islands. Nowhere else in the world can this conjunction of plants—some typical of arctic conditions and some of subtropical habitats—be seen.

Or consider the maidenhair fern *(Adiantum capillus-veneris)*, its delicate, fan-shaped leaflets tremulous within a shadowy fissure in the limestone pavement. This is an essentially Mediterranean–Old World subtropical species of fern, almost at the northernmost limit of its range. Yet peep above the rim of the fissure and there, roots clenched into a pocket of peat in the bare rock pavement, may be the twining stems of bearberry *(Arctostaphylos uva-ursi)*. This little arctic shrub is now rare on Ireland's hills, but in the Burren it grows on the highest summits and down almost at the sea. The lemon-flowered hoary rock-rose *(Helianthemum canum)* hugs the rocks in much the same way, yet its affinities are with the rock-roses of the Pyrenees and Spain.

It is not hard to imagine the origins of the arctic and mountain plants, given the history of glaciations in Ireland, but it can be much more difficult to fit their postglacial survival into a picture of a Burren cloaked for millennia with pine and hazel and then with vigorous grasses and herbs. The answer may be that the northwestern hills of the Burren, their cliffs and screes, were never forested and thus served as an open refuge for postglacial plants until new pavement habitats were bared by soil erosion to the east and south. There are no convenient lakes to offer pollen cores that might confirm this. But the plant community typical of the high limestone pavements and summit plateaus at the northwestern corner is dominated by mountain shrubs such as mountain avens, bearberry, crowberry, and juniper, and its dense mats of stems and leaves all but crowd out the smaller herbs and mosses. To some botanists this seems to be a vegetation never colonized by trees, perhaps because of the fierce storms that sweep in from the Atlantic in winter.

This does not explain why arctic plants should creep down to colonize rock and turf directly beside the sea or thrive so strikingly in temperatures that also suit a Mediterranean orchid. Nowhere else in northwest Europe does mountain avens grow so profusely, and its approach to the sea is paralleled in Scotland only on the north coast, a further five degrees toward the North Pole.

The spectacular abundance of mountain avens is almost equaled by the sudden brilliance underfoot, between mid-April and mid-June, of the blue stars of spring gentian—again, from the summits to the sea. The gentian has the image of an alpine, conjuring the rocky peaks and fleecy clouds above Europe's high mountain pastures, but it also grows at quite low levels in central Germany, and, indeed, on the karst of northwest Yugoslavia. In Ireland, the flower is shared among the Aran Islands and the shores of limestone lakes in Galway and south Mayo. But since it was reported, some time before 1650, "in the Mountains betwixt Gort and Galloway, abundantly," it is to the Burren, its stronghold, that "the Gentian of gentians" has drawn its admirers. In his book on the Burren flora, Charles Nelson explored the plant's reproduction, given that frost is necessary for the successful germination of its seeds. Freezing temperatures are not unknown in County Clare, even by the sea, and enough gentian seed does germinate to permit some variation in the color of the flowers. But mature plants also produce underground stolons that sprout new rosettes at their tips, and these proliferate further, so that "by vegetative multiplication the spring gentian could persist for ever." In a paper on landscape heritage, John Feehan pictures "hordes of blue gentians" invading the abandoned limestone pastures of the Bronze Age in a healing wave, as part of a possible golden age of Ireland's natural diversity.

Problems surrounding the arrival of the Burren's southern plants are part of what Nelson has called "an unfinished enterprise that obsesses Ireland's botanists and geographers." He pointed to the particular challenges set by the dense-flowered orchid that has a few other stations in Ireland but otherwise keeps to Mediterranean latitudes. Like all orchids, its life cycle is intimately bound up with pollinating insects and symbiotic soil fungi, which, as Nelson argued, could not have been present in the severe tundra conditions of 15,000 years ago: this orchid has to be a postglacial migrant, with all the questions this invites.

On the other hand, the dust-fine spores of fern, flung out into the wind, comfortably explain the northward dispersal of maidenhair fern to limestone cliffs and crevices on the mild western coasts of Ireland, Cornwall, and Wales. In the Burren, it penetrates even into Aillwee Cave, growing by the artificial lights switched on for tourists, and must regularly be scraped off the rocks with toothbrushes. Conditions in the deep fissures of the limestone pavements are almost cavelike in their shelter, humidity, and constancy of temperature. Within a few hundred yards of the sea, the black stems of maidenhair fern spring up through the shadows to set a fluttering green line of fronds where the crevice widens to the light.

Most of the Burren's limestone is almost flat-bedded, and its level pavements are exceptionally strange to those who walk them, as if rock were in some way acting out of character. On some, the stone is pillow-smooth, every edge rounded, and the long, deep crevices that mark out its jointing are perhaps a foot (30 cm) wide and 3 feet (1 m) or more deep (geographers reared on the limestone topography of England call the flat pavements "clints" and the deep crevices "grykes," but in Ireland the soft Gaelic word *scailp*, for a cleft in a rock, seems more natural and sympathetic). On other pavements, the surface is entirely shattered and covered with angular blocks, some of which shift disconcertingly beneath one's feet with a harsh and hollow clunk. (Others have been said to "quiver in the breeze with a tinkling noise," but I have yet to meet them.) Either sort of pavement may be littered by shoulder-high erratic boulders of limestone or of granite from Connemara, carried along and abandoned by the ice sheets of the Midlandian period.

Some of the finest pavements are on the high shoulders of the hills, where masses of rock already sectioned and loosened by the percolation of water were swept away by the glaciers, leaving the slopes tiered in cliffs and terraces. In the lowland Burren, to the east, pavement stretches away from the scarps in bare, shattered rock sheets, fretted with cracks and hollows deepened by water. A dark brown organic soil has collected in many of these crevices, initiated perhaps by algal crusts, the wind-blown droppings of goats, and the growth and death of mosses. Eventually, higher plants can grow, including even fragmentary, tight-woven rugs of grass. The limestone of the high plateaus crumbles readily into a light soil for grassland, and thin bands of chert (a flintlike form of quartz) at these levels trap enough water in springs to let cattle drink. Otherwise, the upland karst is dry, and falling rain is lost at once through the maze of clefts and fissures.

The pavement and its crevices support a great range of plants, but few of them belong specifically to this world of stone. Maidenhair fern and a few lime-loving mosses certainly choose the shady fissures almost exclusively, and the scrambling red stems of madder *(Rubia peregrina)* grow more abundantly along the coastal pavements of the Burren than anywhere else in Ireland. The benevolent chaos of the shattered pavement seems especially to suit wood sage *(Teucrium scorodonia)* and the creamy, five-petaled burnet rose *(Rosa pimpinellifolia)*. The red helleborine *(Epipactis atrorubens)* also raises its mid-summer spires reliably in the midst of the rattling shards of stone.

But the microhabitats of the pavement are also miniatures of larger habitats elsewhere: little bits of grassland, rock crevices, woodland floor, peaty fen or heath, so that most plants have simply translocated from more typical communities away from the pavement. It is easy, however, to see in these wayward specimens, cupped in the gray rock or peering over a windswept parapet, some special distillation of their style and beauty. Never was the blackthorn *(Prunus spinosa)* a more tense and spiky structure than when dwarfed to a bonsai in a fissure's stony throat; never was the brassy carline thistle *(Carlina vulgaris)*, its tap-root jammed in a crevice, a more pagan, indestructible emblem of the sun.

Classification of the Burren's grasslands has been difficult, as the common pasture species are mingled with much more unusual plants. Under many south-facing slopes of the Burren, for example, the more diverse plant communities are dominated by a slow-growing, drought-tolerant alpine grass of transient beauty, blue moor-grass *(Sesleria albicans)*. This grows in the mountains of continental Europe but in the Burren is a glacial relict, like the mountain avens that share its abundance, and is grazed by cattle, goats, and hares. It is known simply as blue grass, and its stumpy flower spikes, as they rise in spring, are clusters of a striking, steely hue that fades when the anthers emerge.

But just as the apparent chaos of the Burren rock has slowly yielded to geologists, geographers, and hydrographers, the region's six hundred–odd flowering plants and ferns have challenged ecologists to establish order in their associations and habitats. In the 1960s, at the prompting of the Royal Irish Academy, two British botanists, Robert Ivimey-Cook and Michael Proctor, sampled the species and their soils to establish the main plant communities. In a paper titled *The Plant Communities of County Clare*, regarded as a classic of the method, these were defined by the presence of certain key plants— inseparable companions, as it were, in the shifting patterns of vegetation.

For example, the association of mountain avens and slender St. John's wort *(Hypericum pulchrum)* defines the Burren's most common and characteristic limestone grassland community, spread across the rock on mats of dark, organic soil. The mountain avens, ling heather, and large lime-loving mosses often dominate the community so thoroughly that "grassland" becomes a misnomer. The Burren's diversity of microhabitats makes it very rich in species: forty to fifty in a sample of 5 square yards (4 m²) is not unusual, and they may include the Burren's rarer plants—spring gentians, for example, and Irish eyebright *(Euphrasia salisburgensis)*.

This community also demonstrates a process that allows plants of acid soils, notably ling heather, to flourish over limestone, side by side with calcicole (lime-loving) species. A black organic soil accumulates rapidly under the mat produced by mountain avens and the mosses, and this begins to leach at the surface, allowing the heather to root in a thin layer of lowered pH. Heather acidifies the soil from decay of its own fallen leaves and flowers, and one by one the lime-loving species in the community disappear. The ultimate result is a heath, and a new kind of community—one marked, perhaps, by the silvery rosettes and white and velvety cat's-paw flowers of mountain everlasting *(Antennaria dioica)* or, on the higher pavements and plateaus, by a new partnership of mountain avens with bearberry and juniper.

In the natural succession of plants in the Burren, hazel is both a bold pioneer and penultimate woodland canopy. Its scrub already fills many of the deeper hollows of drift soil, especially in the broken country at the center, and presses up against the cliffs below Cathair Chomáin in a rustling sea of leaves. Hazel also roots into the deeper fissures of the bare pavements and twines into gnarled, subterranean groves with blackthorn, common and Irish whitebeam *(Sorbus hibernica)*, yew, and ash. When the twigs reach the pavement's surface, they are trimmed smoothly, like topiary, by the browsing of feral goats. These wild, watchful troops, with ragged coats and swept-back horns, clatter over the stone as the Burren's eerie familiars: a willful yet precise constraint absorbed within the ecosystem. From time to time, a gang of small-time rustlers with a muddy truck corners a flock of goats and whisks it away to some distant slaughterhouse, but the total disappearance of the animals would be lamentable. Like the cattle, which also include the pavement tree-tops in their browsing, the goats are a check on the hazel's ambitions. Without herbivores, hazel would weave an almost limitless thicket across the bare stone—might cover, indeed, most of the Burren and then, on the deeper soils, slowly mix with ash as the dominant tree.

In relation to many other plants, however, the Burren's large mammals have an altogether opposite role: rather than limit their spread they act as prime agents in seed dispersal. In such a windswept habitat, one might expect that eolian (wind-borne) transport would be far the more significant, but except for a few well-designed aerial species, most seeds are blown no further than a few yards from the parent plant. After the wind, "epizoochory" (seed transport by adhesion to people or animals) is the most important means of wide dispersal, and about half of the 460 flowering plants of the Burren are well-suited to it.

Sheep, with their fine-meshed fleeces, are well-known to carry seeds in calcareous grasslands. In the Burren, it is the shaggy-coated goats that take on this role; also the hares that travel long distances each day and groom their coats in sheltered places. Cattle also play a part, but are probably better at spreading seeds by an alternative means, known as "endozoochory": in one experiment, 662 seedlings of twenty-four common species were germinated from a single cow pat. One conclusion from a study of the Burren's dispersal ecology is that the goats should keep their right-of-way and the feral shagginess of their coats be maintained by culling any short-haired domestic animals that may find their way to join them.

The hazel woods today may, on deep soil, have a canopy 20 feet (6 m) high, but on the scrub over broken ground and rock-strewn slopes the effect is that of an intimate, three-dimensional maze, muffled in bright green mosses and dappled with sun. The hazel's branches, often rotten and brittle and breaking at a touch, are checkered with bark lichens and sleeved with a community of epiphytic mosses, such as *Ulota vittata* on the higher twigs and branches, and rich muffs of *Neckera complanata* and *Isothecium myurum* on the larger boughs below. The rocks, too, are cushioned and rounded with mosses and ivy, but woodland flowers grow up through them, and the hazel copses are a special retreat of helleborines. At Tobar MacDuach, on the limestone's eastern fringe, the tall, green-flowered, broad-leaved helleborine *(Epipactis helleborine)*, in the words of Nelson, "crowds every footfall within the wood," attended by the pollinating wasps of late summer.

The orchids of the Burren—twenty-two species at least; more if one yields to the siren-song of variation and possible subspecies—blossom to a calendar that begins in late April and continues into September. The spikes of the early purple orchid *(Orchis mascula)* are the first to light up the hillsides, but the variation of color that is possible between one bold flower-head and the next—from crimson, through rosy purple, to pinks and even pure white—gives fair warning of variants and hybridizations waiting to tax the orchid lover later in the spring. The *Dactylorhiza*, the spotted orchids, in flower from June through August, include among their forms one with white flowers and unspotted leaves that commemorates the remarkable Patrick O'Kelly. In the late 1800s this local farmer's son had taught himself enough about the Burren's flowers and ferns to set up as plantsman and nurseryman, and his salesmanship for the striking white orchid he listed as *Orchis maculata alba* eventually secured his fame in the modern formulation *Dactylorhiza fuchsii fuchsii okellyi*, or O'Kelly's orchid. When the spotted orchid crosses with the fragrant orchid *(Gymnadenia conopsea)*, the hybrid is a *Dactylorhiza* endowed with scent, and as the Burren's fragrant orchids are themselves a subspecies (densiflora), this is likely to be a hint of cloves.

The fragrant orchid is among the species notorious across the British Isles for appearing in hundreds or thousands in some years and meager dozens in others. Like the bee orchid *(Ophrys apifera)* and other monocarpic orchids (bearing one fruit and then dying), it takes several years to reach maturity, then spends all its energies on flowering and producing prolific numbers of seeds. The seeds of all orchids are minute, a few cells without any food reserve, and in the Burren the odds on landing on bare rock or gravel are high. To survive at all, the seedling must strike up the right relationship with a soil fungus: a mycorrhizal marriage. The raspberry-pink pyramidal orchid *(Anacamptis pyramidalis)*, which throngs the Burren's coastal sand dunes at the town of Fanore in June, is fed by its root fungus for the first four years, forms leaves the following spring, and waits another year or two before flowering. Autumn lady's tresses *(Spiranthes spiralis)*, an exquisite miniature, takes thirteen years to mature, but the appearance in September of its spiral of tiny white flowers is an event worth waiting for. "To savour the fragrance," Nelson noted, "you must, appropriately, prostrate yourselves before the plant."

The orchids of the Burren and these islands add their own refinements to pollination mechanisms patented by the worldwide Orchis family. In their most highly evolved devices—as in the early purple orchid, for example—a visiting bumblebee has the whole pollen mass stuck to its head with something like superglue: this then pivots in such a way that the bee must deposit lumps of it on the stigmas of several plants in succession. In other orchids, fertilized by long-tongued moths and butterflies, sticky discs of pollen are clamped around the insect's proboscis: the pyramidal orchid was singled out by Darwin for its perfect adjustment of this mechanism. A few species go in for mimicry, imitating female insects to encourage attempts at copulation, but where the system is too specifically synchronized it can also be easily upset. In the Burren, for example, the fly orchid *(Ophrys insectifera)* attracts to its flower lips in late May the males of just one species of small, burrowing wasp *(Gorytes mystaceus)*, and this only in the brief interval before real female wasps emerge. A spell of bad weather at the wrong time can be disastrous for seed production. Indeed, the Burren's annual parade of orchids of all kinds is a spectacular celebration of chance.

BURREN WILDLIFE

A wilderness of limestone—warm, dry, calcareous, full of caves and holes of every dimension—seems so distinctive a habitat as to command a special fauna. In such a rocky setting in a zoo, for example, one might look for brown bears, sagebrush lizards, or rattlesnakes. In the Burren, rewards are more modest and elusive. One evident link between animal and habitat is the host of snails that creep abroad on the rock and its plants after a good shower of rain (notably the bright and stripy *Cepaea nemoralis,* most variously patterned of Europe's spiral snails). They glide across the fossil ghosts of their marine ancestors, building their shells from others laid down in the ocean more than 300 million years ago.

Other affinities are not so obvious. Why, for example, should open stretches of the Burren attract the pine marten, of all the small mustelid predators? *Cat crainn,* or tree cat, is the Irish name. "A rank wood is their province," advised Arthur Stringer in *The Experienced Huntsman* in the 1700s, "for they breed in the tops of hollow trees and continually lie in such places in the day time," sometimes curled up in an empty bird's nest. The martens' main refuge in the Burren, is, indeed, a wooded nature reserve at Dromore Village (see Appendix for list of nature reserves) dotted with small limestone lakes, where about ten of the animals keep to dense cover, traveling regular pathways on the ground. But they also live and breed on the limestone pavement and cliffs, which is where most human glimpses occur, and they will den in the roofs of deserted farm cottages. Study of the few remaining martens in Britain suggests that, whatever habitat they choose, good vertical elements such as crags or cliffs are essential.

Since the Ice Age, the Burren's martens have known their terrain both with and without pines, so their survival in near-wilderness as an adaptable "relict"

Mike Hartwell, Department of the Environment Northern Ireland

Giant's Causeway, County Antrim

Matthew Parkes, with permission of the Geological Survey of Ireland

Footprints, older than 385 million years, in purple siltstone on the rocky northern coast of Valencia Island in County Kerry

Billy Clarke

Natterjack toad

Declan McGrath

Coumshingaun Corrie in the Comeragh Mountains, County Waterford, formed by the eroding action of a glacier

Bog-cotton

Ring fort at Lisadian in Northern Ireland

Erica × *stuartii*

Sundew

Reconstruction of the Ballyglass Neolithic house in the
Ulster History Park at Omagh, County Tyrone

Blanket bog on uplands

Merlin

Irish mountain hares

David Herman

Aspens growing on a cliff ledge on
Mweelrea Mountain, County Mayo

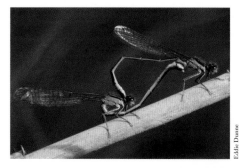

Eddie Dunne

Common blue damselfly

Eddie Dunne

Marsh fritillary butterfly

Cliffs of Inishmore, Aran Islands, showing the stone-walled fields of the island

Typical Burren landscape with Mullaghmore Mountain in the background

Bord Fáilte — The Irish Tourist Board

E. C. Nelson, from *Wild Plants of The Barren and the Aran Islands*, 1999

O'Kelly's orchid

E. C. Nelson, from *Wild Plants of The Barren and the Aran Islands*, 1999

Eddie Dunne

Tompot blenny

Red helleborine

Doolough Lake, part of the Delphi salmon
and sea trout fishery in southwest Mayo

Salmon Angler on the Delphi Fishery,
County Mayo

Blasket Islands off County Kerry

Gray seals

Typical rocky shoreline on the western coast

Bord Fáilte — The Irish Tourist Board

Royal Society for the Protection of Birds Northern Ireland

A western sandy shore edged by dune system

Wintering waders on an estuary in Northern Ireland

Farming landscape

Hawthorn

Fuchsia

Wildflower meadow

King Oak on Charleville Estate, Tullamore, County Offaly

Dúchas, The Heritage Service

Entrance to Connemara National Park,
County Galway

Daniel Kelly

Mossy oakwood at Killarney

Lakes of Killarney surrounded by woods

Dúchas, The Heritage Service

Mossy interior of Reenadinna Yew Wood
at Killarney

Mike Hartwell, Department of the Environment Northern Ireland

Mourne Mountains in County Down, designated as an
Area of Outstanding Natural Beauty

Paudie O'Leary

Dúchas, The Heritage Service

Glenveagh Castle on lake in Glenveagh National Park, County Donegal

Dúchas, The Heritage Service

Diamond hill in Connemara National Park, County Galway

Wooded island, safe from the surrounding overgrazing, in Tawneyard Lake, County Mayo

Billy Clarke

Stoat

Pine marten

Billy Clarke

feed on legumes typical of short, calcareous turf near the sea. The great variety of wild grasses nourish the larvae of the brown butterflies, and among the Burren's four fritillaries—all tawny, handsomely inscribed, and fast-flying (and thus hard to chase after on the broken pavements)—is one that seems to link with the region's tree-covered past. The pearl-bordered fritillary *(Boloria euphrosyne)* is a woodland species elsewhere in Europe and lives nowhere else in Ireland but the Burren. Its caterpillars feed on the shoots of the common dog violet *(Viola riviniana)* and hibernate for winter immediately within its curled-up leaves.

A butterfly that can suggest a special affinity with the cryptic chaos of limestone is the self-effacing grayling *(Hipparchia semele hibernica)*. It flutters abroad in late summer, feeding at thyme *(Thymus serpyllum)* or heather on heathy grassland, but is never caught with outstretched wings (which are brown and eye-marked). It rests while leaning over, almost flat to the ground, as if to hide its shadow, and in this posture its gray-smudged hind wings are a crust of lichen, blurred and indiscernible against the rock.

Among the Burren's moths, the most entomologically famous is also quite brilliantly colored (lime, jade, and malachite have all been used to describe its striking shade of green) and this makes it all the more remarkable that it remained undiscovered until 1949. The Burren green *(Calamia tridens occidentalis)*, about 1.6 inches (4 cm) long, inhabits mainland Europe but in Ireland it is found only in this one location, and nowhere in Britain. In a special expedition mounted in 1950 to confirm its Irish residency, light traps set up near Black Head attracted onlookers from miles away to see (in an English entomologist's anecdote) "the great lights and the gintlemen catching flois." The noctuid moth, which emerges an hour or so after dark, was duly found fluttering in the grass, and both male and female were trapped. It is a distinct Irish subspecies, marked by with gray hindwings, with its nearest relative on the Iberian Peninsula. The moth's food plant is thought to be the Burren's abundant blue grass *(Sesleria albicans)*, which flowers in striking, steel-blue spikes in April or May but is drained of color as the season wears on. The moth is widely spread along the coastal areas of the Burren, and the towns of Ballyvaughan and Ballyryan are two locations where it may be found resting on grass stems.

The eastern scrubland of the Burren, on the other hand, is the best place to watch, on a sunny day in spring, for the rapid-flying male of the ornate emperor moth *(Pavonia pavonia)*. The eyes of its hind wings are flushed with gold and its elaborate antennae aquiver for freshly emerged females (which, however, fly only at night). Of the other day-flying moths of the Burren, the most striking in summer are the burnets, their gray wings splashed with carmine. The six-spot burnet *(Zygaena filipendulae)* is the commoner of the two, laying its eggs on bird's foot trefoil *(Lotus corniculatus);* the other, the transparent burnet *(Zygaena purpuralis)*, while quite a rarity in Britain, can be locally abundant on the Burren slopes where its food plant is wild thyme.

TURLOUGHS: THE BURREN'S VANISHING LAKES

Just as the hare's form pressed into heather or grass will conjure up the absent animal, the grassy, rock-rimmed hollows of the Burren turloughs insist, in summer, on evoking all the water that is not there. The Irish is *turlach*—"the place that dries up" (more usually, and wrongly, defined as "a dry lake"), so that, once again, the old language defines a landform for geographers. A turlough basin, flooded from autumn to spring, may have been scooped out by ice or erosion by river or stream, but often has sunk above the collapse of an underground cave carved by water in the limestone.

Although the turloughs are so strongly identified with the Burren, they extend through much of the low-lying western limestone regions of Ireland, notably in Clare, Galway, Mayo, and Roscommon. Many have been drained through surface channels dug by farmers, but an inventory in the 1980s found sixty that still flood in winter to an area of at least 25 acres (10 ha), and there are perhaps four or five times as many smaller ones. The largest—at Rahasane, County Galway—often drowns almost 250 acres (1 km²) of pasture. Most turloughs are never much more than 6 feet (2 m) deep, but some large ones near Gort, north of the Burren, reach more than 16 feet (5 m) in midwinter and the deepest, at Peterswell, commonly 48 feet (15 m). Nearby, at Caherglassaun near Kinvara, is the oddity of a turlough in which water rises and falls by 3 feet (1 m) or so twice a day, in time with the tides in Galway Bay, a few miles distant.

The eccentric flooding rhythms of turloughs reflect the tortuous underground drainage of karst country, and land features similar to turloughs are found in karstified landscapes elsewhere in the world: *les lacs mystérieux* of eastern Canada, for example, or the *poljes* of Slovenia, which leave pastures under water in winter. What is characteristically Irish is a high karstic water table that brings moving groundwater close to the surface. Unmediated by any substantial covering of glacial drift or surface river network, this high water table reflects the capacity of subterranean fissures to carry water off to deep springs. Between October and late April, the system overflows, and water in the turloughs rises and falls in response to rain—often simultaneously, in turloughs across a wide area. A long, dry spell, on the other hand, may even empty many of the hollows. Turloughs in the Burren uplands, generally much smaller than those of lowland pasture and sometimes only a few dozen yards in diameter, can flood at any time of year within hours of heavy rain (much to the discomfort of incautious campers tempted by a sheltered, grassy bowl), and empty again a few days later.

Occasionally in modern Ireland, a limestone lake that was never commonly regarded as a turlough will suddenly be drained of its water as some subterranean bath-plug of glacial silt gives way. In 1996, for example, Lough Fuinshinagh in County Roscommon, 7 feet (2 m) deep and almost 2,000 acres (800 ha) in area, emptied almost overnight, leaving thousands of perch and pike flapping in the mud. Thus the geological processes of ages can still spring surprises.

In a typical Burren turlough in summer, a rim of black moss on rocks and holly trees sets an upper contour to the vanished winter flood, and the plant's name, *Cinclidotus fontanaloides*, rather nicely suggests the tinkling and babbling of a stream. But the welling up and sinking down of water is largely a silent affair (at least above ground), with the water gushing up, crystal clear, from a boulder-filled slogaire or swallow hole, that may, in summer, be as obvious as the drain at the bottom of a bath (the townland name of "sluggary," wherever it occurs, is a good topographical guide to the phenomenon). The welling water brings up lime, which precipitates out as a white marl, itself a sealant against drainage. Some turloughs do not empty completely in summer but keep enough water to float white and yellow lilies; in others, after a spell of warm weather, the empty sink holes may be draped with a filamentous freshwater alga, *Oedogonium*, which dries out in pale, feltlike sheets of algal paper.

Coming to the turloughs in the early 1930s, the first botanist to give them proper attention, Praeger, found their summer sward so closely grazed as to be "usually nibbled to the last leaf—often more closely shorn than could be done with a lawn-mower." But enough leaves remained for him to judge the apparent indifference to drowning of plants inhabiting the margin of the turlough, between winter high-water mark and summer low-water level. He was surprised to see how many plants would put up with several months of submergence. Dewberry *(Rubus caesius)*, for example, a creeping, sparsely fruited kind of blackberry, would grow 10 feet (3 m) below the flood limit; creeping cinquefoil *(Potentilla reptans)* and silverweed *(Potentilla anserina)* spread their net of stolons almost twice as far down: indeed, silverweed is the summer's dominant herb in many turlough basins, its ferny leaves gleaming in wind-blown ripples like those of the absent water.

Writing on turloughs and their vegetations in *Wetlands of Ireland,* Roger Goodwillie noted that comparatively few terrestrial plants are adapted to survive unpredictable months underwater: once the soil is fully flooded and trapped oxygen is used up, they lose the respiration of their deeper roots and must seek oxygen dissolved in the water at the soil surface and refreshed by the riffling winds. The stolons of *Potentilla,* lacing the drowned surface layer, survive to shoot up again even if some of the roots have died, a strategy common to some other plants with rhizomes (perennial underground stems) and ground-level buds that wait out the winter.

The turlough or fen violet *(Viola persicifolia),* with round-faced flowers of pale china blue, seems to flourish only on short-cropped, damp, and limy turf that is flooded in winter, a rare habitat in modern farmland and one that now confines the fen violet in Britain, for example, to just two protected sites. It can be found on the River Shannon floodplain, however, and in the Burren turloughs it chooses a band of turf about 3 feet (1 m) lower and damper than the zone enjoyed by the heath dog violet *(Viola canina),* and sometimes, in such proximity, the two violets hybridize.

There seem to be no plant species that grow exclusively in turloughs, and Goodwillie found it useful to think of them as ecotones or boundary zones, "the shores of underground lakes or the callows of underground rivers" that each year flood to a different depth and duration. The unpredictability is shown by the woody plants that spread into a turlough basin in years of normal winter floods but then are widely killed by exceptional inundations and have to reinvade.

The really flamboyant plant of the turlough flood margins is the fifth and largest representative there of the *Potentilla* family, the shrubby cinquefoil *(Potentilla fruticosa).* Along with silverweed and creeping cinquefoil come the trailing yellow stars of tormentil and the entirely disparate-seeming, upright, and plum-flowered marsh cinquefoil *(Potentilla palustris)*. The bright yellow flowers of shrubby cinquefoil are identical with those of the prostrate silver-weed but are borne aloft on twiggy stems up to 3 feet (1 m) high. At one or two turloughs, the shrubs interlace so densely as to make a barrier, and by the peak of flowering, in July or August, they set a complete circle of gold around the hollow. But the plant has known even more spectacular days. In Praeger's *Tourist's Flora of the West of Ireland*, published in the early twentieth century, was a photograph of a great carpet of flowering *Potentilla*, covering several acres of flat limestone heath near Ballyvaughan. Nelson, seeking out the site with photograph in hand at the end of the 1980s, found only sleek cattle grazing on artificially created grassland, its bland green lawn quite innocent of any flowers.

Turloughs shelter and feed substantial flocks of migrant wildfowl. The one at Rahasane, in south Galway, covers some 618 acres (250 ha) or more and is a prime winter birding location, attracting ducks, whooper swans from Iceland, and Greenland white-fronted geese. In summer, as the water retreats and the turf dries out, cattle move in to graze lush grass that the birds have helped to fertilize. At turloughs from east Mayo to the Shannon Estuary, a yellow-flowered cress sprawls across patches of moist, cattle-trodden mud, often near the central swallow hole. This is northern yellow cress *(Rorippa islandica)*, listed as a rare plant in the Irish *Red Data Book*. It occurs in Greenland, Norway, and Britain, as well as in the Alps, but is especially common in Iceland and abundant there around Lake Myvatn, a noted wildfowl site. The unpredictable water levels of turloughs seem to suit an exceptionally adaptable plant.

Compared with plants, the aquatic animals of the turloughs have had little attention. A lake of crystal-clear and temporary water, most of it over grass, or a bare bed of marl does not seem to promise much interesting life, even though frogs and newts may spawn there. It can also be hard to find general patterns in the faunal ecosystem of turloughs, when they vary so much in their flooding regimes, food sources, and enrichment of their water. Julian Reynolds, a freshwater biologist who has made a special study of turloughs, finds them harsh and precarious environments, their uncertainties symbolized by wind-rows of dead snail shells or rafts of caddis-fly (*Trichoptera* spp.) cases left high on the shrubby banks.

The turloughs' main food source for animals is decaying vegetation, but the types of animals present are strongly linked to the presence or absence of a single fish species, the stickleback (*Gasterosteus* spp.). Where the fish are absent a grassy bottom will often teem with water fleas *(Cladocerans)*, freshwater shrimps *(Amphipoda)*, and mayfly nymphs *(Ephemeroptera)*, but abundant sticklebacks (which may find refuge in the swallow hole) leave the bottom life dominated by beetles and snails. The actual species present can be unpredictable, because each drying-up of a turlough interrupts the struggle for dominance and survival. Many of the dominant forms are insects that leave the water at some stage. "Thus," wrote Reynolds in *The Irish Wildlife Book*, "the battle that starts again has been subtly or savagely altered, and a victory once gained is not a permanent feature in the evolution of the water body."

There are typical turlough water beetles, rare or unknown elsewhere in Ireland or Britain. Among them are *Hygrotus quinquelineatus, Coelambus impressopunctatus,* and *Graptodytes bilineatus* (known also from mossy temporary ponds in Spain and southern France). The flightless beetle *Agabus labiatus,* very common in the turloughs, is also found in temporary waters in ancient locations in the bogs.

The discovery of thousands of the "fairy shrimp" *(Tanymastix stagnalis)* in temporary summer pools at Rahasane and other turloughs has been especially intriguing—the first freshwater *Anostraca* to be found in Ireland and a species absent from Britain though common in continental Europe. It is an odd and pretty creature in its own right, a transparent, iridescent crustacean, up to half an inch (12 mm) long, that swims incessantly on its back, fanning itself through the water with rows of leaflike limbs on its thorax. How could it have passed unnoticed until the 1970s?

Accidental introduction is almost certainly the answer. The shrimp is confined to a life in temporary pools, which lack the risk of being eaten by fish. Observations at Rahasane in 1974 suggested that many female shrimps carry eggs in the brood pouch until they die and decay in the lowest crevices of the mud. This improves the chance that, when the pool fills again, it will be deep enough, and last long enough, to sustain the next generation. The thick-shelled eggs of *Anostraca* can survive both drought and repeated immersions until conditions for hatching are right. They can also survive in mud that sticks to the feet of migrant duck and waders, the hooves of cattle, or the boots of wildfowlers—all, quite possibly, involved in the arrival of the fairy shrimp in Ireland and its spread from one pool to another.

Turloughs are almost peculiar to Ireland, and their conservation is given priority under the European Union Habitats Directive. As ecotones—transition zones between aquatic and terrestrial systems—they create special problems in deciding what to conserve. Drainage is a continuing threat, but of greater concern, very often, is the danger of water pollution through interconnected systems. Some turloughs show a distinctive zonation of plants, others are valued wintering sites for birds, still more have important invertebrates. The conservation of turloughs demands a balance of measures almost as subtle as the hydrology that sustains them.

THE BURREN UNDERGROUND

The Burren's geological fascination is deepened by the youth of its landforms, because these bare limestone hills, with their slopes of silvery scree, their terraces of polished fissured pavement, are at least partly the product of soil erosion stemming from early human settlement. But the natural, karstic topography, too, is young in geological terms. Present rates of corrosion suggest that even the largest of the Burren's depressions—closed valleys of sometimes unsettling aspect—probably started as swallow holes when the limestone's shale cover was breached in the late Pliocene, less than three million years ago.

The karstic dynamic continues, as streams of acid bog water, flowing off impermeable shales at the margins of the Burren, dissolve new swallow holes and run on underground toward the sea. They are sluicing out the glacial silt that blocks many older caves and forming and enlarging new ones, sometimes quite rapidly, at the modern shale edge in the west and south Burren. Exploration by Irish and British cavers is adding all the time to the known network of hydrological connections. "For many cavers," according to Philip Chapman in *Caves and Cave Life*, "the sinuous, underground canyons [of the Burren] with their glistening, water-sculpted walls, echoing the sound of rushing water, represent the nearest thing to paradise on earth."

The longest cave is Poulnagollum, "The Cave of the Doves," high on the flank of Slieve Elva. It has almost 9.3 miles (15 km) of galleries in which, as Chapman described, "red peaty water burbles along, ankle-deep, in clean-washed T-shaped canyon passages with scalloped walls." In the Coolagh River Cave, on the other hand, reckoned the most exciting by experienced cavers in wetsuits, the main passage may fill to its roof in a flood. Good judgment would seem necessary also for exploration of the Burren's many "green holes"—caves that exit, fathoms deep, in the surging swells of the Atlantic. For the family tourist, a stroll through the caverns of Aillwee, above Bally-vaughan, with its dry walkways and spotlit stalactites, may seem a sufficiently novel experience. For the dedicated crawler, the great cavern of Pol-an-Ionain eventually offers a 21-foot (6.5 m) stalactite, one of the longest known in Europe—a wonder that may yet be made more generally accessible.

LIFE IN A LABYRINTH: THE ARAN ISLANDS

Moored in Galway Bay at a tangent to the Burren and carved from the same mass of Carboniferous limestone, the three Aran Islands of Inishmore, Inish-maan, and Inisheer are so obviously fragments of the mainland as to set up expectations of sameness. The rock underfoot (and there is a lot of it) retains much the same physical habits as the Burren, and, allowing for the extra force and saltiness of wind, a good many plants and insects are indeed shared. But there are great differences in the experience of exploring the Burren, on the one hand, and the islands on the other.

On Aran (the three islands condense into a single, evocative word) the interface of rock and ocean takes on an even grander, more elemental pres-ence, and the structure of the limestone cliffs and pavements is dramatized in extraordinary ways. But the landscapes of the island interiors are intensely the work of human subsistence over four millennia. Away from cliff edge and storm beach there is scarcely one stone on another that has not been placed by human hands, scarcely a cupful of soil that has not been stirred—even, in places, directly manufactured—by human intervention. Ashore, in the Bur-ren uplands, there is often a sense of wilderness, of a great historic distance between the landscape and its early farmers; on the islands the intimate coex-istence of people, rock, and wild plants is as current and tangible as the near-est stone wall.

On Inishmore, the largest island, some 1,000 miles (1,600 km) of walls en-close about 14,000 tiny fields, some scarcely as big as a generous suburban lawn. This intricate labyrinth of stone is repeated on Inishmaan, the middle is-land, and Inisheer, smallest of the three. The open mesh of many of the walls, a lacework of balanced slabs and wedges, whiskery with lichens, can lend them a deceptive fragility, so that a visitor lifting a single rock is invariably shocked by its weight. The heft is enough to give sudden extra meaning to the smooth, fissured pavements within the walls and the great mass of rock beneath them.

The limestone of the islands is tilted down toward Galway City at the northeast corner of the bay, so that the highest cliffs on Inishmore face out, southwest, into the full force of the Atlantic. The terrible energy of the winter storm waves, distilled in Robert Flaherty's cinematic classic *Man of Aran* (1934), has built great ramparts of jumbled boulders, up to 16 feet (5 m) high, along the tops of cliffs almost 200 feet (60 m) above the sea. These "storm beaches" have been created by a progressive hurling-back of loosened slabs by as much as 30 feet (10 m) from the edge of the rock.

The highest cliffs are usually vertical, following the natural pattern of jointing in the rock, and undercut by the horizontal chiseling of the waves. At intervals, another enormous block of stone slips down, to serve, for a time, as a shield at the base of the cliff. The intersection of joints—a cuboid, almost mathematical lattice of weakness within the rock—helps the sea to make caves, both by its own power and that of the masses of air it drives into crevices. Eventually, as the caves penetrate inland, blocks collapsing from the roof may open a shaft to the surface, to create puffing holes. At Poll na bPéist ("the hole of the serpents") on the southern shore of Inishmore, a similar process has opened an exactly rectangular abyss, like a dark swimming pool more than 330 feet (100 m) long, in the wave-cut terrace at the foot of the cliff.

Aran's terraces, like those of the Burren, are carved along the thin beds of shale that interleave the limestone. The five highest and most obvious can be traced the length of the islands, giving Inishmaan, in particular, a sharp flight of steps up the slant of its profile. Separating the level treads of the terraces, in many places, are shallow cliffs up to 20 feet (6 m) high; elsewhere the scarp is weather-worn or blurred by broken stone. Much of the original glacial debris was scoured from the islands by storms in the early postglacial era of higher sea levels, but more than enough loose rock remained to build the homes and cashels (the fortified, circular enclosures) and great ceremonial forts of the earliest people, as well as the subsequent maze of field walls. Most of these were built to clear and enclose rocky pastures, often little more than bare pavement interspersed with shreds of grassy turf. On Inishmore there are spectacular expanses of flat pavement where one can walk for more than 30 yards (30 m) without once crossing the fissure of a *scailp*. At the other extreme are fields so closely dissected that one stride could span two or three parallel clefts, each with its linear, sunken garden in which to peer after maidenhair fern. But loose stone could also be jammed into these fissures and seaweed and sand laid down on top to create an organic compost. By supplementing in this way the natural pockets of soil, some 700 acres (284 ha) of a total 11,000 acres (4,450 ha) in the islands were growing crops by the early nineteenth century, principally potatoes, with rye for fodder and thatching.

Some plots of rye *(Secale cereale)*, tall and tough-fibered, are still grown in Aran, still pulled by hand for thatch, and the grain threshed out against a rock to save for next year's seed. This relictual agriculture has secured the survival of arable weeds long scarce or even extinct in mainland Ireland and Europe. Among them are darnel *(Lolium tementulum)*, a tall grass with seeds that once gave a toxic taint to rye flour, bristle oat *(Avena strigosa)*, a primitive grain crop and thus genetically valuable, and the beautiful but transient blue cornflower *(Centaurea cyanus)*.

Aran gives refuge to another plant that has never been recorded on the Irish mainland. The purple milk vetch *(Astralagus danicus)* is thought to have arrived in the early postglacial period, in the grassy interval before the trees spread north. Its flowers, like loose heads of violet clover, blossom in May on the cliff tops of Inishmore, west of Dun Aengus, and also in Inishmaan.

To many botanists, any scrutiny of island wildflower species now offers a chance to apply the theory of island biogeography developed by Americans Robert MacArthur and Edward O. Wilson in the 1960s. This milestone in evolutionary biology followed Darwin and Wallace in appreciating that islands, with their isolated communities of plants and animals, are the ideal laboratory in which to sort out some of the rules of nature's biodiversity. MacArthur and Wilson had noticed that the faunas and floras of islands around the world show a consistent relationship between the area of the islands and the number of species living on them. They also explored laws of population biology, which control the balance of new arrivals of species with extinction of the existing ones. Their theory is now a basic tool of ecology, especially in trying to judge the potential for biodiversity in national parks, wildlife reserves, and similar "islands" of conservation.

The tiny but floriferous archipelago of Aran nicely demonstrates that the diversity of flora increases with island size: Inisheer, with 1.5 square miles (4 km²), has 289 species; Inishmaan, with 3.5 square miles (9 km²), has 352 species; Inishmore, with 12 square miles (30 km²), has 417. The many studies of the Aran flora over the past 125 years also offer evidence for the idea that an island's plants are in a state of continual flux and that the species list shows a balance between extinctions and new arrivals. "It is striking," says botanist Cilian Roden in *The Book of Aran*, "how species appear and disappear from the flora lists as decades pass." The transience of seashore plants is familiar enough: a sudden appearance, a brief flourishing, and then decline, as happened with frosted orache *(Atriplex laciniata)* in the 1970s. But wild sage *(Salvia verbanaca)*, so rare in Ireland as to merit legal protection, suddenly appeared in 1984 along a well-traveled road on Inishmore, more than 50 miles (80 km) distant from its nearest mainland colony.

Even in its enduring Burren flora there are mysteries: why, for example, has mountain avens never made the crossing when its Mediterranean partner, the dense-flowered orchid, can be found on all three islands? But in general the flowers of the limestone crags of Aran repeat, sometimes even more vigorously, the vivid communities of similar habitats ashore. Early visitors to Dun Aengus, the massive cashel at the cliff edge on Inishmore's southern skyline, find the turf inside the outer wall "profusely set with the harsh blue jewels of the spring gentian," as Aran's distinguished interpreter, Tim Robinson, described it, and little drifts of Irish saxifrage *(Saxifraga rosacea)* lie in the rocky pastures like startling patches of snow. By midsummer the rocks and pavements are lit up with bloody cranesbill, its furious magenta echoed more gently in pyramidal orchids in the dunes and the tall spires of tree mallow *(Lavatera arborea)* brought in from the cliffs to flourish in island gardens. Yellow is the interweaving color of the pastures, in kidney vetch *(Anthyllis vulneraria)*, yellow rattle *(Rhinanthus minor)*, and hoary rock-rose, the limestone's yellow-flowered shrub.

On islands innocent of herbicides and pesticides, and where motor traffic is confined to the few surfaced roads of Inishmore, the abundance of wayside insect life is a measure of what has been lost in Europe generally. Among half a dozen species of bumblebee is one of Aran's few claims to an island race: a dark-haired subspecies of a mainland carder bee *(Bombus muscorum)*, which the islands share with some Scottish islands and the Scilly Islands, off Cornwall. The number of gray Aran moths (and their degree of grayness) invites speculation on natural selection in this limestone setting. The islands also share with the Burren the paler, grayer version of the Irish race of the grayling, one of some twenty resident island butterflies. But there is no dulling touch of gray in the bright blue butterflies, often in myriads on the grassland of Inishmore, the common blue and the small blue, or in such strikingly colored day-flying moths as the black and red burnets and the turquoise forester *(Adscita statices)*.

The small animals of Aran present familiar puzzles of island biogeography: why are some species there and not others? Thus, while the islands have no frogs, which are widespread and common in the Burren, the viviparous lizard is so frequent that twenty have been seen on one summer walk across Inishmore. The Burren naturalist Gordon D'Arcy guessed that they arrived on Aran, moribund in hibernation, among the boatloads of peat sods brought as fuel from the Connemara mainland.

But in any exploration of the islands' natural history, the most intriguing creatures are ultimately those conjured by rain, wave, and sea spray from the Aran rock itself. In cliffs and beach cobbles, in the limestone masses and their interleaving bookmarks of shale, the remains of marine life in a subtropical coral sea emerge by the millions, often as gleaming, shelly hieroglyphs but also in forms weathered to a full relief in the matrix of gray stone. Most numerous are the brachiopods or lamp-shells, primitive filter-feeders that look superficially much like the cockles and scallops and other bivalve mollusks of today. They belonged, in fact, to another phylum, with the upper valve humped above the lower, rather than in matching left and right halves, and with internal coils of ciliated organs to waft water through the shell. The larger brachiopods, such as *Dictyoclostus*, were anchored in the mud of the seabed, perhaps 300 feet (100 m) down, and their shells, etched out to the finest spines, now stud the sea-washed lower terraces of Aran. Smaller brachiopods lived in dense colonies, anchored by stalks to rocks and reefs. The teeming thousands of species of these lamp-shells crowded the seas as far back as the Ordovician and Silurian periods, but today they survive in fewer than three hundred forms, in habitats as different as the deep water of Norwegian fjords, the mud of Malayan beaches, and in Galway Bay.

The fossils of Aran and the Burren attest to a rich ecosystem, probably close to that of today's Caribbean. Beach cobbles are polka-dotted with the calcite of colonial corals, such as *Lithostrotion,* or discs from the stems of crinoids, the branching "sea lilies" allied to feather-star echinoderms. In cliffs and boulders, sometimes in stone walls, embedded spiral forms are those of Carboniferous gastropods or of early ammonoids, called goniatites. These are an extinct class of cephalopods, active, free-swimming hunters and carnivores but with their bodies coiled in an external shell. Modern cephalopods, such as cuttlefish and squids, have internalized or lost their shells, and the early form survives only in the beautiful pearly nautilus of tropical seas.

Limestone is a rock of considerable character, transformed by the fall of light or the flow of water. Whether as the matrix of ancient fossils, the armature of mystical hills, or the cradle of dazzling wildflowers, it makes memorable landscapes of the Burren and the Aran Islands.

SELECTED REFERENCES

Aalen, F. H. A., K. Whelan, and M. Stout, eds. 1997. *Atlas of the Irish Rural Landscape.* Cork: Cork University Press.

Asher, J., M. Warron, R. Fox, P. Harding, G. Jeffcoate, and S. Jeffcoate, eds. 2001. *The Millennium Atlas of Butterflies in Britain and Ireland.* Oxford: Oxford University Press.

Chapman, P. 1993. *Caves and Cave Life.* London: HarperCollins.

Curtis, T. G. F., and H. N. McGough, eds. 1988. *The Irish Red Data Book. 1: Vascular Plants.* Dublin: Stationery Office.

D'Arcy, G. 1992. *The Natural History of the Burren.* London: Immel.

Nelson, E. C., and W. Walsh. 1991. *The Burren: A Companion to the Wildflowers.* Kilkenny, Ireland: Boethius Press.

O'Connell, J. W., and A. Korff, eds. 1991. *The Book of the Burren.* Kinvara, Ireland: TirEolas.

Otte, M. L., ed. in press. *Wetlands of Ireland.* Dublin: University of Dublin Press.

Pilcher, J., and V. Hall. 2001. *Flora Hibernica.* Cork: Collins Press.

Robinson, T. 1995. *Stones of Aran: Labyrinth.* Dublin: Lilliput.

Waddell, J., J. W. O'Connell, and A. Korff, eds. 1994. *The Book of Aran.* Kinvara, Ireland: Tir Eolas.

— *Chapter 9* —

Rocky Shores and Sandscapes

At the southwestern corner of Ireland, where rocky peninsulas thrust out into the long Atlantic swells, any fair summer sky finds a rich blue reflection in the deep, salty waters of Lough Hyne near Skibbereen in West Cork. Its beauty is exceptional, framed in woodland and rocky promontories, and so is its morphology, because its only link to the ocean is a brief and narrow channel called the Rapids. Here the water changes direction four times a day, flowing with glassy force and a thrilling surface rush of some 10 miles per hour (16 kph). At the turn of the tide there is a pause, a silence. Long, dense fronds of kelp on the bed of the creek stand up clear of the water, hold their pose for a moment, then bow the other way.

The dramatic contours of Lough Hyne, with its deep internal trough and sheltered basins, are thought to have their origins in a glacial plucking-out of rock from the old red sandstone of Munster. Until about 4,000 years ago it was a freshwater lake, but then rising sea level crept over its sill. This invasion occurred when Neolithic people were active in the area, and some geologists, looking at the neat, almost artificial, look of the rapids' shallow channel, have wondered if human skills were engaged in its making, perhaps with the promise of seafood in mind. Today the lake's attraction for scientists is as a pristine aquarium, a temperate tidal pool .62 miles (1 km) long and .46 miles (.75 km) wide. Here live several thousand species of marine animals and plants in an almost complete range of exposed and sheltered environments.

The stillness of the lough, stirred only by its replenishing tidal pulse, has encouraged an extraordinary biodiversity: seventy-three kinds of sea slug; more than one hundred kinds of sponge; twenty-four kinds of crab; eighteen kinds of anemone. At the rapids, beneath the long brown tresses of the kelp forest, jewel anemones share the rocks with fan-worms, corals, and sea-squirts, and the lough's vertical underwater cliffs are embossed with the same brilliant throng of marine life. Even in its amphipods—sand hoppers and the rest—the lough is richer in species than any location so far examined in Ireland. Shelter and warmth have made the lake particularly hospitable to southern, even Mediterranean, species, and three Lusitanian gobies (*Gobius* spp.) are among its seventy-odd inshore fish. Some biologists suggest that such species may have accumulated during favorable swings in climate—one of which, indeed, coincided with the first incursion of the sea in Neolithic times.

Lough Hyne's attractions for research have developed over a century and earned it protection as the Republic of Ireland's first marine nature reserve. Marine scientists have not disdained studies of the everyday, bucket-and-spade species of the seashore, and one especially conjures up the contrasts in exposure of Ireland's 1,800 miles (2,900 km) of rocky shores, from bleak Atlantic headlands battered by some of the world's most violent surf to sheltered inlets hung with dense curtains of seaweed.

This is the dogwhelk (*Nucella lapillus*), a predatory sea snail that is the special scourge of mussels and barnacles, drilling through their shells or finding the joints in their armor. It extends not only along European coasts but also, by way of Iceland and Labrador, to shores as far south as New York. It suffers a name change on the way. As *Thais lapillus*, America's eastern dog whelk, it is linked to the closely related *Thais* species of the U.S. west coast and thus to the mollusk's evolutionary home in the Pacific. The dogwhelks belong to the Muricidae, a family that includes the highly decorative spiny whelks of the tropics. They are known to historians for a valuable dye, imperial or Tyrian purple, once prepared from the toxic secretion purpurin, which some species use as a predatory weapon. Tyre, in the eastern Mediterranean, supplied dye to Roman nobles, and middens of broken dogwhelk shells at sites along the

Irish west coast are evidence of purple dye manufacture and trade in Early Christian and Bronze Age periods.

Ecologically, the interest of this voracious gastropod lies in its intense variability. It occurs in a wide range of colors, not always easy to relate to diet or habitat, and shell form can vary considerably, even along a short distance of coast. Some of the variation may be influenced by rate of growth or genetic grouping, but research at Lough Hyne has demonstrated graphically how shell proportions can change when selected to fit differences in physical conditions. Out on the nearby headland of Carrigathorna, exposed to the full crash of the waves, the dogwhelk shell is squatter, and its wide aperture has room for a larger foot to keep a better grip in rough seas. In the shelter of the rapids, *Nucella* shells are taller and much thicker and have a narrower aperture, the better to resist attacks by crabs. The rough-sea whelks are harder to pull from the rocks and they hang on more tenaciously in currents. But their thinner shells and larger front doors yield far more readily to crustacean pincers. Thus selection matches the dogwhelk's shape to its location in the intertidal zone and punishes any tendency to stray from its niche.

Ireland's twice-daily tides are generated at a distance, as masses of water in the great expanse of the Atlantic respond to the pull of moon and sun. As this energy reaches the continental shelf, the tides surge northward up the west coast and around the top of Scotland, while part of the same Atlantic "wave" veers along the south coast to flood into the Irish Sea and the English Channel. Twice a month, when the moon is reduced to a ghostly arc and at the full moon, the spring tides (from the Saxon *sprungen*, with its sense of brimming) reach into the farthest corners of creeks and estuaries and coastal lagoons. Irish tides do not match the great extremes to be found, say, in Nova Scotia's Bay of Fundy, but the spring tides of the west coast, with an average amplitude of about 15 feet (4.5 m) can, on the ebb, expose more than 300 feet (100 m) of gently sloping shore.

At the outermost Atlantic headlands, where dark planes of rock tilt down into deep, clear water, the tide falls back from successive bands of a few resilient species. The wide, tar-black selvedge of *Verrucaria* lichen contrasts starkly with the studded white armor of barnacles, and the rock wall darkens again nearer the water with a mat of mussels, small and densely packed. Finally, at the furthest dip of the waves comes a ragged hem of purple laver *(Porphyra umbilicalis)* and a swirl of tough brown dabberlocks *(Alaria esculenta)*.

Here, at the first fringe of the kelp forest, a pink glow beneath the weed marks the crust of coralline red algae that decorates so much of the sublittoral rock.

The acorn barnacles that stud the rock in such astronomical numbers (half a million or more per square meter have been counted) are actually crustaceans that have adopted a sedentary lifestyle, whisking plankton into their mouths as the surf surges over them. Of the two dominant species, *Chthamalus stellatus* and *Semibalanus balanoides*, the former chooses the higher reaches of rock and so courts a longer exposure to air and sun.

Semibalanus belongs to the north-temperate or boreal fauna that rings the top of the Earth, adapting to oceans cooled by glaciers and drifting pack ice. Most of the seaweeds and marine animals on the shores of Ireland and Britain are duplicated along the coast of New England. But arctic currents dip down through the boreal zone to establish some species of their own: Among the *Laminaria* seaweeds, for example, a big arctic kelp grows along the coast of Maine but is missing from the shores of the eastern North Atlantic. In Britain, currents flowing down into the North Sea bring arctic-boreal species from the Scandinavian coast, but these have not extended to the warmer coasts at the west of the island.

Ireland, too, lies in the boreal Atlantic region, but in a zone of transition between warm-water Lusitanian species and those of the arctic north. The thrust of the Gulf Stream up the Irish west coast extends the range of colorful southern species, which are missing from the cooler waters of the east coast.

Sea temperature is crucial, both to the establishment of species and to the seasons at which they breed. The northern acorn barnacle *(Semibalanus balanoides)* breeds in midwinter at the southern shores of Ireland and Britain, at the southernmost limit of its range, but leaves it until later in the year in colder waters farther north. *Chthamalus*, however, is a southern species at the northernmost limit of its range, and here it breeds in midsummer. It is one of many species carried north by a warm, salty current, flowing from the mouth of the Mediterranean, following the slope of the European continental shelf and bringing Lusitanian plankton up the western side of Ireland and into the English Channel. Along much of the European coast, this ocean current is held out from the land by a rim of less saline water mixed by the outflow of big rivers, but along Ireland's western shores the ocean surges in unchecked, enriching the coastal boreal fauna with species more common to Brittany or the northwest of Spain.

Among the more striking animals are the crawfish or spiny lobster *(Palinurus elephas)* and the fan mussel *(Pinna fragilis)*, which grows up to 12 inches (30 cm) long. Southern anemones, sponges, and seaweeds have all found their niches on rocky Irish shores, and the little red-mouth goby *(Gobius cruentatus)* is among the rockfish that have taken up occasional residence. Perhaps the most obvious of the exotics is the purple sea urchin *(Paracentrotus lividus)*, sometimes found in thousands in tide pools from Cork to Donegal but in general greatly overexploited for export to the fish markets of France. On the exposed limestone coast of the Burren and in pools on the Aran Islands, these urchins have bored deeply into the soft rock, using both teeth and spines to grind out their holes. Here they remain, sometimes immured for life in their globular cavities and absorbing food directly from the water.

In the shelter of Lough Hyne, on the other hand, the flocks of sea urchins roam abroad to strip the rocks of seaweed. A particular pattern of predation seems to have changed the behavior of *Paracentrotus*. On its home ground in the Mediterranean, it comes out to graze by night, reducing the risks from attack by crabs and predatory rockfish. In the warm shallows of Lough Hyne, however, seagulls are the top predators. To avoid them, the crabs have taken to prowling by night, while the larger urchins, adapting to a new balance of risks, often come out to feed by day.

Even the most exposed of Ireland's rocky coasts have a diverse and animated sublittoral world. The starfish, sponges, and carpeting anemones are all the more intensely colorful for living in water free of misty sediment, and the steep, cathedral light allows kelp beds to continue to a depth of 98 feet (30 m). As the fathoms pile up, this undulating *Laminaria* forest, patrolled by foraging wrasse *(Labrus* spp.) and pollack *(Pollachius pollachius)* gives way to a dense sward of leafy red seaweeds and, scattered among them, the grapefruit-sized globes of Ireland's largest sea urchin, the common or edible sea urchin *(Echinus esculentus)*. So severe is the grazing of this animal that large patches of the rock are left coated only with a glowing, rose-red crust of coralline algae *(Hildenbrandia rubra)*, the same living substance that makes jewel boxes of so many tidal pools.

The ragged meadow of red seaweeds grades into a turf composed entirely of animals: delicate, plantlike colonies of hydroids and bryozoans, filtering the water with flowerlike tentacles. Anchored here and there among them are bonsai growths of sea-fan coral, the lacy puppet gloves of dead-man's-fingers

(Alcyonium digitatum), and the sculptured masses of sponge. These rocky deeps are also the territory of the lobster *(Homarus gammarus)*, and where cliffs plunge sheer into the Atlantic the local currach fishermen venture in fine weather to set their pots in the shadow of the rock. Crawfish are less willing to enter baited pots, but have been heavily fished with tangle nets, in some places almost to extinction.

For the scuba diver, the exceptional clarity of Ireland's west-coast deeps creates some memorable undersea perspectives, and popular bases for the sport have been developed at a number of centers from Donegal Bay to Kerry. Where the underwater cliffs are of limestone, fretted into horizontal ledges and caves, marine life is correspondingly enriched. Indeed, one of the most intriguing sites for marine study is a limestone habitat at a quite different and unexpected corner of the island.

In the narrow channel at the top of the Irish Sea, where the tide rushes out between Fair Head and the Mull of Kintyre, Rathlin Island sits on a base of fine-grained white chalk, worn into underwater terraces. On the northwest coast, the lowest of these terraces drops away to a depth of almost 600 feet (180 m), creating the tallest underwater cliffs in Ireland or Britain ("Diving from these cliffs to 50 meters," wrote Bernard Picton in *Sea Life of Britain and Ireland*, "is like being on the outside of a skyscraper, with a vertical wall disappearing into blackness below.") The pitted and fissured stone is densely covered with dead-man's-fingers, hydroids, plumose anemones, the red sea-squirt *(Dendrodoa grossularia)*, and many sponges. Caves and arches in the chalk shelter unexpectedly delicate species, such as the luminescent sea-pen *(Virgularia mirabilis)* and burrowing brittle stars of the Ophiuroid order. Above sea level, most of Rathlin's rocks are basalts. These break up underwater into angular boulders and they shelter communities of animals and plants quite different than those of the chalk. Such a striking contrast of habitats and communities, sharing the same fierce tidal environment, has earned Rathlin the promise of conservation as a marine nature reserve.

On the intertidal rocks in the shelter of Ireland's bays and inlets, the underwater zoning of seaweeds is echoed even more elaborately. Its vertical sequence of species is predictable on sheltered shores right around the coasts of Ireland and Britain and is true also, in general terms, for temperate American Atlantic coasts.

The highest seaweed on the Connemara shore, for example, begins just below the line of the gray-green lichen *(Ramalina siliquosa)* and the brilliant orange lichen *(Caloplaca marina)*. It is the short, branching channelled wrack *(Pelvetia canaliculata)*, often reduced to a crisp, black stubble by exposure to air and sun. It manages to absorb its food from a sea experienced more often as spray than as submerging water, and swells to an olive-green suppleness in the wash from each spring tide. On the coast of Maine this European wrack is missing, but a related plant, the spiral wrack *(Fucus spiralis)*, climbs almost as far up the shore and spends nearly three quarters of its life out of water.

The spiral wrack, with flatter fronds, forms its own distinct zone on the upper Irish shores, just below the channelled wrack. Then comes the broad slope of the middle shore, sometimes covered by a mixture of two kinds of wrack but more often by one or the other. The knotted wrack *(Ascophyllum nodosum)* and the bladder wrack *(Fucus vesiculosus)* are both lifted in an incoming tide by the gas-filled swellings in their fronds, so that what seems like a dense blanket of weed becomes a more spacious forest, the long fronds of its canopy combed all one way by the current. Below this, in turn, comes the wide fringe of toothed wrack *(Fucus serratus)*, often covered with tiny animals—hydroids, bryozoans, and coiled, spirorbid worms, all trusting to a quick return of immersion in the sea.

Differences in tolerance of drying sun and wind help to explain the marked zonation of seaweeds. Those at the lowest fall of the tide, such as the broad fronds of dabberlocks *(Alaria esculenta)* and oarweed *(Laminaria digitata)*, flop down into the water even when their holdfasts are exposed. Just below them, *Laminaria hyperborea* holds its stem erect and sticks up out of the water at the bottom of a spring tide, exposing its growing point, or meristem, to the sun. Indeed, any of the truly sublittoral kelps can be damaged by a single exceptionally low tide on a day of scorching sun.

Tidal factors do not altogether explain zonation: the growing habits of the plants are important, too. Carrageen moss *(Chondrus crispus)*, known as Irish moss in America, from North Carolina to Labrador, wins its zone on the lower shore because its holdfasts fuse into a crust that the spores of *Fucus* species cannot colonize. Some plants outlive and ultimately dominate their competitors, and some tough-fronded species may eliminate their rivals by continued brushing of the rock surface. Most grazing animals feed on a film of microalgae and tiny juvenile seaweeds, and selective feeding by sea snails such as limpets and periwinkles may play a big part in deciding which seaweeds grow where.

Some of the most sheltered western shores of Ireland have a remarkable density and diversity of seaweeds. In south Connemara the coast is so intricately fretted that, on some inlets, only the thick golden fringe of knotted wrack distinguishes the seashore from a freshwater lake. The region has no fewer than 336 recorded species of seaweed, a dozen more than the U.S. North Atlantic coast and about 70 percent of Ireland's total. A mild, cloudy climate and the pure, highly aerated oceanic water help the plants to use a wide range of habitats, and the deep pools of Connemara's granite shores extend their diversity still further. Between rocky promontories in Mannin Bay and Cashla Bay, dead fragments of calcareous seaweeds *(Lithothamnion corallioides* and *Phymatolithon calcareum)* have built up into coral strands, their skeletal remains, originally red, now bleached a pale pink in the sun. This is *maerl,* a word from the coast of Brittany in northwest France, where mining the limy gravel for the dressing of acid grassland is a long-established industry. This was also the case in Ireland, at a more modest and local scale, until the coming of ground limestone, and proposals now exist for dredging huge drifts of living *maerl* that lie offshore, notably in Galway Bay. But these banks shelter marine animals in spectacular density and variety and biologists see them as an exceptional and little-studied ecosystem, too valuable for casual destruction.

Few seaweeds in Connemara are in any way special to this coast, and of southern species only the brown wrack *(Bifurcaria bifurcata)* that thrives in the pools is at all common. What is striking is the diversity of species on a single stretch of shore and their exceptional robustness. A botany professor in Galway University sent some specimens from Connemara for illustration in a marine algal flora: they were rejected as being "too luxuriant to be typical." Their vigor and abundance have, at times, helped to shape Irish social history, and the Galway peninsula and the neighboring Aran Islands offer powerful examples of this.

The islanders know about thirty species, each with its own value or disadvantage as fertilizer for crops or as raw material for the kelp industry. Up to a century ago, tons of wrack were cut from the lower rocks for potato planting in March and April; carried up in baskets and mixed with sand, it created the humus of hundreds of acres of made soil in fields of virtually bare limestone.

Indeed, the affinity of the potato for the potash-rich nutrients of rotting seaweed, and the abundance of knotted wrack in sheltered bays, help to explain the huge surge in population on Ireland's western coastal fringes from the late eighteenth century onward. The wrack allowed human settlement to increase on poor, boggy, often mountainous land within carting distance of the shore, and the constant addition of shell sand, seaweed, and manure, repeated over decades, created infields of artificial soils. So precious was the seaweed that families staked out their rights to collect it on particular stretches of shore—rights that were jealously guarded, inherited, and sometimes violently contested. The passion of such disputes can linger even today, when some 350 part-time cutters harvest knotted wrack in the creeks of Connemara, Mayo, and Donegal. Like their counterparts in the Scottish Islands, they tow it to shore in rafts of tethered bundles—no longer to fertilize potatoes but for the extraction of alginic acid, the viscous, gumlike acid that serves as an emulsifier and stabilizer in food products such as low-fat spreads and mayonnaise.

The industrial harvesting of seaweed on the Atlantic shores of Ireland and Scotland began with the use of seaweed ash in making soap and glass, and continued into the early decades of the twentieth century for the extraction of iodine; it was a strenuous seasonal occupation of coastal farm families for the better part of 200 years. Most of the harvest was of *Laminaria* species, chiefly *hyperborea* and *digitata,* and it relied on the stems of the kelp (called in Ireland "sea rods") cast up on the shore in huge, slippery banks after storms. Well into modern times, poorer shoreline families competing for subsistence would wade into rough seas at first light to drag the stems ashore. They were dried on low walls and then fired in stone kilns beside the shore—a slow, melting process that raised great clouds of smoke along coasts from Aran to Rathlin Island. It yielded a residue that set as hard as stone and formed the raw material for iodine.

The gel-like properties of some seaweeds are familiar from traditional use of carrageen moss in jellied fruit desserts or blancmange. This is one of several edible species associated with coastal traditions in western Europe and now becoming fashionable as health-promoting food ingredients, rich in vitamins, iron, and iodine. The flat fronds of dulse *(Palmaria palmata),* waving in crimson hands along the fringes of low tide pools, dry out as a salty, chewable substance that has had many devotees. In the early era of the Boston Irish (who knew it as *dillisk)* the immigrants bought it from street-corner pedlars;

it is harvested commercially in North America, notably in the Bay of Fundy, and cooked like spinach or added to soups and fish dishes. Another of the membranous red weeds, laver *(Porphyra purpurea)*, clings to exposed rocks like shreds of party balloon: in the classic Welsh recipe for laverbread it is boiled, rolled in oatmeal to make patties, and then fried in bacon fat.

The fact that "exotic" seaweeds typical of Japanese cuisine are also common along both shores of the North Atlantic has underlined the shared ecological patterns of temperate marine life around the world. But enough differences remain between regional Pacific and Atlantic regimes to make any transfer of species an ecologically risky process.

In its native Japanese waters, *Sargassum muticum* is an unremarkable seaweed. Arriving in Western Europe in the 1970s in shipments of the Pacific oyster *(Crassostrea gigas)*, it rapidly became an aggressive, invasive plant, displacing some native seaweeds at sites along shores from the Mediterranean to Norway (it has also colonized, in a very similar way, the west coast of North America, from Mexico to Canada). In 1995 it was found on oyster bags in Strangford Lough, on Ireland's east coast, and has since been found in Kerry and Connemara.

Despite the increasing vigilance of the island's Marine Institute, Ireland seems bound to succumb eventually to additional troublesome species, especially those that can hitch a ride in the shellfish trade. An omnivorous Japanese crab *(Hemigrapsus sanguineus)*, already achieving devastating densities in France and Spain, could arrive any day in half-grown oysters from Brittany. Imported American oysters have yielded live slipper limpets *(Crepidula fornicata)*, which already smother great areas of some bays in the English Channel, and egg cases of the American oyster drill *(Urosalpinx cinerea)*.

Strangford Lough in County Down sets parameters of marine life almost without equal in Ireland or Britain and they have made it an obvious choice as a marine reserve. This 19-mile (30-km) inlet, an arm of the Irish Sea sheltered within the Ards Peninsula, holds at high tide some 60 square miles (150 km²) of water. At the rocky narrows between the fishing villages of Portaferry and Strangford, the tidal rush is like that of Lough Hyne's rapids hugely magnified: a spectacular surge of water flowing in at up to 11 miles per hour (18 kph) and setting up dramatic whirlpools and eddies. This forceful renewal of plankton-rich sea guarantees a year-round mixing of water, unlike the summer layering that deoxygenates the deepest trough of Lough Hyne. But the essentially placid regime of the inner lough and the absence of inflowing rivers has allowed an exceptional sorting of sediments, from cobbles near the entrance to reaches of mud so fine that even birds may hesitate to walk on it.

This great variety of habitats and water movements supports more than two thousand kinds of marine animal, almost three quarters of the species found on the entire northern coast. "Not a single spot remains unoccupied," in the enthusiastic words of ecologist Bob Brown in his book *Strangford Lough.*

In some areas it is impossible to see the bottom because of the tangled carpet of animal life. Brittle starfish may number over 500 to the square meter, extending a forest of thin arms into the current to feed. Dense mussel beds may themselves be almost smothered by the animal life growing on their shells. Vast meadows of anemones, hydroids and sea squirts sway silently back and forth in the flow.

The ecological star of the lough has been the horse mussel *(Modiolus modiolus)*, its colonies building solid reefs in soft mud as new, long-lived generations attach to the dead shells of their predecessors. More than one hundred other species use the reefs as anchor or shelter, among them the variegated scallop *(Chlamys varia)* and a large range of bryozoans, sponges, and tunicates. Bottom trawling for scallops, however, has caused severe damage to the mussel reefs, and a survey in 2002 found no living specimens. The fine silty mud of Strangford also shelters a small, burrowing lobster *(Nephrops norvegicus)*, more familiar on restaurant menus as the "Dublin Bay prawn" and also trawled intensively in the mud of the western Irish Sea. A good representation of the fish and bottom-dwelling fauna of Strangford is offered by the Exploris Aquarium in Portaferry.

Ireland's burgeoning aquaculture industry, which now dots almost every sheltered bay with tanks, rafts, and barrel lines, has prompted even more welcome initiatives to restore Ireland's natural beds of shellfish. There is reseeding of the great scallop *(Pecten maximus)* in bays from Bantry in west Cork to Mulroy in County Donegal, and greatly depleted beds of the native flat oyster *(Ostrea edulis)*, at the head of Galway Bay, are also being revitalized.

More by luck than judgment, Ireland's marine ecosystems are still largely in a healthy state, but waste nutrients pouring out from the rivers in runoff from farming fertilizers and sewage are feeding coastal phytoplankton, sometimes including new and often toxic dinoflagellates introduced to Europe in ships' ballast water. Harmful algal blooms can be nourished by perfectly natural upwelling of nutrients offshore, but some small, enclosed sea inlets are suffering the kind of eutrophication that can happen in freshwater lakes, and plumes of enriched water, surging out from the estuaries, are carried along the coast by local currents.

Toxic algae, carried ashore in wind-driven water, have halted the tightly controlled marketing of shellfish from Ireland's most productive bays, sometimes for long periods. At Killary Harbour, for example, the 9-mile (15-km) fjord that opens to the Atlantic at the border of Galway and Mayo, hundreds of tons of mussels are grown suspended on ropes to avoid predation by starfish. The steep slopes of the fjord are sparsely populated, and the great inflow of fresh water from rainy hillsides and two salmon rivers is relatively low in nutrients, yet toxic blooms sometimes close off mussel exports for many months on end.

Killary Harbour's dramatic setting, with Mweelrea, Connacht's highest mountain, standing guard at the entrance, speaks directly for the action of ice in shaping the Irish shoreline. A glacier, surging through the valley, bulldozed all the rocks on its floor and swept them westward. At what is now the mouth of the fjord, where rock outcrops became islands, the debris was left in a sill, so that today's seabed rises gently from the thirteen muddy-bottomed fathoms of the inner waters.

SANDSCAPES

On Ireland's sandy shores, the glacial impact is more remote and mediated by water and wind. Along the northwest coast of Donegal, where sea and mountain are interfolded in a maze of dunes and beaches, the origin of the sand in an outwash of eroded and pulverized quartz and granite is obvious. Elsewhere, longshore currents have carried glacial sediments to make beaches far from their source.

Almost 466 miles (750 km) of Ireland's 4,660-mile (7,500-km) shoreline (a journey lengthened by many indentations and islands) are lined with dune systems, spaced out around the island between stretches of shingle and rock. There are no sandscapes quite on the heroic scale of the Great Beach of Cape Cod, but many long strands (as the Irish call their beaches) dwarf a human figure between breakers and dunes. The beaches of the west and north are wide and flat, dissipating the great energy of far-traveled swells; on the east coast they are narrower, steeper, and more elaborately molded by waves jostling within the Irish Sea.

At the retreat of the ice, great drifts of sand and gravel bordered the bays and estuaries and spilled out across slopes of dry seabed. As sea level rose rapidly again, much of this material was sucked away and sorted under the waves. About 6,000 years ago, the tides began to move the sand ashore again, where wind snatched it up from beach ridges to build massive fields of mobile dunes. Without the right plants at hand to stabilize it, the sand peaked and flowed in bare crests above the shore.

The earliest dated dunes are on the Bann estuary on the north coast, which, like many others in Ireland, sheltered human occupation from Neolithic times. Nearby, at Magilligan, a great triangle of sand jutting into the mouth of Lough Foyle displays the dynamics of dune development in a raw plain of beach ridges. At this northern end of the island, still tilting up on the rebound from a dense weight of ice, a slight withdrawal of the sea allowed huge amounts of sand to accumulate onshore—one of the largest deposits in Ireland or Britain.

Today the sand circulates in a cycle that erodes the older dunes in storms and creates new beach ridges from the recycled sand. At the southeast of Ireland, on the other hand, the rapid erosion of soft cliffs of glacial debris continues to supply the sediment for dunes at a distance along the coast. For almost every dune system there is a slightly different story, a variation in the complex interactions of sediment, wind direction, sea level, and wave power.

Most of Ireland's dunes are losing, rather than gaining, sediment, and where they are eroding at the seaward face, exposed layers of black peat may testify to ancient littorals buried by storms or the sudden migration of sands. These movements have often been hastened by human activity. In Donegal, for example, the stocking of dunes with rabbits for food (common in pre-Famine Ireland) led to a massive sand-blowing that buried the village of Rosapenna, together with the mansion of the local earl. Overgrazing, and the cutting of marram grass for thatch, fodder, or bedding, led to other catastrophic episodes right up to the twentieth century.

Marram grass *(Ammophila arenaria)* is the toughest, most resilient of beach grasses, with long, spiky leaves in-rolled against the wind. It grows best in a constant inflow of sand, putting out new roots beneath the surface as it rises, so that a single marram plant can accumulate and bind a dune many yards high. Its stabilizing role led to widespread planting by coastal landlords, notably to fix the drifting sandhills of County Sligo early in the nineteenth century. (On Cape Cod, too, where the colonists' cattle devastated dunes in the 1700s, the inhabitants of Truro were out planting "beach grass," spring after spring, as a duty set by local law.)

Most Irish dunes are extremely well-vegetated, but often lack the scrub and heather characteristic of sand systems elsewhere in Europe. The spiny sea buckthorn *(Hippophae rhamnoides)*, which forms dense thickets in the dunes of continental Europe, is absent as a native shrub in Ireland. It was, however, planted at the well-preserved dune system at Murlough, County Down, where the enormous crop of orange berries now feeds flocks of thrushes arriving on winter migration across the Irish Sea. The Murlough dunes, a nature reserve, have been protected for two centuries and are unique in having a growth of heathers to match that of similar landscapes across the sea in Scotland. Botanists speak of the "dramatic abuse"—overgrazing, burning, and cutting—which has all but obliterated heather from Irish dunes elsewhere. Severe selective grazing by rabbits has sometimes allowed lichens (especially *Peltigera* and *Cladonia* species) to cover older dunes with a mantle of gray. But moderate grazing by cattle or sheep can preserve floristic diversity. Without it, many east-coast back dunes would grow a thick, dense sward of the prickly burnet rose, a beautiful but ungenerous neighbor.

Dune plants vary greatly between the coasts of the east and west. Along the Irish Sea the dunes are drier and sunnier and warm up quickly in spring. This makes them a refuge, especially at the southeast corner, for plants of Atlantic-Mediterranean affinity that are at the limit of their range in Ireland. One of the rarest, the woolly cottonweed *(Otanthus maritimus)* survives precariously at a single location in County Wexford, protected by a ring of boulders against the incursions of bikers.

The warm, mossy sward of the eastern fixed dunes is characterized botanically by the presence of a small wild pansy *(Viola tricolor,* spp. *curtisii)*, which mingles yellow and purple in its flowers. Orchids are plentiful, too—bee and pyramid orchids *(Ophrys apifera* and *Anacamptis pyramidalis)* and the pink or purple spikes of marsh and spotted orchid hybrids. But much of the ecological interest lies in plants that grow in the eastern dunes but are missing from the west, among them the sagelike wild clary *(Salvia verbenaca)* and plants in the borage family—the small, early forget-me-not *(Myosotis ramosissima)* and hound's-tongue *(Cynoglossum officinale)*. The dunes of the southeast, under great pressure from recreation and development, hold the last few sites of the wild asparagus *(Asparagus officinalis)*, and another traditionally edible plant, the fleshy sea-kale *(Crambe maritima)* is actually flourishing at many sandy and shingle beaches around the island now that its spring shoots are no longer valued as a potentially succulent vegetable.

On the west and northwest coast of Ireland, from Galway Bay to Malin Head, the frequency and power of winds from the Atlantic have shaped a dune landform shared only with the islands of Scotland, further north. It is called by a word from Scots Gaelic, *machair,* meaning "plain," and resembles, in its most perfect examples, a great green lawn of close-cropped grassland spread out behind the shore. There is often a shallow barrier of dunes on the seaside and a chain of shallow, sandy lagoons on the landward margin. In the working model for the development of machair, an earlier system of ridged dunes has been flattened down to the water table, eroded by frequent strong winds throughout the year and heavy grazing and trampling by sheep and cattle. Only on Ireland's machair coast does annual average wind speed reach 21 feet (7 m) per second, with a top storm speed of 164 feet (50 m) per second. Machair also coincides, both in Ireland and the Scottish Outer Hebrides, with the last communities of Gaelic-speaking farmers ("crofters" in Scotland), subsisting on small holdings of land and treating the dunes as commonage for livestock to graze.

Along the west coast in general, beaches are rich in fragmented sea shells, and these lighter particles form much of the sand that arrives on the machair in a constant wind-blown sift of grains. Where this ready supply of calcium feeds into microhabitats with distinctive wetland plant communities, the range of terrestrial snails is often remarkable. The dune system at Dooaghtry, below Mweelrea Mountain in Mayo, is a notably rich sanctuary for mollusks, with perhaps up to sixty species, but overgrazing by sheep now appears to have extinguished two of the rarer wetland snails *(Catinella arenaria* and *Vertigo angustior).*

The core plant species of machair blend fescue grass *(Festuca rubra)* with a tight weave of plantain *(Plantago lanceolata),* daisy *(Bellis perennis),* and bright yellow bird's foot trefoil. But the shell sand brings machair soil to an average pH of 7.8 and nourishes calcium-loving plants such as squinancywort *(Asperula cynanchica).* Indeed, its influence often reaches far inland from the machair, so that peaty hillsides growing crowberry and juniper may also find a berth for the rare hoary whitlow-grass *(Draba incana)* and, even in Donegal, the dense-flowered orchid. On the machair itself, the grazing pressure keeps vegetation extremely short, but even here, in August, the grass of Parnassus may lift a flush of delicate white cups above the sward.

A habitat as unusual as machair has special significance for birdlife. In winter it attracts large flocks of migrant golden plover, small parties of snow buntings *(Plectrophenax nivalis)*, and occasional small flocks of barnacle geese visiting from offshore islands. In summer it offers nesting grounds to waders such as lapwing, dunlin *(Calidris alpina)*, and ringed plover *(Charadrius hiaticula)*. But perhaps its most distinctive avian function is as a feeding ground for a small and engaging crow, with glossy blue-black plumage and bright vermilion bill and legs.

The chough *(Pyrrhocorax pyrrhocorax)* was originally a crow of mountain valleys where its buoyant, acrobatic flight, playful use of updrafts, and ringing contact call of "chee-ow! chee-ow!" were born of a rocky landscape. They echo now on coastal cliffs and islands, where the choughs nest in remote caves and crevices. The bird has also adapted to feeding at the margins of long-established pastoral agriculture, notably on seaboards where natural grassland is closely grazed by sheep and cattle. Here it probes with its curved bill for larvae hidden in the thin, salty turf or flips over cow dung for beetles and grubs.

The chough has known steep declines in the past, often to do with human persecution. It is extinct in England, and its steady decline in Europe has left Ireland as its most secure stronghold. More than nine hundred pairs survive, with a particular concentration along the cliffs of Kerry and Cork. But at the other end of the island, along the coast of Antrim, the last few choughs are seriously threatened by the kind of agricultural change (plowing and reseeding of grassland, fencing out of sheep from grassy cliff slopes) that has extinguished their habitats in Britain and Europe. The chough needs its small, playful bachelor flock as a school for young birds, and the loss of the flock is the beginning of extinction.

The insects of old, permanent pasture are disappearing over much of Europe, and Irish machair may well be a last refuge for some of them. The click beetle *(Selatosomus melancholicus)*, for example, with larvae that hunt in the soil beneath cow dung, has its closest European relatives in the Vosges Mountains of France and in the Alps and Pyrenees. But the island's dune systems in general have accumulated insects now scarce in Europe. As one example, the robber fly *(Machimuus cowini)* is found only at scattered sites between Hungary and the Atlantic coast, and the east coast dunes of Ireland would now seem to be its headquarters. The dunes of the south and east also shelter insects in solitary populations remote from more usual, much warmer, European habitats. They include two species of large, flightless grasshopper *(Conocephalus dorsalis* and *Metrioptera roeselii).* The occurrence of such dry grassland insects reinforces the regional difference between these dunes and those on the northwest coast, where cold-climate insects probably arrived in an early, postglacial colonization.

A few dune systems in the west of County Kerry shelter Ireland's one species of toad and thereby add another mystery to the island's biogeography. The natterjack toad belongs more typically to the warmer coasts and heaths of continental Europe and is the only vertebrate member of the so-called Lusitanian fauna of southwest Ireland (see Chapter 4).

The security of the Kerry natterjacks became an early conservation issue in the recent development of dune systems as golf courses, chiefly for tourism: Ireland now has almost half of all the sandy golf links in the world. For many small-farm communities on the Atlantic coasts, the local sandhills, often grazed as commonage, seemed an obvious resource for development, and planning control was inadequate to halt the change of use. In many early developments, dunes were reshaped with bulldozers, slacks drained, vegetation changed by rolling with heavy rollers and fertilizing. As conservation protest mounted, course designers became more ecologically sensitive, attempting compromises with the original dunescape. Many golf courses now have fine displays of wildflowers in the roughs (flowers that rarely appeared under the former pressure of grazing) and this can seem reassuring. But the loss in biodiversity has been inexorable. Plants and animals specially adapted to the natural conditions of the dunes have disappeared or have been relegated to isolated fragments of dune. The pressure for yet more golf courses is unremitting. By the end of the twentieth century there were fewer than ten intact dune systems remaining in Ireland of the 190 originally existing along the island's 466 miles (750 km) of sandy coast.

ALONG THE TIDELINE

A beachcombing naturalist will take to the tideline on the long sandy beaches of Ireland's Atlantic coasts with considerable expectation. Prevailing southwesterly winds, the clockwise swirl of the Gulf Stream, a shoreline open to the full fetch of an ocean: all the factors seem promising. Even the well-washed innocence of most drift material, in general free of condoms, syringes, and tar-balls if not of ubiquitous plastic debris, seems to augur happily for the occurrence of what were once called "natural curiosities."

Chief among them, but by no means frequent or common, are tropical drift fruits and seeds, the peregrine disseminules of jungle vines and other pod-bearing plants growing in the West Indies and Central America. Their arrival on European coasts has been discussed for at least four centuries, with the first Irish record of the sea heart *(Entada gigas)*, the most prominent and robust sea bean, dating to 1696. "It is very easie to conceive," wrote Hans Sloane, founder of Britain's famous Chelsea Botanic Garden in *Philosophical Transactions*, "that growing in Jamaica in the woods, they may fall from the Trees into the Rivers, or by any other way conveyed by them into the Sea." He posited a westerly trade wind "for at least two parts of three of the Whole Year, so that the Beans being brought North by the Currents from the Gulf of Florida, are put into these Westerly Winds way and may be supported by this means at last to arrive in [Ireland and] Scotland."

Sloane's theory, written when the currents of the oceans were scarcely understood, is strikingly accurate. Perhaps a score of species of tropical plants have peregrine seeds capable of staying afloat in salt water for fifteen months, the least time that it takes a small object to drift across to Europe. The glossy skin of the sea heart, like the maroon Morocco leather of antique desktops, is so tough that snuff-boxes have been made of it; no less durably, it served as a teething ring for many a coastal baby. The 2-inch (5-cm) seed is spilled from pods more than a yard long, growing on woody rain forest vines.

Western and northwestern Irish beaches have produced about eight drift species of tropical American (mostly West Indian) origin. Largest by far is the coconut *(Cocos nucifera)*, which arrives still in its husk and sometimes barnacled and bored by marine mollusks. Along with sea hearts come half a dozen smaller seeds, notably the dark-brown horse-eye bean *(Mucuna sloanei)*, named for its conspicuous black hilum; the similar sea-purse *(Dioclea reflexa)*; and the dove-gray, acornlike nickar nut *(Caesalpinia bonduc)*. The smallest transatlantic traveler, not from the tropics, is North America's sea pea *(Lathyrus japonicus)*. This is a rare coastal plant in Ireland but common on the Canadian east coast and on the shores of the Great Lakes. It seems probable that several thousands of its seeds arrive each year on Irish beaches in the west and south, and that they account for the plants occasionally found growing in such locations as Inch Strand, County Kerry, and in sandy coves in County Cork.

To rear plants from the tropical seeds has been an intriguing challenge to professionals and amateurs alike. The tough seed coats should be nicked with a hacksaw and allowed to soak in warm water for forty-eight hours before planting in a sterile compost doused in fungicide. The sea heart is notoriously reluctant to germinate, and even a successful emergence of the radicle will often be followed by fungal attack. But experiments in the National Botanic Gardens at Glasnevin, in Dublin, have reared vigorous plants from the horse-eye bean and the sea purse (both vines) and from the nickar nut, a spiny shrub with foliage like that of the rose family.

Drift seeds are usually delivered at the top of the tide, at the line of seaweed fronds and plastic granules. Here, too, is a landing for the buoyant marine organisms that have lived at the mercy of the wind.

One of the most beautiful to human eyes is the violet-blue seasnail (*Janthina* spp.), a pelagic gastropod that drifts beneath a raft of silver bubbles trapped in mucus. The animal's shell is translucent and so fragile that it seldom floats ashore intact. Rachel Carson waited for years to find one, so that my discovery, on the beach below my house, of two *Janthina* at once must make up for never winning the lottery.

The snail's own survival is unusually governed by chance, because its most substantial prey is also randomly adrift on the ocean. The by-the-wind-sailor *(Velella velella)* is a colonial hydrozoan, an organism in a class with the big Portuguese man-o-war *(Physalia physalis)* but miniature in scale and differently rigged. While the Portuguese man-o-war drifts beneath a gas-filled float the size of a party balloon, the by-the-wind-sailor depends on a low, stiff triangle of a sail, mounted diagonally on a deep-blue, oval float an inch or two across. In a further and subtle refinement, the sail is set northwest to southeast on some floats and northeast to southwest on others, so that in the same wind some will sail leftward and the others to the right, either of them veering as much as sixty degrees from the wind's direction. Both models of the by-the-wind-sailor occur in all the big oceans, where their two divergent paths disperse the species. In the northern hemisphere, where the ocean winds twist clockwise, it is most often the left-sailing one that ends up at the Irish shores, where the little discs and sails, often prettily rainbowed, may be scattered along the tideline or even piled in continuous drifts and numbered in millions, as they were in 1992. *Janthina* occasionally arrives with them (having, perhaps, not quite finished chewing), but snail and jellyfish first meet far out on the ocean by bobbing blindly against each other, like toy boats on a park pond. In some years, prolonged southwesterly winds that herd by-the-wind-sailor ashore may also bring shoals of Portuguese man-o-war into the waters off the south of Ireland, but this notorious stinging power pack is rarely cast up in one piece. The common jellyfish of Irish beaches, shared with coastal waters all over the world, is the transparent and innocuous moon-jelly *(Aurelia aurita)*. The much larger compass jellyfish *(Chrysaora hysoscella)*, marked with brown rays like an art deco lampshade, is similarly harmless. There are no regular stinging jellyfish in Ireland's inshore waters, but two potentially harmful pelagic species, the *Cyaneids*, sometimes called the lion's mane jellyfish—the blue *(Cyanea lamarckii)* and the large brown *(Cyanea capillata)*—may sometimes be driven inshore. Shoals of another oceanic drifter, the small, amber-colored and luminescent *Pelagia noctiluca* were cast up on the west coast in the fall of 1998.

TURTLES

The annual abundance of jellyfish in Irish waters must help to explain the regular migratory appearance, between July and October, of one of the ocean's most remarkable animals, the leatherback turtle *(Dermochelys coriaceae)*, which feeds heavily on them. This is the world's biggest turtle and heaviest reptile, growing to some 6.5 feet (2 m) in total body length: an exceptional specimen, washed ashore at Harlech in Wales in 1988, measured 9.5 feet (2.9 m) and weighed 2,018 pounds (915 kg). Its appearance is also strikingly different from that of the hard-shelled species. It has no horny carapace but is covered with what looks and feels like smooth, black rubber, and the flattened barrel of its body has ridges running along the back, like seams or struts, to hold it rigid. The leatherback's form was shaped in the Triassic, and it can live a hundred years or more.

This turtle was once thought rare, and those captured by Ireland's coastal fishermen were often towed ashore for public exhibition and photographs in the local paper. Leatherbacks become entangled in fishing gear, notably in the ropes of lobster-pot buoys that they may mistake for jellyfish. One entangled in a lobster-pot line off Crookhaven, County Cork, in 1993 was full of lion's mane and compass jellyfish. In the 1980s a zoologist, Gabriel King, made a personal crusade of informing fishermen about the leatherback. They responded well, and have sometimes taken great personal risks in releasing entangled turtles.

Records collected by King and others show a seasonal migration to the Atlantic waters off Ireland and Britain that echoes similar movements elsewhere in the world (a study in the eastern Pacific, using a satellite, tracked a female leatherback for 1,732 miles, or 2,780 km, over eighty-seven days). The turtle is regularly recorded as far north as Norway, Newfoundland, and Alaska, a passage into cold made possible by its ability, unique among reptiles, to generate body heat internally and conserve it through insulation by the thick, cartilaginous shell and a vascular system that minimizes heat loss. Its deep-core warmth has been measured at as much as 64°F (18°C) higher than the temperature of the surrounding water.

The origin of Ireland's leatherbacks is still unclear, because there are breeding rookeries both in West Africa and on the tropical coasts of the western Atlantic. The thrust of the Gulf Stream makes a birthplace in the west more likely, and in 1997 an animal tagged in French Guiana was found dead on a beach in Wales.

Ocean currents bring two more species of turtle to Ireland, but these arrive involuntarily, with no return ticket, and most are young animals no bigger than the leatherback's head. The origin of one species is known for sure, as almost the entire world population of the little Kemp's ridley turtle (*Lepidochelys kempii*) nests on a single beach in the Gulf of Mexico. There have been half a dozen records of strandings of the ridley, usually freshly dead in winter, but many more of the common loggerhead (*Caretta caretta*), which has rookeries on both the American coast and in the Mediterranean. A typical stranding of loggerheads followed a run of westerly storms in the late winter of 1992. Along with several dead turtles were two that reached Kerry weak but alive and were taken into an aquarium.

A closer acquaintance with the ocean's more exotic species is part of Ireland's new and long-overdue awakening to the sea. Unprecedented state investment in marine science is directed mainly to "sustainable" exploitation, but the advent of aquariums such as those in Dingle and Galway, the rapid growth of boat ownership and marine sports, and the excellence of underwater films on television are all engaging an island people with the new wild world surrounding their shores.

SELECTED REFERENCES

Brown, R. 1990. *Strangford Lough: The Wildlife of an Irish Sea Lough.* Belfast: Queen's University.

Guiry, M. D., and G. Blunden, eds. 1991. *Seaweed Resources in Europe: Uses and Potential.* Chichester, UK: Wiley.

Moore, P. G., and R. Seed, eds. 1985. *The Ecology of Rocky Coasts.* London: Hodder & Stoughton.

Myers, A. A., C. Little, M. J. Costello, and J. C. Partridge, eds. 1991. *The Ecology of Lough Hyne.* Dublin: Royal Irish Academy.

Nelson, E. C. 2000. *Sea Beans and Nickar Nuts.* London: Botanical Society of the British Isles.

Quigley, M. B., ed. 1991. *A Guide to the Sand Dunes of Ireland.* Dublin: Trinity College.

Wood, E. ed. 1988. *Sea Life of Britain and Ireland.* London: Immel.

— *Chapter 10* —

Wild Islands and Their Seas

"They were made to be plucked and eaten,
and we might as well pluck and eat them as any others.
Where would the people get feather beds to give
their daughters when they get married,
only for the rock-birds?"

—Tom O'Flaherty, *Cliffmen of the West*

For several centuries, the presence of people on the bleaker, more remote island outposts of Europe depended substantially on the summer harvest of birds for winter food, warmth, and lamp light. On St. Kilda and on Scotland's northern isles, men climbed cliffs at night to lasso sitting gannets with nooses dangled from rods or lowered themselves on ropes to snatch fulmars *(Fulmarus glacialis)* to squeeze for oil. On Ireland's Aran Islands, villagers stretched fishing nets up the face of the sheer limestone cliffs, flapped them to frighten the guillemots *(Uria aalge)* off their eggs, then heaped the tangled birds into currachs (small boats with a skin of tarred canvas) dancing in the foam below. Split, salted, and dried in the wind, the *forachan* were eaten "every day of the week except Friday" (then the Roman Catholic day of fasting).

Young gannets the size of farmyard geese are still taken to eat in Iceland, the Faroes, and Scotland's Outer Hebrides, and the Faroese still hook plump shearwater chicks from their burrows. Up to the late 1970s, scores of thou-

Storm-washed islands off the west coast of Ireland.

sands of guillemots were shot annually on both sides of the North Atlantic by the hunters of Newfoundland and of the Faroes and southern Norway. In Ireland, however, the hunting of seabirds never became a mere sport, and from the early twentieth century, the crowded tiers of the island's cliff colonies were left to their own clamorous excitements.

Ireland rarely approaches the awesome assemblages of nesting seabirds to be found a little further north: St. Kilda has bigger colonies of gannets, fulmars, and puffins *(Fratercula arctica)*; the north of Scotland has larger concentrations of guillemots and razorbills *(Alca torda)*. Even the great Irish colonies of storm petrels *(Hydrobates pelagicus)* are a fact to be taken largely on trust, because the birds visit their nests underground, at night, on uninhabited islands. But precipitous rocks and islands off the tips of the Kerry peninsulas are still among the most spectacular and important seabird colonies in western Europe, and other Irish islands lend refuge to particular species or have a strong geographical role in patterns of migration.

Breeding on small islands a good distance from shore makes particular sense for birds designed better for swimming and diving than escape from rats or foxes. Sitting tail to the wall on a narrow ledge of a cliff or fluttering in at midnight to the entrance of a deep, dark burrow are extreme but effective defenses.

The cluster of crowded bird islands at the southwest corner of Ireland also speaks for an exceptional food supply. Indeed, while northern Scotland may have larger colonies, the highest densities of birds at sea are found off Cork and Kerry. Here strong ocean currents lift cold water from the seabed at the edge of the continental shelf, and the upwelling of mineral nutrients nourishes summer plankton fronts. These, in turn, feed zooplanktonic crustaceans and shoals of fish such as herring, sprats *(Sprattus sprattus)*, and mackerel *(Scomber scombrus)*. Seabirds and cetaceans alike are drawn to these concentrations of food, just as they are at the strong shelf-line upwellings off California or southwest Africa. At the Irish approaches, gannets plunge among white-sided dolphins *(Lagenorhynchus acutus)* and Manx shearwaters *(Puffinus puffinus)* skim the waves among minke whales *(Balaenopera acutorostrata)*. More stirring and mixing goes on as the currents meet the long headlands and the islands, and mineral enrichment again supports a great burgeoning of phytoplankton. In a study at the Scilly Isles, at the southwest corner of England, plankton productivity was shown to increase dramatically over an area of sea twenty times the size of the islands.

The dramatic scale and intensity of seabird movement in Ireland's southwestern waters have made Cape Clear Island, off Baltimore in West Cork, one of the best vantage points for sea watching in Europe. From a perch on precarious cliffs, often slick with rain and spray, viewers can observe prodigious numbers of birds in endless procession. They begin in February, with the first passage of gannets toward their big Kerry colonies from winter quarters further south. The migrations gather pace in March, with notable movements of Manx shearwaters and storm petrels, and by April, razorbills and other auks are streaming westward past the island in passages approaching ten thousand an hour. July brings the year's largest movements of Manx shearwaters, eastward at dawn and then westward at dusk, when the birds pouring back to their Kerry breeding islands may number almost thirty thousand an hour. Perhaps the best period for sea watching begins in August and continues through the autumn. There are heavy passages of gannets, many young, dark birds among them, and foggy, drizzly days bring these migrants so close inshore that they turn their heads in passing to take note of the watchers on the cliff. Wet and windy weather brings remarkable seabird movements within sight of land, not only of Cape Clear but of many headlands and islands along Ireland's Atlantic coasts: in the long processions of birds are great numbers of southern (sooty, great, and Balearic) shearwaters *(Puffinus griseus, Puffinus gravis; Puffinus mauretanicus)* and northern (great and arctic) skuas *(Stercorarius skua; Stercorarius parasiticus)*, and such rarities as Sabine's gulls *(Larus sabini)* and Leach's petrels.

On misty mornings in autumn, especially in early October, birders gather in high excitement at the gardens, bushy hollows, and bogs in the interior of Cape Clear. The island, a mere 5 miles (8 km) long, seems to act as a magnet for vagrant birds, not merely from Europe but even from Asia and America. "It is almost impossible," wrote Tim Sharrock in his 1973 *Natural History of Cape Clear,*

> to stay on the island during this period and not see at least one real Irish rarity, such as the spotted crake *(Porzana porzana)*, wryneck *(Jynx torquilla)*, Continental coal tit *(Parus ater ater)*, bluethroat *(Luscinia svecica)*, Blyth's reed warbler *(Acrocephalus dumetorum)*, barred warbler *(Sylvia nisoria)*, subalpine warbler *(Sylvia cantillans)*, greenish warbler *(Phylloscopus trochilloides)*, Bonelli's warbler *(Phylloscopus bonelli)*, Richard's pipit *(Anthus novae see-landia*, also *A. richardi)*, tawny pipit *(Anthus campestris)*, woodchat shrike *(Lanius senator)*, red-backed shrike *(Lanius collurio)*, hawfinch *(Cocco-thraustes coccothraustes)* or yellow-breasted bunting *(Emberiza aureola* or *L. chrysophrys)*. . . . Early October is the time when American birds are most likely to appear—dowitcher *(Limnodromus* spp), white-rumped sandpiper *(Callidris fuscicollis)*, yellow-billed cuckoo *(Coccizus americanus)*, olive-backed thrush *(Catharus ustulatus)*, red-eyed vireo *(Vireo olivaceus)*, American redstart *(Setophaga ruticilla)* and rose-breasted grosbeak *(Pheucticus ludovicianus)*.

Such a remarkable annual roster of wandering birds made Cape Clear an obvious choice for a bird observatory—a counterpart to Fair Isle, off the north of Scotland, already a mecca for British ornithologists. It was set up in 1959 and recorded more than 240 species in its first decade. But its more serious value lay in the regular, methodical sea watches that established the huge scale of seabird movements and began to make sense of the often mysterious passages east and west. The movements of the birds seem often to be guided by the barrier of the coast, and the westerly processions may often gather up birds that have drifted east in gales and deliver them back to the ocean.

Most numerous of the seabirds that nest around Ireland and Britain are the guillemots, fulmars, and kittiwakes *(Rissa tridactyla)*, and many thousands of these join the gannets and puffins on the cliffs of Great Saltee Island, off Kilmore Quay in County Wexford at Ireland's southeastern corner. This was where the first Irish bird observatory was established, in 1950, and during the decade or so of its existence a total of 209 species were seen there. Great Saltee lies on the main migration route of passerines from continental Europe to Ireland, and from April to mid-May the island gives temporary refuge to hundreds, even thousands, of warblers, swallows, and martins.

The chance of rarities in spring and autumn and the huge summer concentration of seabirds have made the island a popular destination for Irish birders, even though, unlike Cape Clear, it has no ferry and landing by chartered trawler on to a boulder beach needs a fairly calm day. The island has not been farmed or inhabited since the 1940s, and an unexpected delight in early summer is to find field after field densely carpeted with bluebells *(Hyacinthoides non-scriptus)*, more commonly a woodland plant. Such abundance is echoed in a number of windswept (but humid) maritime locations, and bluebells bloom in drifts, along with woodland primroses *(Primula vulgaris)* on the exposed granite cliffs of southwest Britain.

Unlike the bluebells, the gannets of Great Saltee are relatively recent settlers, part of a steady increase that has brought its population in the eastern North Atlantic to almost 274,000 pairs. In a progress typical of the species, the birds first prospected the Great Saltee cliffs in the 1920s, took thirty years to become established, and then increased their numbers rapidly, reaching one hundred nests by 1965 and more than one thousand by 1990. In long-established colonies, growth is slower, until, sitting virtually wing to wing, the birds have occupied every usable patch of rock.

The success of the North Atlantic's gannets has brought them closer to people: the youngest of Ireland's five colonies, discovered in 1989, is on Ireland's Eye in Dublin Bay, on the doorstep of the capital city, and already numbers upward of thirty nests. But the largest and oldest colony, second biggest in the northern hemisphere, occupies the peak of a drowned headland jutting up from the sea about 11 miles (18 km) off Kerry: a stark and precipitous mass of sandstone, polished at wave level by the ocean surge.

Landing on Little Skellig needs calm conditions rare even in summer. Yet many people have achieved it, and from the most fragile of boats. The gannet colony was noted as long ago as 1700, and in the mid-eighteenth century there were an incredible number of birds. Despite the hazards of setting foot, there was local exploitation of the gannets for food: an exceptionally low level of about thirty pairs in 1880 may have reflected the hungers of the mid-century Famine. Yet within twenty years, thousands of pairs were reported again. During this period, the gannets were sharing the rock with large colonies of puffins, which burrowed into the hummocks of sea-pinks *(Armeria maritima)*. Eventually, this skin of vegetation and peat soil was so undermined it disintegrated, dispossessing the puffins entirely and providing new sites for the gannets' nests.

Today the nest sites and guano of 26,850 pairs of gannets whiten the cap of the island, and a closer approach by fishing boat nudges into a clamorous traffic of birds. Thousands are sitting on their eggs; thousands more wheel in the sky around the rock or plunge with folded wings to smite the sea. In this plummeting dive, from as high as 130 feet (40 m), the gannet does not spear its fish but seizes it underwater and swallows it whole. Little Skellig is a nature reserve owned by BirdWatch Ireland, and the colony's hectic spectacle is recorded every summer by boatloads of birders and other tourists, cameras braced to the rise and fall of a powerfully rhythmic sea.

From the mainland, the island's white-capped silhouette contrasts sharply with the dark double-pyramid of its close neighbor, Skellig Michael (formally the Great Skellig). This is Ireland's most dramatic holy island, named for the saint of high places, and a haunt of hermits, monks, and pilgrims since early Christian times. An open cut-stone stairway climbs steeply from the landing to a platform more than 560 feet (170 m) above sea level, where a cluster of dry-stone beehive huts and the ruins of a ninth-century church look out to inspiring infinities.

There are no gannets on Skellig Michael, but many thousands of Manx shearwaters and even vaster numbers of its diminutive cousin in the Procellariidae (tube-nosed) family, the storm petrel, smallest of all Atlantic seabirds. When lightkeepers were stationed at the island, their usual computation of petrel numbers was one word, "enormous." No greater precision could be brought to the petrel colonies of Puffin Island, closer to land, or the Blasket Islands off the Dingle Peninsula, or the empty, windswept islands ranging north to Donegal.

Now that lighthouses are unstaffed and run by resident computers, even fewer people see the petrels and shearwaters fly in from the ocean as darkness falls. This nocturnal strategy (to avoid attack by gulls), and the habit of nesting in burrows and crevices on often precipitous terrain, has tantalized ornithologists wanting firm population figures. Perhaps three quarters of the world's storm petrels breed in Ireland and Britain, but estimates of their total have remained wildly uncertain, ranging from 70,000 to 250,000 pairs. The total of Manx shearwaters in burrows beside them on the breeding grounds is computed at about 300,000 pairs, more than 90 percent of the breeding population.

Why "Manx" shearwater and why *Puffinus puffinus* for a bird that is not a puffin? There are few, if any, shearwaters on the Isle of Man (in the Irish Sea), but in the seventeenth century the species' plump nestlings were harvested in thousands from a nearby islet, the Calf of Man, and sold widely for food as "puffins." Ornithology borrowed from the vernacular, but popular confusion between birds that nested as neighbors later transposed "puffin" to the more eye-catching species, which had formerly been known by local names such as "sea parrot." Puffin Island off Kerry is, as it happens, the major Irish stronghold of the Manx shearwater, with perhaps ten thousand pairs, and it also holds thousands of storm petrels, but it is the large colony of puffins (some seven thousand pairs) that gives the island its name.

The petrels and shearwaters have many habits in common, using the darkness of the islands for courtship, mating, and contacts at the nest. For both, the breeding season is a long one (the shearwater incubates its eggs for fifty days or more), and each of the parents in turn may spend several days at sea. But the colonial lives of the two species are quite distinct, and that of the Manx shearwater includes a daily ritual still puzzling to biologists.

Each evening, an hour or so before sunset, the birds gather in rafts on the sea close to their islands. At the major breeding sites off southwest Wales and Ireland, these assemblies can number thousands of shearwaters. If the sea is rough, they skim to and fro in long flocks over the water. Otherwise they float, waiting for darkness, preening or bathing, and perhaps paddling slowly toward their adopted cliffs. As night falls, the rafts break up and the birds disperse to their burrows in the dark mass of the island.

What is gained from this idle gathering—time stolen, as one might suppose, from fishing? A theory proposed in the 1960s suggested that the evening assembly might allow the shearwaters to assess their current population level and to adjust their breeding effort to match the prevailing supply of food and nesting burrows. This idea has been rejected, however, as an evolutionary heresy, because natural selection would seem to demand that each shearwater make the best possible effort to hand on its genes. On the other hand natural selection may favor genes for social behavior that in turn favors survival of all.

The storm petrels keep no such sunset rendezvous but flicker straight in from the ocean after dark. Their calls in flight are soft and vibrating—a modest vocalization, like the muffled crooning, churring, and hiccoughing the petrels utter underground. The shearwaters' brand of spookiness, however, can become distinctly strident as the breeding season nears its peak, and the wild crowing of birds on the wing and on the ground joins the crooning that drifts from the burrows. The chorus has provoked some naturalists to shockingly violent behavior. In 1888 the respected Irish ornithologist, R. M. Barrington, camped on Skomer Island, off Wales, which today holds some 100,000 pairs of shearwaters. He wrote in *The Zoologist:* "We lay down to sleep, but it was a mockery, for as the night wore on, the noise became worse and at times awful. Unable to sleep, we determined to go out, and either frighten or kill some of the shearwaters. Armed with a stick each, we killed all we could carry—forty to fifty— in half-an-hour. Our midnight raid had no effect in quieting the birds, and we got no sleep until after two in the morning."

The many seductive questions on the lifestyle of the midnight petrels has drawn a succession of Irish and British ornithologists to the birds' breeding islands and sea stacks, whether for a few snatched and hazardous nights on banding expeditions or for long-term studies demanding great personal commitment. As mentioned earlier, the veteran Irish ornithologist, Robert F. Ruttledge, seeking the rarely encountered Leach's petrel spent a night on every island off Galway and Mayo. These included the rocky pinnacles of the Stags of Broadhaven, where his tent was blown away in a storm.

The counting of nesting shearwaters and storm petrels can be exceptionally difficult and dangerous and hard to carry out accurately. Even where the colony can be reached with safety, one burrow may lead to more than one

nesting chamber (or, just as likely, to a rabbit or a puffin). Shearwaters sometimes nest in boulder scree, and storm petrels seek out every sort of dark
cranny, from spaces under boulders on the shore and holes under hummocks
of sea-pinks or heather, to crevices in scree slopes poised above the sheerest of
cliffs. A more confident counting of petrels has waited on new sampling
methods, using tape recordings of the male petrel's purring calls to elicit re-
sponses from birds sitting tight under stones and hummocks. The totals these
samples produce when fed into a computer model will always be a rough fit
for the realities of the ocean wilderness, but some sort of best-effort baseline
has seemed worthwhile. Increasingly, the isolation and grandeur of Ireland's
uninhabited islands make them a target for recreational adventure and summer disturbance, and conservation strategies need to be able to measure the
impact of change.

Inisnabró and Tíreacht islands from Inishvickillaune, the outermost Blasket Island.

ISLANDS OF THE BARNACLE GOOSE

Some of the western islands are so forbiddingly bleak in winter that only an adequate supply of *poitín* (moonshine), one might think, could ever have sustained their human settlement. The illicit distillations of alcohol on the Inishkeas, off the Mullet Peninsula, in County Mayo, were indeed of famous quality, being mashed in copper stills that were family heirlooms, lowered out of sight on ropes in the caves of the western cliffs. But a tragedy of 1927, when the islands' currachs were overtaken by a sudden storm and ten young men were drowned, proved beyond solace, and within a decade the islanders moved to the mainland.

Today the ruined villages of north and south islands make a broken frieze above the shore, and storms litter the fields with brightly colored buoys from

The Great Blasket from Inishvickillaune, the outermost Blasket Island.

distant fishing fleets. For much of the winter, indeed, the unkempt and sodden appearance of the islands give them the feel of freshly thawed arctic tundra, so that large flocks of barnacle geese, which breed on such terrain in the Arctic, have taken to them as their special home in Ireland. Between 2,500 and 3,000 of the birds arrive each October, by far the largest concentration among the 8,100 barnacles that winter along the west coast.

These beautifully marked geese, with black necks and white faces (like a smaller, silvery version of the Canadian goose, *Branta canadensis*) derive their common name from the early Old World belief that they hatched from goosenecked barnacles. The truth is scarcely less remarkable. The geese nest in colonies on cliff ledges in east Greenland, and their goslings are led by their parents into jumping into space at less than forty-eight hours out of the egg. This free-fall descent, wing-stubs whirring, may end in fatal collision with a spur of cliff or on a ski slope of soft snow. The parents wait on the scree at the foot of the cliff to gather up what is left of the family (if a patrolling arctic fox has not reached them first) and lead them on a trek to the safety of the nearest lake.

David Cabot

Barnacle geese *(Branta leucopsis)* flying in to Inishkea Island off County Mayo.

This extraordinary sequence has been filmed in high arctic latitudes of East Greenland by David Cabot, who has made a long-term study of the barnacle geese of Inishkea. Families end up year after year in favorite corners of particular fields on the islands, but the proportion returning with young is always low; one or two breeding seasons in the long lives of the birds is enough for replacement. Cabot has banded almost nine hundred of the geese, and his arctic expeditions were made in hope of finding them on their nesting grounds. Only a handful of the banded Mayo birds were recovered in Greenland, however, and most of them have turned up subsequently at Islay, an island at the southwest of Scotland, where rich grassland and stubbles attract huge flocks of wildfowl. On Inishkea, the geese feed incessantly, their small, stubby bills cropping a close maritime turf. Cattle follow them around to feed on their droppings, still high in nutritional value and resembling green cheroots: a grazing flock of two hundred to three hundred birds will produce up to six thousand droppings every hour.

In the diverse achipelago of seabird islands, spectacular colonies belong, for the most part, to the operatic grandeur of ocean fortresses such as the Skelligs and Blaskets. But at the eastern side of Ireland, in the Irish Sea, comparatively small rock islets and even tinier, sandy ones in coastal lagoons, are crucial refuges for migrant terns. Among them is a species brought almost to extinction in Europe and North America in the nineteenth century, and even now remains the scarcest seabird of the temperate North Atlantic.

The roseate tern *(Sterna dougalli)*, with long tail streamers and a delicate, salmon-pink flushing its underparts, became, like the white-plumed egrets, an object of plunder for the millinery trade. As it grew scarce, the ruthlessness of egg collectors threatened to eliminate it altogether on its breeding grounds, among them a number of islets off County Dublin. By 1900 the tern had vanished from Ireland, returning decades later only after protection in Britain had helped to raise its numbers.

Its breeding population in Britain, Ireland, and France reached a peak of 3,812 pairs in 1968 and then declined almost continuously over the next two decades, drifting down to a low point of 561 pairs. One reason was certainly the persecution on its wintering grounds along the coastline of Ghana in west Africa, where the children of fishing communities trap it with baited hooks buoyed up by bladders, as a sport. But in Ireland, too, its misfortunes were extraordinary: its best breeding island (a sand bar in Wexford Harbour) was washed away by storms; its next colony was attacked by rats; another important island was sprayed with herbicide by overzealous lighthouse keepers. In Britain and France, too, colonies were devastated by rats and gulls. Today more than half the European roseate terns nest on the small island of Rockabill, off north County Dublin, where close wardening has brought the number of nests to more than five hundred. In the shade of the sprawling tree mallow no longer sprayed as a weed, more chicks are reared per pair here than at any other colony under study.

The rocky cliffs of the Irish islands, while dominated in summer by seabirds, extend the nesting habitats of other significant species. The chough, flamboyant red-billed crow of the Atlantic seaboard is not a colonial breeder, but on undisturbed islands may share out the crevices and caves at a sociable density of several pairs per mile. Peregrine falcons and ravens, which make such fractious neighbors along mountain crags, pursue their traditional squabbles high above the island surf. The ravens are attracted by sheep and seabird carrion and the lack of competition from foxes.

"PEOPLE OF THE SEA"

The notable mammals of the islands, indeed, are those at home in the sea. Of Ireland's two resident seals, it is the large gray seal *(Halichoerus grypus)* that makes the remoter island coves its own, raising a doglike head from the sea to inspect a human through coal-black eyes. The common or harbor seal *(Phoca vitulina)* is the round-headed pinniped of estuaries and sand bars and only common in some locations (Strangford Lough on the east coast has about one thousand individuals), but the gray seal is numerous enough to populate coastal folklore as "the people of the sea" and to invite an unassuageable feud with fishermen.

Ireland is at the southwestern edge of the gray seal's distribution in the northeast Atlantic: the really big breeding colonies are at the huge "seal cities" at islands off western Scotland and along undisturbed coasts in Wales. With protection since 1914, the UK population has recently been rising at a steady 7 percent a year, to an estimated 108,500 animals in 1994. No up-to-date estimate for Irish waters exists, leaving a twenty-year-old figure of 2,000 to 2,500 still serving as a very uncertain minimum. The production of pups in autumn at Ireland's main westcoast breeding colonies seems steady: about 150 a year at the Inishkea Islands, about the same at the Blaskets, and another 150 at the

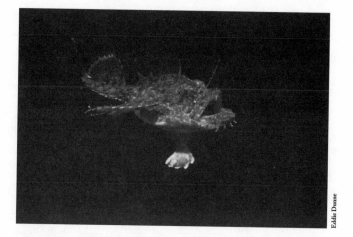

Angler fish *(Lophias piscatorius)*, whose tail, renamed as "monkfish," is a prime Irish restaurant delicacy.

Saltees and smaller east coast islands. But the migration of gray seals from Scotland and Wales substantially boosts the numbers in Irish waters. In the molting season early in the year, as many as 2,200 seals may be hauled out on the rocks of the Inishkeas, growling and crooning in a restless kind of repose.

The appetite of the average gray seal has been calculated variously from 11 to 33 pounds (5 to 15 kg) of fish per day—sufficient at any level to incur the general hostility of trawlermen. A sampling of the seals' diet off the Mayo coast, during the spring cod gillnet fishery, found them taking the smaller specimens among commercial whitefish and noncommercial species such as poor cod *(Trisopterus minutus),* the wrasse species of rocky seabeds, and substantial numbers of squid. A similar study in the Irish Sea confirmed the seals as opportunistic hunters, their prey varying according to seabed type and time of year.

Particular seasonal predations, notably of spring-run salmon caught in drift nets off the Mayo coast, have brought retribution in the form of brutal and illegal culls by local fishermen at island breeding grounds. Some juvenile seals die today off the south coast, drowned in tangle nets of near-invisible nylon. These are laid on the seabed to catch the remarkable angler fish *(Lophias piscatorius),* whose tail section, renamed as monkfish, is a prime Irish restaurant delicacy. Despite evidence that the most valuable fish stocks are not the principal prey species of the gray seal, the fishing industry continues to favor management by culling. This crude measure is resisted by Irish scientists who want the seal considered as just one dynamic element in a complex marine ecosystem.

WHALES AND DOLPHINS

Old sea bones, like mossy stone sculptures, decorate my garden in Mayo: a sperm whale's vertebra dug from the dunes; the skull of a beaked whale salvaged from a lonely gravel spit; dolphin vertebrae nailed to the shed in a wedge of flying forms. The big sandy beach at the foot of the hill is a burial ground for cetacean corpses drifting in from the Atlantic. Now and then it is also a joyous place for watching the animals alive: bottlenosed and Risso's dolphins *(Tursiops truncatus; Grampus griseus)*, circling the reefs beyond the surf.

Ireland has grown more fully conscious of its cetacean wildlife, even declaring the Republic's coastal waters a sanctuary for whales and dolphins. Enthusiastic study, much of it amateur, tries to put numbers and patterns on the comings and goings of the animals offshore. This is research done for its own sake, but also with the hope of adding Ireland to the world's prime whale-watching tourist destinations.

The island's wide undersea margin of continental shelf is certainly one of Europe's best locations for cetacea, with a great diversity of marine habitat and rich feeding grounds. To date, twenty-three species have been recorded, from the common harbor porpoise *(Phocoena phocoena)* to eight species of dolphin and fourteen kinds of whale. Much of the information is from records of strandings, both live and dead, and from Ireland's brief chapter of commercial whaling, but extensive modern surveys and sea watching by the Irish Whale and Dolphin Group have been building up a picture of relative abundance and seasonal movement.

A year-long study funded by the oil and gas industry used bottom-mounted hydrophone arrays to monitor the songs of the large baleen whales to the west of Ireland and Britain, from the Faroe-Shetland area down to the Bay of Biscay. The pulsing moans of the fin whale *(Balaenoptera physalis)* rang through the sea at every latitude, with the highest counts of individuals between October and January. The massive blue whale *(Balaenoptera musculus)*, 79 to 92 feet (24 to 28 m) long, was also there, with a peak in numbers off western Ireland in November and December. Nothing in the pattern of calls for either whale matched the accepted picture of a regular migration between wintering grounds off northwest Africa and feeding grounds at the Arctic Circle. Only among the humpbacks *(Megaptera novaeangliae)*, the whale least encountered, was there some picture of southward movement in early spring.

Telescope views of groups of fin and sei whales *(Balaenoptera borealis)* feeding over several days off the Old Head of Kinsale, County Cork, in winter months have been among more recent excitements of Irish whale-watching. The minke *(Balaenoptera acutorostrata)*, smallest (23 to 33 feet, 7 to 10 m) and most common of the rorquals, is the one most likely to be glimpsed, mainly in July and August, from islands and headlands on the coast. The killer whale *(Orcinus orca)* also travels the west and south in small pods, often pursuing herring and salmon, and its recent appearances included a seven-strong escort for a tourist ferry in the Blasket Islands in August 1998.

Some cetacean species such as the harbor porpoise and bottlenosed dolphin live in coastal waters throughout the year, while other dolphins move inshore in summer in pursuit of shoaling sprats, herring, and mackerel. In particular, the common dolphin *(Delphinus delphus)* is widespread west and south of Ireland, its numbers peaking in July as the dolphins move into shelf waters with their calves. The far heavier Risso's dolphin, with blunt head and distinctive scars, is a scarcer visitor, feeding mainly on squid, but its inshore behavior can be memorable. The naturalist Tim Sharrock was sitting watching seabirds at the tip of a rocky point on Cape Clear Island, as recounted in *The Natural History of Cape Clear Island*, when

> three Risso's dolphin began to tear through the water directly towards me, stopping just in time to avoid crashing headlong into the rocks, then turning and repeating this "game". Frankly, it was a somewhat unnerving experience for a sea-level observer to have three 13-foot monsters, each weighing a third of a ton, charging head-on to within a few feet.

Sharrock's book confirmed that this small island, already outstanding for its bird observatory, was also a peerless vantage point for spotting whales and dolphins. Schools of killer whales are frequent in August and September in family parties, as Sharrock suggested, "for the male usually leads the way but frequently turns back and rounds up any laggards." Scores of solitary large whales, passing far out to sea, have been caught in the telescopes of cliff-perched ornithologists, among them humpbacks in classic breaching behavior. Dolphin schools may reach 150 animals, often feeding among diving flocks of shearwaters, terns, gulls, and gannets.

A few peninsulas north of Cape Clear, at Dingle Harbour in County Kerry, a famously sociable bottlenosed dolphin, named Fungie, became the focus of intense tourist interest after his first approaches to swimmers and fishermen in 1983. By 2000 he was being visited by some 150,000 people a year and still interacting vigorously with boats, divers, and swimmers. His home range in the mouth of the harbor was a mere 1.5 square miles (4 km²), unusually small for a resident dolphin, and his most sociable behavior, including touching, was reserved for a chosen few who regularly spent time swimming with him.

Small populations of bottlenosed dolphins roam many bays in the west of Ireland—indeed, this coast has probably the highest concentration of the species in Europe. Fungie has not been alone in his befriending of humans. In 2000, at a creek near the village of Doolin in County Clare, an immature female dolphin began to swim regularly with people, and zoologists from the Irish Whale and Dolphin Group quickly engaged with the community, devising a practical management plan to let the animal choose how much contact it wanted to make.

A science-led development of commercial cetacean-watching has been pioneered in Ireland in the Shannon Estuary, where bottlenosed dolphins in May, June, and July may total some eighty animals. They like to chase salmon in the strong tidal races, sometimes in gatherings twenty strong. The more usual group, however, is five to ten, and on a good day for watchers, the animals will demonstrate speed-swimming, somersaults, fluke-slapping, breaching, and bow-riding—all, apparently, for their own amusement. Tour boat operators, cooperating with cetacean scientists based in Kilrush, refrain from crowding the resident animals too closely and avoid their resting places. Photographs have made more than half the dolphins familiar by patterns of nicks, notches, and scars, and their identification cards now help in continuing study of behavior. The scientists hold the advisory initiative, and in exploratory cruises west from the River Shannon across the continental shelf have been testing potential for yacht-based encounters with deep-water whales and dolphins.

Professional and amateur cetacean enthusiasts, acting together through the Irish Whale and Dolphin Group, successfully achieved the 1991 government declaration that all Irish territorial waters are a whale and dolphin sanctuary.

This was the first time a European country had made such an unequivocal statement on whale conservation, and it carried important implications internationally.

It also drew a final line under Ireland's own desultory adventures in whaling. These ranged from the local opportunistic capture and butchery of inshore schools of dolphins and pilot whales *(Globicephala melaena)*, going back over centuries, to commercial ventures off the west coast in the early twentieth century. The modern Irish operations were part of the expansion of Norwegian whaling through the use of small and powerfully fast steamers armed with a harpoon gun mounted in the bow. Norwegian entrepreneurs set up two whaling stations on the Inishkea Islands, at the northwest of County Mayo and in nearby Blacksod Bay. Between 1908 and 1922 at least 818 of the great whales were killed at sea and stripped of blubber ashore. The record season was in 1920, when the Blacksod company caught 101 fin whales, twelve sperm *(Physeter macrocephalus)*, nine blues, and three sei. Norwegian boats continued to hunt around Ireland, killing minke whales on a small scale, up to 1976, when Ireland's new Wildlife Act reinforced earlier whaling prohibitions. It protected all cetaceans from "wilful interference" within the state's exclusive fishery limits (up to 200 miles, or 320 km, from the coast).

The 1991 declaration of a sanctuary was a statement of political sentiment. It responded to no specific threat, and added no new legal protection. But the very idea of a sanctuary brought with it responsibilities and set new actions in motion. It stimulated a national stranding and sighting scheme, and prompted spending on research into cetacean food resources, contamination of the Irish marine environment, and the by-catch of dolphins and porpoises by trawlers and gill-netters.

As public interest and sympathy intensified, the Irish Whale and Dolphin Group organized a network to provide trained help with live strandings, and television news coverage of seashore rescue efforts have become increasingly familiar. The refloating of eight pilot whales in Cork Harbour in 1995 and of seventeen common dolphins in Blacksod Bay in 1997 have been among the more successful episodes. Sperm whale, killer whales, northern bottlenose whales *(Hyperodon ampullatus)*, a minke whale, and Risso's, white-sided, and striped dolphins were among other species beached alive in the 1990s.

Mass strandings are still an enigma, but the belief that cetacean herds are sometimes beached in an act of group support or altruism is held by many marine scientists. In New Zealand, where mass strandings, especially of pilot whales, are a regular occurrence, it has been suggested that females having problems in giving birth may bring the rest of the herd ashore: a still-born calf, only 5 feet (1.55 m) long, was found after the stranding of pilot whales in Cork Harbour. Support for a sick member of the herd has also been suggested at times in a long-term program of autopsies on stranded sea mammals that began at University College, Cork, in 1995. Pleurisy with pneumonia, heart problems, septicemia, calving problems, and colic have all been fatal conditions. Capture in fishing nets is another common cause of death in animals washed ashore in storms.

How realistic is it to hope to see whales or dolphins from Irish cliff tops? An hour or two of continuous sea watching with binoculars, preferably in calm weather, is probably the least that is worthwhile, but a good day in the right place may yield more than one party of dolphins or porpoises and even a distant blow of a fishing minke. Where bottlenosed groups are resident, watching from a boat can improve the odds. Casual close sightings of dolphins, on the other hand, as a simple matter of luck, may reward any visit to a shore with deep water at hand.

THE "SUNFISH": GENTLE LEVIATHAN

One summer evening, perched high among the sea-pinks of the Bills Rocks in Clew Bay, County Mayo, I tracked the high dorsal fin of a basking shark *(Cetorhinus maximus)*, cruising the sea toward the mountains of Achill Island. Its motion was almost imperceptible, a black triangle in the empty plain of the ocean, yet the whole of that huge vista seemed charged by its presence. In other waters these sharks may be seen in shoals of fifty at a time, gathered at some rich ocean eddy of copepod plankton. But a solitary animal, in that particular bay, spoke for the innocent grandeur of the second largest fish in the world, outranked only by the whale shark *(Rhincodon typus)* of the Pacific.

For most marine scientists, the true "sunfish" is *Mola mola*, the big, disc-shaped ocean wanderer that turns up in sparse numbers at the Irish and other European coasts. But the basking shark was elected the sunfish, *liabhán greine*, for a long period of Irish folk history (or occasionally, as in Kerry, *ainmhíde na seolta*, "the monster with the sails"). Under either name, it made determined hunters of the fishermen of Ireland's Atlantic coast.

"They are taken in the Hot Season in the Months of June and July," wrote the Rev. William Henry in an unpublished natural history (1739) as he described the shark hunt off Killybegs, County Donegal.

As they sleep on the Surface of the Sea, they are Discovered by their Fin, which being extended above the Water, resembles the Sail of a Boat. They lye in this posture, til the Fishermen, making up to them, strike them with their Harpoon Irons. Whereupon they dart down to the Bottom and, rolling on the Ground, work the Harpoon deeper into the wound. Then, being irritated, they rise again to the Surface and shoot away with an incredible Velocity, Dragging the Boat after them . . . and they bear away to Sea sometimes for Leagues; til at last Dying, they Float on the Surface till the Fishermen come along their side, and cut out the Liver, which affords several Barrels of Oyl. In this Dangerous War with these smaller Leviathans, it is necessary to have 100 Fathoms of small Cord fixt to the End of the Harpoon, to give it play: and for a man to stand by the Gunnell of the Boat, with an Hatchet, to cut the Rope in Case of any stop of its running off, or the Fish's emerging too suddenly; either of which Accidents might overset the Boat.

The basking shark lives in temperate seas in both hemispheres, foraging north to the edge of arctic waters, but only in the northeast Atlantic is there any long history of exploitation. Not until 1950, after at least two centuries of shark hunting, did scientists explore in any detail its anatomy and biology. A study of sharks caught off the Scottish Hebrides was quick to address traditional exaggerations of its size, scaling down the reputed 40 feet or more (12 m) to a rarely exceeded 34 feet (10.5 m). But the shark's other vital statistics remain impressive. Swimming steadily at about 2 knots with its cavernous mouth agape, it can filter more than 2,000 tons of seawater an hour, the volume of a 164-foot (50-m) swimming pool. The zooplankton trapped by its long, bristlelike gill-rakers nourishes a liver of enormous size, often weighing a ton, and charged with oil rich in squalene, a low-density hydrocarbon. This gives the shark buoyancy, so that, in swimming forward, the tilt of the big pectoral fins is enough to lift it from the seabed or hold it at the surface. The basking behavior is thought to have a twofold purpose. The sharks congregate in areas—often off headlands—where currents bring plankton to the surface and concentrate it into drifts, and the act of gathering may itself help them to find a mate.

In 1950, in an article for the Zoological Society of London, L. H. Matthews and H. W. Parker described the basking shark as "clearly an indiscriminate plankton-feeder" and thought it unlikely that it seeks out particular kinds of plankton-rich water. More than fifty years later, the sharks have been tracked in the western English Channel and shown to be selective filter-feeders that do indeed choose the richest and most profitable plankton patches. They forage along thermal fronts until they discover, rounded up by currents, dense concentrations of their chief copepod prey, the reddish *Calanus* crustacea. Here they swim around on a convoluted path for a full day or more, steadily filtering the water. Then, as plankton falls below a threshold level, they move on, swimming perhaps 12 miles (20 km) in a straight line between one plankton patch and the next. Occasionally a plankton front, swept inshore by winds and currents, attracts spectacular assemblies of the sharks. In May 1998 they appeared in some hundreds off the Lizard Peninsula in southern Cornwall, bringing crowds of people to the cliffs to marvel and to try to count their fins.

Among the acute and sometimes mysterious perceptions of sharks in general is a sensitivity to electrical fields, and this in the basking shark could extend even to electroreception of copepod muscle activity. The animal may also be guided by the scent of dimethyl sulphide, the gas produced by phytoplankton when grazed by zooplankton and used as a foraging cue by some tube-nosed seabirds, such as petrels and shearwaters. The sharks' foraging areas in the Channel study were shared with large shoals of mackerel and other fish; gannets dived among them. *Calanus* species are an important food for many commercial species, including herring, so that the keen nose of this shark for plankton fronts in the ocean could make it extremely useful in monitoring key plankton species. One approach was suggested by a Scottish experiment in the 1980s, in which a basking shark was fitted with a radio transmitter and tracked by satellite for seventeen days.

New knowledge about the species' behavior helps to interpret the dramatic swings of fortune in the story of the sunfish hunt around Ireland. This hunting was concentrated off the coast of Connacht, where the sharks were traditionally in greatest evidence in April and May. From the early 1700s, oil extracted from the liver was used to fuel lamps, waterproof wool and timber, soothe burns and bruises, and rub into aching muscles. It fueled lighthouses around the coast and filled the flickering "lanthorns" in the streets of Dublin and other Irish cities; in purity and versatility, it was second only to oil from the sperm whale. But there could be drastic fluctuations in numbers of the sharks, and the farmer-fishermen of the west coast were sometimes tempted to great risks, of the kind memorably restaged in 1932 for Robert J. Flaherty's film *Man of Aran*. In this gripping "documentary" (no basking shark had been caught from Aran in a full generation), the shark was hand-harpooned from currachs, the fragile, lath-and-tarred-canvas craft powered by oars that are still the general workboats of the islands. This portrayal was accurate enough as reconstruction. The islanders had boiled the livers in iron cauldrons and the oil was sent to the mainland in barrels or burned in the little cottage lights called *muiríni*—a pool of oil in a scallop shell with the pith of a rush as wick.

Most of the mainland shark hunters, fishermen from the Claddagh in Galway or from Connemara, sailed out in late spring in open, carvel-built boats

Fishing currachs on Caher Island, County Mayo, with Ballybeg Island and Inishturk Island in the background.

(known as hookers), heading for the "Sunfish Bank" off the west coast of Connacht. Here, on two days at the start of May in 1815, they killed between one hundred and two hundred sharks. Such a massacre was part of a wider boom in fishing that helped support a new and dense settlement of people along the Connacht coast. The main catch was herring, with the basking shark close behind in value, and at the height of activity in the early 1800s, ten thousand fishermen might converge in summer on Inishbofin, County Galway, the nearest island to the Sunfish Bank.

Great hauls were frequent, but both quarry species were notoriously unpredictable. The disappearance of herring from the west coast was just one of the blows to the Connacht community in the depression that followed the Napoleonic Wars, and the gradual deterioration of the basking shark fishery as the century wore on was a minor note in the general tragedy of the Famine and emigration. Paraffin oil was distilled from coal in 1850, and within a decade it was capturing the market once held by shark and whale oil. The Sunfish Bank, so far from harbor, finally ceased to lure. In 1873 a shoal of sharks was seen off Inishbofin, and two boats put to sea. One of them was capsized by a harpooned shark and five men drowned. For the rest of the century, few basking sharks were taken anywhere off Connacht.

Although they are usually hunted in deep water, the sharks show little fear of shallows and will follow where plankton leads. At the tip of Achill Head, County Mayo, which sets a high promontory of rock into the northward sweep of the Atlantic, the currents swirl into a cove below the cliffs, called Keem Bay. Here, in crystal water above a slope of pure white sand, basking sharks were tangled in nets and lanced to death for almost thirty years after World War II.

As the war ended, there was an acute shortage of industrial oils. The sharks were blundering into salmon nets stretched out from the shore and ruining them. A local entrepreneur put the two facts together, made stronger nets of manila, and recruited local fishermen in currachs. The approach of a shark was signaled by a lookout man, perched high on the cliffs, and the currachs took up stations at the nets. At the peak of the operation, in the 1950s, more than one thousand sharks were being killed annually—as many as 1,800 in 1952. But, typical of such ventures around Ireland and Britain, the fishery had virtually collapsed by the mid-1970s and disappeared soon after.

It was easy to suppose that the cause was simply overfishing. This assumed a local, resident, slow-breeding population that spent the winter somewhere offshore, in deep water, and swam in to the coast in spring. But it was equally possible that the sharks were part of a greater and more complex migration, responding to shifts in plankton distribution and abundance. For so little to be known about such a huge animal at the end of the twentieth century was really remarkable.

Both in Ireland and Britain, sighting schemes have involved fishermen, ferry operators, mariners, and the wider public in an effort to find out more. In 1993 records were made of 425 individual sharks around Ireland, with the greatest concentrations off the east, southwest, and northern coasts. They were scarce off the west, the traditional hunting ground, and unexpectedly plentiful in the Irish Sea: the Isle of Man, indeed, is now a prime location for them, and the abundant sharks off the coast of County Dublin probably belong to the same population. The sightings peaked in June and September, with April almost a blank, so that here, too, there were departures from past patterns. The abundance of *Calanus* copepods and the timing of their increase in the spring seem strongly influenced by the North Atlantic Oscillation. This is an atmospheric mechanism, comparable to El Niño in the Pacific, which controls the strength of west winds, sea temperature, and mixing of surface water, all relevant to the abundance and distribution of the basking shark's favorite plankton.

It is the only big shark that adopts the technique known as ram-filter feeding (the whale shark and the megamouth, *Megachasma pelagios*, are mainly suction feeders). It takes energy to forage in this way—perhaps more, at times, than the density of plankton supplies. The animal seems to live on an energetic knife edge: when should it stop swimming and close its mouth? The idea that it could not feed at low plankton densities because it would not achieve a net gain has been broadly accepted for half a century, but it was not explored in the field until the shark-tracking study off Plymouth on the English Channel already described. This found it foraging for long periods in prey densities of less than $\frac{1}{25}$ of an ounce per cubic yard (1.36 g per m^3) of water. Indeed, as David W. Sims reported to the Royal Society of London in 1999, the sharks would start to feed at less than half that—about 400 copepods per cubic meter. This threshold for an energy gain is much lower than biological modeling would have suggested.

This relates sharply to another long-standing mystery: what happens to the sharks in winter? It has been widely accepted that, when coastal zoo-plankton falls back in autumn, they migrate into deep water and rest on the bottom, suspending all activity. Historically it had been thought, from the pattern of sightings in spring, that the sharks migrated northward from a single population wintering off the Moroccan coast of Africa. A more general distribution has since seemed probable, with deep-water hibernation and a spring revival as water temperatures rise and plankton multiplies, starting off in Portugal and spreading up to the west of Scotland. This theory has seemed to be supported by the fact that some basking sharks caught in winter have shed their gill rakers and are developing a new set. But others have been caught with gill rakers intact and zooplankton in their stomachs. Significant amounts of *Calanus* plankton do in fact overwinter in deep water off the northeast Atlantic continental shelf, so life may go on year-round for the great gray torpedoes.

Indeed, the offshore migration may have a lot to do with their breeding cycle. Those caught at the surface have been almost all female. Some at the start of summer have had recently healed mating scars, but only one pregnant female has been caught, off Norway in 1936: it gave birth to five living young while being towed into harbor. The mating of the sharks has been observed just as rarely: a pair were seen mating in the northern Irish Sea, early in June 1999, after some four hours of courtship. There is still speculation on the pe-riod of gestation, and it may be that, as with many other shark species, the fe-males store sperm for delayed implantation, waiting until autumn to grow their young for live delivery months later.

The lack of knowledge about the basking shark's numbers and breeding ecology has moved marine conservationists in Ireland and Britain to seek pro-tection for the species, still being caught for their liver oil when the market price is worthwhile and also for their fins and tails to serve the East Asian ap-petite for shark-fin soup. Norwegian fishermen have been the major commer-cial catchers in the northeast Atlantic, ranging as far as the south and west of Ireland, and to Iceland and Faroe, in small vessels with a harpoon gun mounted in the bows. Until the late 1990s their catch was limited by an an-nual quota set (at 100 metric tonnes of livers) by the European Union, in agreement with Norwegian fisheries. In 1998, however, Britain added the sunfish to its protected species of wildlife, an action not yet followed in the Republic of Ireland.

SELECTED REFERENCES

Cabot, D. 1999. *Ireland: A Natural History.* London: HarperCollins.

Fairley, J. 1981. *Irish Whales and Whaling.* Belfast: Blackstaff.

Foster, J. W., ed. 1997. *Nature in Ireland: A Scientific and Cultural History.* Dublin: Lilliput.

Lloyd, C., M. L. Tasker, and K. Partridge. 1991. *The Status of Seabirds in Britain and Ireland.* London: Poyser.

McNally, K. 1976. *The Sun-Fish Hunt.* Belfast: Blackstaff.

O'Flaherty, T. 1935. *Cliffmen of the West.* London.

Robinson, T. 1986. *Stones of Aran: Pilgrimage.* Dublin: Lilliput.

Sharrock, T., ed. 1973. *The Natural History of Cape Clear Island.* London: Poyser.

Chapter 11

The Shining Waters

The television maps of Ireland's weather forecasters leave plenty of room at the left of the island to accommodate the low-pressure systems sweeping in from the Atlantic. At the fall equinox, their contours bunch up into tight whorls, packed with wind and rain. From October onward, a procession of two or three gales is a regular punctuation of winter; an inch of rain may fall in a day, for several days together.

On the midland plain of Ireland, in fields beside the River Shannon, water begins to rise slowly, as ecologist Stephen Heery described: "first as shining lines picking out the rectangular pattern of drains; then as an advancing floodline whose true position is hidden, because the water lies just at the surface between the tufts of grass and sedge." By Christmas in most years, the creeping flood has drowned the riverside meadows in a shallow, silvery lake some 25 miles (40 km) long, and the Shannon's raised banks are left as islanded ribbons of grass, thronged with migrant wildfowl and wading birds.

Europe's last big undrained river flows all too gently to the ocean: at its central section it drops by a mere 39 feet in 25 miles (12 m in 40 km). In summer much of its water is stored in bogs, lakes, and fens, but winter flows can spread the Shannon across a sinuous floodplain of some 241,100 acres (100,000 ha).

By kind permission of the *Irish Times*

The floodlands of the River Shannon in winter.

Generations of farmers have watched the wildfowl paddling where their cattle should be grazing, and early native governments made extravagant promises to "drain the Shannon," reluctant to acknowledge that economics and topography would probably never allow it. Eventually, faced with spiraling estimates of cost, they turned to new environmental subsidies for farmers supported by the European Union. These now help to protect a wildlife habitat unique in Europe: a highly fertile floodplain at the margins of modern agriculture, sanctuary not only for migrant birds but for plants in a grassland vegetation left largely unchanged in a thousand years.

Few in Ireland talk of "floodplain," or even, as in England, "water meadows," but rather of "the callows," from the Irish word *caladh*. What are meant, specifically, are the fields of the middle Shannon, between Athlone and Portumna, which flood in winter but dry out sufficiently in summer for the cropping of hay or for grazing. At Clonmacnoise, a ruined monastery founded by St. Ciaran in the sixth century looks out to fields that have not been plowed, drained, resown, or artificially fertilized since the monks worked there. The soil is enriched by silt and the droppings of waterbirds.

The monastery grew up on the ancient Pilgrim's Road, which ran east-west across Ireland on the ridges of the sinuous glacial eskers that offered high, dry passage through the bogs. The eskers fixed the crossing points at the Shannon and, in time, the location of its few road bridges. Between them, the callows wind as a wide green ribbon, bordered for much of the way by somber, brown horizons of peatland.

The callows are too wet for hedgerows, so that open drains or ditches mark out the fields into long strips running down to the river's edge. Most of the six hundred small farmers who share the meadows cut a single crop of hay in late June or July, and in a year of heavy summer rains, perhaps once in a decade, the water rises before even this can be harvested. In an averagely benign year, with a fine late summer, the grazing of the "aftergrass" that follows the hay crop actually helps to maintain the mixture of species by keeping down the growth of grasses.

The callows vegetation is thus seminatural, a mosaic of plant communities in which Ireland's native grasses grow with a variety and affinity long lost elsewhere. Most of Ireland's cattle pastures and silage meadows have been re-sown with a single dominant species, rye grass: perennial rye-grass *(Lolium perenne)* and Italian rye-grass *(Lolium multiflorum)* in their many artificially bred variations. The rye grasses, nourished with artificial fertilizer, were held to produce more meat per animal than the old permanent grasslands.

In European grasslands in general, pockets of land where five or six wild species still wave together are thought well worth conserving. On drier stretches of the Shannon callows, as many as ten grasses can be found within a couple of paces, gold and silver oat grasses mingling with fescues, meadow grasses, and species with names from the past: timothy *(Phleum pratense)*, sweet vernal grass *(Anthoxanthum odoratum)*, or Yorkshire fog *(Holcus lanatus)*. Their shining flowers and fleecy seed heads are a revelation in a countryside where most grass is mown for silage while it is still young, soft, and green. Flowering between them, among the eyebright and fairy or purging flax *(Linum catharticum)* are orchids modern farming would scarcely acknowledge: lesser butterfly *(Platanthera bifolia)* and twayblade, fragrant orchid, a whole flush of pyramidal orchids so deeply, vividly pink as to merge with the dense red clover *(Trifolium pratense)*.

The flora of the callows is rich rather than rare, although the very encounter with such brilliant wildflower meadows is itself increasingly rare in Ireland. On such botanical holy ground, the impulse is to walk on tip-toe, as if one's weight could find harmless points of space between the plants. Stephen Heery, a plant ecologist who lives beside the callows and has written their natural history in *The Shannon Floodlands*, has found 216 plant species in habitats ranging from open drains and marshy backwaters to peatlands and dry grasslands. In the hay meadows, many plants have their origins in the clearings of ancient fen woodland: angelica *(Angelica sylvestris)*, common valerian *(Valeriana officinalis)*, or meadow-rue *(Thalictrum flavum)* with its haze of creamy yellow flowers. In patches of wet grassland along the river, and most abundantly at Clonmacnoise, grows the rare marsh pea *(Lathyrus palustris)* with its large, deep-magenta flowers. The floods have spread another marshland rarity, the summer snowflake *(Leucojum aestivum)*, which may raise its white bells in the secrecy of a dense reed bed.

The lush, damp summer vegetation of the callows, humming with insect life, gives food and cover to one of Ireland's largest concentrations of breeding waders (there are others on banks and islands of the Erne, in County Fermanagh, and at Lough Neagh). To stand at the edge of the callows at twilight in early summer is to hear the strange aerial "drumming" of displaying snipe from many points in the sky (a sound created by the rush of air through the outer tail feathers of the diving bird). Lapwing, curlew, and redshank *(Tringa totanus)* also thrive on the invertebrate food so abundant in undisturbed soil.

Along with the snipe's eerie vibrations, the nighttime call of the corncrake *(Crex crex)* is the most evocative sound of the meadows, intriguing—even conceivably annoying—tourists in their bunks in cabin cruisers and holiday barges moored along the river. This secretive land-rail, the size of a big thrush, is rarely seen at any time, since it arrives by night at the end of its spring migration from Africa and seldom flies again until the return journey, preferring to run at great speed, even through dense grass, if danger threatens.

Crex crex closely describes the rasping call of the male, repeated emphatically and continual, hour after hour, from the denser shadows of the grassland. *"Ratulae vero raucae at clamosae infinitae,"* noted Giraldus Cambrensis in the twelfth century, and one hardly needs to know Latin to sense which Irish bird he was discussing. In the mid-1900s ornithologists learned how to call the male birds into sight in daytime by rasping notched sheep ribs together. But a sound once inseparable from summers in rural Ireland, often unceasing from dusk to dawn, is now confined to a few precious localities.

The corncrake migrates between southeast Africa and Europe, where changes in grassland farming methods have steadily reduced the breeding habitat it needs. Arriving in late April and May, it seeks tall vegetation for nesting cover—typically, in Ireland, old, moist hay meadows of the sort once mown late in summer and with the scythe. The bird's decline began in the late nineteenth century with the introduction of horse-drawn mowing machines, and gained pace with the spread of tractors, the change from hayfields to silage cut earlier in the summer, and the drainage and reseeding of meadows with rye grass. Progressively, nests, chicks, and adults were destroyed in the tines of the mower, and traditional permanent meadowland was whittled away.

The decline of this species in Ireland was first noted in the east of the island about 1900 and moved progressively westward, intensifying decade by decade. Densities of one pair per acre, once widespread in farmland, thinned out and melted away. As early as 1948, the Jesuit ornithologist P. G. Kennedy made a remarkably accurate analysis and prediction in *Studies*, the Irish intellectual journal of his order: "There seems to be little doubt that the chief causes of the decrease in numbers have been the introduction of mowing machines and the earlier harvesting of hay. . . . This is especially the case where the grass is cut for ensilage. Hence it is considered likely that in fifty years or less, according to the present trends, the Corn-Crake will become a very rare, if not an extinct, species."

By 1978, when the Irish Wildbird Conservancy made a national survey, the total of birds reported was down to 1,062, most of them in western coastal areas. By the summer of 1993, the number of male birds heard singing in Ireland had fallen to 174, an 81 percent decline in only five years.

The corncrake is now one of the few bird species of Western Europe considered to be in danger of extinction: only in Sweden and Finland have its numbers been sustained. In the British Isles, at the close of the twentieth century, the corncrake entered what was feared to be a terminal collapse at its last few refuges in northwestern Ireland and the Hebridean islands of Scotland.

Even at such a late stage, ecologists would still venture only that vegetation changes offered the most plausible explanation for the decline. But field research produced conservation strategies aimed at persuading Irish and Scottish farmers to change their mowing methods (many, indeed, had already written to the Irish Wildbird Conservancy—now BirdWatch Ireland—asking what they could do to save the corncrakes in fields they were about to mow). On the Shannon callows and in the corncrake's refuges in Mayo and north Donegal, farmers were paid small grants to delay mowing until August, when corn-crakes have had a chance to hatch the two broods essential to maintaining numbers. The birds are reluctant to cross open ground, so that another pre-mium rewarded the farmers for mowing a corncrake meadow from the center outward, to give the chicks a chance to escape to the safety of the field mar-gins. By 2000 there were encouraging signs of stability in the surviving core areas, and another promising success in coaxing a return of calling males to an uninhabited island off County Mayo (this simply by fencing off a few acres to keep out the sheep).

In the interior of Ireland, however, the Shannon callows are left as an is-land refuge for the corncrake in a sea of inhospitable farmland. Indeed, taking the bird's entire migratory range, perhaps one-fiftieth of Western Europe's corncrakes now rely on these mere 3 square miles (8 km²) of grassland. By the late 1990s the meadows sheltered some sixty pairs, or almost half of Ireland's breeding population. This seemed a precarious number to withstand the nor-mal hazards of summer floods or the ravages of predators, and for a listener at midnight on Banagher Bridge, the faint "crex-crex" can seem almost over-whelmed by the imminence of silence.

In winter, when big windy skies are reflected in the risen water, the picture along the Shannon is one of phenomenal bird abundance. The callows gather flocks of migrant wildfowl and waders on a scale matched only at a few of Ire-land's coastal bays and estuaries. The flooded margins of the Shannon, and especially of a tributary, the Little Brosna, become rippling, busy boulevards of swans, geese, and ducks, while clouds of waders—lapwing, golden plover, dunlin—wheel in tight flocks overhead.

About thirty thousand wildfowl and a similar number of waders arrive for winter from a wide range of the compass: whooper swans from Iceland; Be-wick's swans *(Cygnus columbianus)*—the Russian goose, as locals call them—from the Baltic; white-fronted geese from Greenland; great flocks of wigeon *(Anas penelope)* from Iceland, Scandinavia, and Siberia. Other duck species—teal *(Anas crecca)*, shoveler *(Anas clypeata)*, mallard *(Anas platyrhynchos)*—trace a complicated web of migration from Iceland, northern Europe, and Britain.

For several of these species, the winter numbers on the callows are of in-ternational importance in European conservation. For example seventeen thousand wigeon graze the soft grasses at the edge of the flood; and six hun-dred white-fronted geese feed along the Little Brosna, a substantial flock of this distinctive race that breeds in Greenland and winters entirely in Ireland, Scotland, and Wales. In late winter and spring, the callows gather in a great variety of waders that are building their energies for the 620 bleak miles (1,000 km) of the North Atlantic that lie between Ireland and Iceland. Among them are the handsome black-tailed godwit *(Limosa limosa)*, which shares its life between the two islands, and the whimbrel *(Numenius phaeopus)*, the curlew's smaller cousin, thousands of which use the callows as a staging post on spring migration from Africa to Iceland. Local people, recognizing the seven-note descant of its flight call, welcome it as the "Maybird."

The full, rich spectacle of the Shannon in winter is the privilege of ornithologists making counts from the sky. On the ground, the silvery maze of the flooded valley and the scarcity of riverside roads demand some application in seeking out good vantage points. But these do exist, and the Little Brosna callows, in particular, guarantee large congregations of wildfowl within binocular distance. Flocks of whooper swans can appear on any part of the callows, and the valley's wigeon also move about in large and small flocks, their communal sense expressed in constant, whistling calls. There is something stirringly primal in the glitter of flood water, the restless throngs of birds, yet their grassy sanctuary is shaped by human perseverance in the landscape and a use of the callows that represents a compromise with nature.

The esker ridges of glacial gravel and sand that wind across the callows, providing the only high ground in the region, have a natural history of their own. Their steep slopes have protected them from cultivation, so that, like the callows, they preserve a seminatural grassland that is also rich in flowering herbs. It is, technically, an oat grass–fescue sward on dry, lime-rich soil and echoes the natural vegetation of the chalk hills of southern England or of some Alpine valleys. Of almost one hundred different plants recorded from esker pastures and meadows at Clonmacnoise, only eighteen were grasses: the rest were broadleafed wildflowers. Among them were the legumes—clovers, vetches, and trefoils—that fix nitrogen for the sward, lime-loving plants such as cowslip *(Primula veris)*, the gold-bracted carline thistle, and a wide range of orchids.

The esker grassland appeared after woodland clearance, and without the present mowing or grazing would soon pass into a succession of spiny shrubs, then trees again. Scattered along the eskers are small patches of secondary woodland, dense with hazel, which help to make these ridges a series of wildlife corridors through the midlands (much as, historically, they helped people find a way through the bogs). Unfortunately, the eskers offer convenient supplies of gravel and sand to the construction industry. Most ridges have been scarred by quarrying, and some stretches have vanished altogether. In some quarry pits, migrant sand-martins *(Riparia riparia)* from Africa adopt a transient habitat, scooping holes for their nests in the sandy cliffs.

BIRDS OF THE WINTER ESTUARIES

The winter gathering of migrant waterbirds in the wetlands at the heart of Ireland is echoed by swirling flocks of wildfowl and waders at the many estuaries around the coast. Here, until modern times, the wildfowlers of local communities could find themselves competing with gentleman shooter–naturalists who seemed no less impervious to discomfort.

"A wildfowler's existence is often tinged with melancholy," wrote Sir Ralph Payne-Gallwey, as an easterly gale whistled in the rigging of his boat, "by reason of the broad expanse of waste and shipless water whereon his favourite sport is pursued. . . ." But soon, and typically, he cheered up: "What a scene this is compared to the never-varying turnip-field, the leafless, dripping wood!"

The setting was an Atlantic bay, somewhere between Kerry and Donegal, and Sir Ralph was hunched in the cabin of his cutter to begin his book *The Fowler in Ireland, or Notes on the Haunts and Habits of Wildfowl and Seafowl, Including Instructions in the Art of Shooting and Capturing Them.* First published in 1882, this remains a treasured text in Irish ornithology, not least for its graphic impressions of the island's wildfowl estuaries at a time of extraordinary abundance:

> One of the best day's shooting I ever had was far from shore in a calm like glass: the Wigeon when fired at pitched again at no great distance, and sat so thick on the water that until they were within shot one bird could not be distinguished from another; but merely living black islands of from three to five thousand fowl each.

In the winter before he wrote his book, he traveled the coast in a fishing boat that carried a flat-bottomed punt equipped with a fixed fowling gun in the bow. On this trip he shot 1,500 wigeon.

Whether as early outpost of Europe or as refuge at its farthest fringe, Ireland gathers in wintering wildfowl and waders from breeding grounds around two-thirds of the world. Most migrants to Irish wetlands (which, apart from the Shannon basin, are mainly estuaries, bays, and lakes) are short-haul visitors converging from Iceland and Britain, but others arrive from arctic Canada or Greenland, or, in later waves driven by falling temperatures, from Siberia and Scandinavia. Still more arctic waders use Ireland as a staging post on the way to wintering grounds in Spain or Africa. At Irish lakes and estuaries, callows and turloughs, perhaps a million or more migrant waterbirds seek the comfort of ice-free water and easily penetrable mud. Around the coast, sea ducks and mergansers *(Mergus merganser)* are joined by migrant divers, notably the handsome great northern diver *(Gavia immer)*, which spends the winter months close inshore in the west and south, sometimes in loose parties of up to fifty birds.

Ireland's modest population and late industrialization have left the open character of most of its estuaries largely unchanged. At Belfast Lough, it is true, the growth of the shipyards that built the *Titanic* left little peace for the wigeon, but early harbor works at Dublin Bay actually helped to create Bull Island, a sand spit and salt marsh that today bring great flocks of geese, ducks, and waders to the doorstep of the capital city. At Wexford Harbour, the North and South Slobs drained and reclaimed for farming are shared as a winter sanctuary vital to some twelve thousand of the Greenland white-fronted goose.

The island's influx of migrants begins with modest wisps of waders from the north, arriving on coasts and estuaries at the height of the Irish summer. Little groups of nonbreeding black-tailed godwits appear from Iceland as early as the end of June, and they are followed in July by the first sanderling *(Calidris alba)* and turnstones *(Arenaria interpres)*. Most of the early waders are adults, some still in breeding plumage, and the main flow of juveniles does not begin until August. The early autumn also brings many dunlin of the Iceland race *(Calidris alpina schinzii)* on passage to northwest Africa, but it is the dunlin from Scandinavia and Russia *(Calidris alpina alpina)* that throngs the estuaries of late autumn and mid-winter. Then the soft mud and salt marshes of the broad inner Shannon estuary may hold some twenty thousand of these diminutive waders ("sea mice" in one local name).

The sweeping aerial ballets of great flocks of lapwing and golden plover are among the most affecting pleasures of Ireland's winter birding. The two

Greenland White-fronted Goose.

plovers are the most abundant of the immigrant waders (190,000 lapwings were counted in 1997) and among the most sensitive in responding to hard weather. Most of the lapwings have moved quite short distances, retreating across the Irish Sea from England and Scotland: they use the inland pastures and arable fields of the Irish midlands, as well as the marshes of estuaries like the Shannon. The golden plover are from Iceland, and their highly mobile flocks range widely between marshes, flooded callows, turloughs, and the machair coast and islands of the west. A sharp drop in winter temperature will send them on south, to France and Spain.

Wigeon are still the most plentiful duck of the wetlands, arriving on the west coast with whooper swans from Iceland, and later, as the continental weather hardens, as part of the exodus from Scandinavia and Siberia. How their flocks compare with those Payne-Gallwey knew can only be guessed at. "The companies of wigeon," he wrote, "are so immense . . . that though the wind be from you to them, I have heard the roar of sound transmitted a full mile or more as they rose or pitched." At the end of the 1800s Richard Ussher, writing of the duck's enormous numbers, listed the marine estuaries the wigeon frequented: "the bays and harbours of Kerry, Cork, Waterford and Wexford, Dublin and Dundalk Bays, the Loughs of Strangford, Belfast and Larne, Loughs Foyle and Swilly, Mulroy and Sheephaven Bays, the three bays of Co Sligo, the Moy Estuary and Broadhaven." Even this list quite overlooked the Shannon estuary, which can still muster six thousand wigeon on a good January day.

Throughout the nineteenth century, the numbers of wildfowl shot for food rose greatly with advances in shooting technology. The flint-sparked musket was succeeded by the breech-loading shotgun, one barrel by two, and both by the heavy artillery of commercial wildfowling. On Belfast Lough the greater range and power of the mounted swivel gun made daytime shooting possible and hastened the final departure of the duck; on nearby Larne Lough, a fowler used the swivel gun to kill ninety wigeon at one shot.

An even more formidable cannon was the punt gun, with a recoil that pushed its small boat back through the water. Payne-Gallwey's favorite weapon, mounted in a large, two-man punt, had an oval bore that spread the shot in a lateral swathe ("for general work, twenty ounces is a handy charge"). But even this could be improved on. In his memoir, *The Wildfowler*, Roger Moran described the gun he used to fire on the Shannon in the 1930s, lying in the bottom of the punt and tugging a string at a kick from his father. The gun was double-barreled, almost 10 feet (3 m) long and weighed 1,024 pounds (465 k). Its charge of snipe shot mowed through a flock of golden plover "like a scythe through corn. . . . [Once] we picked up ninety-six golden plover, three curlew and four redshank."

The Morans were cottiers (tenant-laborers), shooting for the market and their own pot (lapwing boiled with turnips). In Wexford Harbour, a far more sheltered stretch of water, guns of a similar size were mounted on open, flat-bottomed punts that were scarcely more than wooden floats. The shooters lay face down and paddled over either side, aiming through the darkness at the guzzling sounds of the birds. Most of the "floatmen" were local farmers and fishermen, and their trade enjoyed a late bonanza in World War II, when the wild geese of the harbor were valued not only as meat but for fluffy down to stuff airmen's jackets. The wildfowling tradition is still vigorous at Wexford, even as the winter flocks flourish in protection on the slobs, but only a few amateur sportsmen take to the floats, and they are lucky to fire a punt gun twice in the season. Some winters still bring five thousand or more wigeon, but the age when they would "blacken the sky" seems irretrievable.

The dramatic reduction of the duck on the Atlantic estuaries in the mid-twentieth century resulted not from overshooting but from a natural calamity. Wigeon are grazers, and their favorite food historically has been species of eel-grass (*Zostera* spp.), marine plants growing as a long-leaved sward on mudflats and salt-marshes. This was the high-energy food on which Payne-Gallwey's flocks of wigeon were feeding, withdrawing as the tide rose and waiting in rafts out at sea.

In the 1930s *Zostera* was afflicted by disease, and great beds of it withered away on both sides of the Atlantic. The disease hit larger *Zostera marina*, and smaller species took over. Its disappearance from the estuaries prompted wintering wigeon to move inland both in Ireland and Britain and to adapt to a wider variety of plant food. Today the main Irish concentration of wigeon is at the flooded callows of the Little Brosna in the Shannon Valley, where fourteen thousand have been counted in midwinter.

The loss of *Zostera* was equally disastrous for the small, shore-feeding Brent goose, which migrates to Ireland and Britain from breeding grounds in northeast Canada and northern Greenland. The birds take their first weeks of nourishment from the plant when they arrive in autumn, and its failure hit them hard. Today, numbers still fluctuate considerably, but recovery of the east coast *Zostera* and protection from shooting since 1960 has secured the Brent as Ireland's most abundant, and increasingly tame, winter goose.

Of the two subspecies of Brent, dark-bellied *(Branta branta bernicla)* and light-bellied *(Branta bernicla hrota)*, it is the small population of light-bellied birds that head for Ireland. These breed on the tundra of Canadian arctic islands and are recognized as a separate population. Their first migrants arrive in the west Kerry bays in late August, but the most important site in autumn is Strangford Lough in County Down, an inlet dotted with drumlin islands, where Brent and wigeon feed on the intertidal *Zostera* side by side.

Small parties of Brent move about widely between the western bays, but the biggest winter flocks (among some twenty thousand birds in the mid-1990s) haunt the estuaries of the Irish Sea coast. Many of the geese that assemble at Strangford Lough in early autumn move south to Dublin Bay, to feed at *Zostera* just over the sea wall from city houses and the coastal suburban railway. More than two thousand at times fly on south to join the Greenland white-fronted geese in Ireland's prime wetland sanctuary, the Wexford Slobs.

Like the light-bellied Brent goose, the Greenland white-fronted goose has a distinct arctic race and population, of modest and precarious numbers, for which Ireland now feels a special responsibility. Its physical differences from the European white-fronted *(Anser albifrons albifrons)* are slight: most birders will settle for the orange-yellow, rather than pearl-pink, bill on an otherwise dark, gray-brown goose (the white front is actually a white forehead). Their total population, breeding on the low coastal fringe of west Greenland, numbers some thirty thousand birds, with only about nine hundred breeding pairs. They winter exclusively in Ireland and Britain, one-third of them at the Wexford Harbour and slobs.

Although surprisingly unrecognized by ornithology as a separate race until 1948, the Greenland white-fronted had been the common wild goose of Ireland, its cackling flight call a traditional sound of the bogs in autumn. The geese feed in the wettest parts, rooting up the bulbs and stolons of plants like cottongrass and beak-sedge, and commute at dusk to roost on lakes remote from people. This, at least, was their common pattern until the early twentieth century, when intensified mining of peat and land drainage of the bogs invaded their winter habitat and disturbance began to fragment the flocks. Displaced white-fronted appeared at the reclaimed estuary land at Wexford Harbour, where large winter flocks of greylag geese *(Anser anser)* were already established. Over the first half of the century the greylag was the dominant goose at the Wexford Slobs, building up to as many as ten thousand birds, but as refugee white-fronted geese began to outnumber them, greylags dwindled dramatically within a few years. Wexford's wildfowlers remained convinced that aggression from the "Russian geese" had driven the greylags away to Scotland.

In the 1950s the total population of Greenland white-fronted was probably fewer than twenty-three thousand, of which five thousand to six thousand birds were using the Wexford Slobs. These could at least be counted, but the rest of Ireland's winter visitors were in small and scattered flocks; many of their remote peatland feeding areas were rapidly disappearing under conifer forestry. When these birds were tracked down by the Wildlife Service, the Irish white-fronted population appeared to have halved between the 1950s and 1970s. A shooting ban followed in the early 1980s, and the number of geese rose thereafter, topping ten thousand at the Wexford sanctuary within a decade. Its 500 acres (200 ha) thus hold one-third of the world's Greenland white-fronted, feeding on the spilt grain of stubble fields and then on grass and fodder beet, often to the annoyance of local farmers. Conservationists, too, are concerned at such a concentration of birds. In 1992 an International Workshop on the Conservation of the Greenland White-Fronted Goose appealed to Greenland, Iceland (where the goose is still shot on migration), Ireland, and Britain to agree on a plan for conserving it throughout its range. Meanwhile, a team of Danish and Irish ornithologists have used satellite technology to track the geese on their spring migration from Ireland to west Greenland. Transmitters harnessed to ganders (stronger and less vulnerable to predators) gave regular broadcasts of their positions at their resting and feeding stopovers in Iceland and east Greenland and at the final summering areas beyond the ice cap. The spring staging posts in Greenland are in areas of intensive mineral exploration, and the skeins of geese have had stressful encounters with helicopters swooping through the sky.

The greylag goose, meanwhile, has recovered from its mid-century rout at Wexford and has built up substantial numbers in winter, notably at Lough Swilly and at marshland sites along the east coast. The origin of the birds, which number more than one thousand at each of the two biggest sites, has been confused, because migrant greylags from Iceland meet up with geese from local greylag flocks, introduced ornamentally at ponds and lakes. The commonest introduced waterfowl in Europe is the big, black-necked Canadian goose, whose explosive feral increase in Britain has caused conflict with landowners. The flocks introduced into Ireland, chiefly on northern lakes, are still comparatively stable, and it is the greylag that now appears to be increasing steadily.

As the Greenland white-fronted grows scarce in its traditional haunts, some of its special aura as a traveler of lonely places has passed to the whooper swan As many as twelve thousand of these arrive from Iceland, mostly during October and often with necks stained orange by the iron-rich streams of that arctic island. They assemble in the big northern sea loughs, Foyle and Swilly, for a further flight to Britain, or move down to Lough Neagh and Lough Beg before dispersing southward in small parties to the shallower lakes and turloughs of the midlands and west. Their bugling calls are among the most stirring of Ireland's wild sounds.

The wedge of lemon yellow on the whooper's bill marks it out at once from Ireland's resident mute swan *(Cygnus olor)* with its black-knobbed orange bill. Both species often feed in the same lake. But up to the early 1900s, the predominant "wild" swan in Ireland was Bewick's, small, dainty, and gooselike, and a winter migrant from Russia. The big increase in whoopers, to outnumber the Bewick's, occurred over a couple of decades in mid-century and was particularly dramatic on the drumlin lakes of Cavan and the wetlands of Galway and Mayo. In a complete swan census in the 1990s, the flocks of Bewick's numbered more than 2,250 birds, the most ever recorded, but their main gathering is now on reclaimed land in County Wexford, with smaller groups at turloughs and Lough Neagh. The swans increasingly dally on migration, detained by the handouts of grain at Britain's managed nature reserves, but Bewick's and whoopers still consort together far afield in Ireland: Inch Lough in Donegal, adjoining Lough Swilly, is one of their favorite haunts in the north.

The 149 square miles (387 km²) of water in Lough Neagh, set in a depresssion in Ulster's basalts, make it the largest lake in Ireland or Britain, but topography accords little drama to its enormous expanse. It is roughly and dully quadrangular, few roads run beside it, and there are no hills around it to offer an encompassing view. Beyond the reed beds, its peaty, relatively shallow waters are nutrient-rich and turbid with phytoplankton, and its margins seethe with midge larvae. This environment is, however, sufficiently welcoming to waterfowl to make the lake, with adjoining Lough Beg, one of the most important wetlands in Europe.

In any season the system holds more duck than anywhere else in Ireland. In winter the surface-feeding ducks, such as wigeon, teal, mallard, pintail *(Anas acuta)*, and shoveler have their largest concentrations at Lough Beg, where clearer water allows much richer aquatic vegetation. Lough Neagh has the internationally important flocks of migrant diving ducks: pochard *(Aythya ferina)* from Russia and eastern Europe (21,000); tufted duck *(Aythya fuligula)* from Britain, Iceland, and Scandinavia (23,500); and goldeneye *(Bucephala clangula)* from Russia and Scandinavia (12,500). The level of the lake has twice been lowered in drainage schemes, and the 78-mile (125-km) shoreline is under constant, piecemeal disruption from farm "improvements." The agricultural grasslands at the fringe, important to the whooper and Bewick's swans in winter, are used by breeding waders in summer, and serious declines in lapwings, curlews, redshanks, and snipe have been monitored in surveys by the Royal Society for the Protection of Birds.

Several of Ireland's urban sea estuaries, like those in most of Europe, have undergone crucial loss of traditional habitat for migratory birds. At Belfast Lough, the modern annexation of mudflats for industry is depriving many species of important winter shelter at this first stop on the flyway down the coast of the Irish Sea en route to Africa: the redshank is typical of visiting waders whose once-substantial numbers have fallen away. Rogerstown Estuary, north of Dublin, while still a successful sanctuary for brent geese and other waterfowl, has been degraded by refuse dumping from the capital and faces a major intrusion by highway construction. At Dublin Bay, by contrast, the sand spit and salt marsh of North Bull Island, in the estuary of the River Liffey, are probably the best-attended bird-watching sites in all Ireland and certainly the most highly protected.

Dublin's dockland is a modest and unpolluted port facility by world standards, and the bay opens out to a recreational suburban shoreline. From late November to early February the tidal mud is thronged with wildfowl and waders, a community sorted by dietary agenda, social habit, and length of bill. Brent geese head for the *Zostera*; flocks of wigeon and teal pack together in the salt marsh; shelduck *(Tadorna tadorna)* and pintail join the waders in harvesting the millions of snails *(Hydrobia* spp.) that graze the mudflat algae at ebb tide. In wheeling clouds of waders (twenty-seven thousand in one winter count), the robust knot *(Calidris canutus)* from Canada and Greenland, and bar-tailed godwit *(Limosa lapponica)* from northern Europe catch the eye, along with silvery hosts of dunlin. Redshanks, abundant in the autumn, are mostly from Iceland, pausing on passage to Africa.

Ecologist Roger Goodwillie, analyzing the available prey and feeding strategies of the North Bull birds, points out that their presence in such numbers indicates "the prodigal nature of the ecosystem." Estuarial mud downstream from a city founded by the Vikings must always have been rich in nutrients, and, until the close of the twentieth century brought funding for more refined arrangements, the bulk of the sewage of Dublin was taken out by ship and dumped into the bay.

Urban sewage and silt brought down by the River Lee have also helped to make an exceptional wetland of Cork Harbour on the south coast. The scale of this near-landlocked estuary and the ramification of its muddy inlets are highly frustrating to bird counters, but its regular support of more than thirty thousand waders now accords the harbor an international importance. There are exceptional numbers of black-tailed godwit and shelduck, along with big flocks of redshank and dunlin.

The estuary's more secluded creeks were also among the first inlets of the south coast to attract the exotic and beautiful white heron, the little egret (*Egretta garzetta*), now established as an Irish breeding bird. The expansion of this Mediterranean species north and west in Europe was one of the most spectacular waterfowl trends of the 1990s, reminiscent of the astonishing global expansion of the Old World's cattle egrets in the mid-twentieth century. After several decades of visits, in increasing numbers and for longer and longer periods, the first breeding of the little egrets in Ireland took place in the sheltered and wooded valley of County Waterford's River Blackwater in 1997, when they added a dozen nests to an existing breeding colony of Ireland's native gray heron. Nesting has since increased, both in Ireland and Britain, and the egret's range is spreading up Ireland's east and west coasts. But the bigger, stronger gray heron, with its slow, flapping flight, is still unchallenged as the familiar sentinel of the island's rivers, estuaries, and coastline. In south Connemara, especially, the herons thrive on the food of sheltered, convoluted bays and nest in small heronries on lake islands in the bogs. Their density has been reckoned the highest of any area of Ireland or the United Kingdom.

Away from the industrial annexations of mudflats, such as those at Belfast and Cork, the biggest changes to estuary contours have been wrought by another alien, this time a plant. For a hybrid species scarcely a century old, the perennial rhizomatous cordgrass (*Spartina anglica*) has had a dramatic impact on both islands. It originated in a cross between *Spartina alterniflora*, a cordgrass endemic to the east coast of North America, and a native European species, *Spartina maritima*. The initial hybrid, which appeared in a southern English estuary in the late nineteenth century, was sterile, but it went on to double its chromosomes in a new, fertile, and rampantly vigorous species called *Spartina anglica*.

S. anglica's rapid spread, and the capacity of its root network to trap new sediment to the seaward side of salt marshes, had clear potential for coastal protection and land reclamation, and it was planted around in Britain and Ireland and as far afield as New Zealand and San Francisco Bay. The first Irish planting was in Cork Harbour in the 1920s, and it now grows in all the estuaries important for waders and wildfowl. It remains quiescent at some, but at others it has spread explosively in response to recent manmade changes in tidal regime.

For ornithologists, the spread of cordgrass has been alarming, seeming to deny waders the burrowing invertebrates of bare mud and threatening the growth of *Zostera*, a favorite estuary food of Brent geese. In conservation areas such as Dublin Bay and Strangford Lough attempts have been made to control *Spartina* over several decades, with limited success. A university project, begun in 1996 to assess the plant's ecological impact in Dublin Bay and to develop a management strategy for its control, soon began to question the traditionally negative view of conservationists. It has found no data to suggest that waders and wildfowl have declined in Irish estuaries because of the spread of *Spartina anglica* and has argued that its stands have actually increased the diversity of intertidal habitats. Even the invertebrate fauna of cordgrass tussocks seems to compare well with that of bare mud or native marsh vegetation such as the glassworts (*Salicornia* spp.). The plants feed the ecosystem with extra organic matter and offer nursery shelter from predators to inshore fish.

In any event, *Spartina* may be losing its initial explosive impetus and finding its equilibrium as a new species: many older stands in Britain, in particular, are receding. The ability of the cordgrass to consolidate sediment and "bind" the edge of salt marsh could take on renewed significance as global sea levels begin to rise. But whatever the future shape of Ireland's coastline, the island seems certain to continue as a vital migration staging post and winter refuge: however much seasonal ice the Arctic may lose with climate change, it will never have a mid-winter sun.

SELECTED REFERENCES

Heery, S. 1993. *The Shannon Floodlands: A Natural History*. Kinvara, Galway: Tir Eolas.

Hutchinson, C. D. 1989. *Birds in Ireland*. London: Poyser.

Moran, R. 1982. *The Wildfowler*. Belfast: Blackstaff.

Payne-Gallwey, R. 1882. *The Fowler in Ireland*. London: Van Voorst. Reprint 1985. Southampton: Ashford Press.

Rowe, D., and C. J. Wilson, eds. 1996. *High Skies—Low Lands: An Anthology of the Wexford Slobs and Harbour*. Enniscorthy: Duffry Press.

Sheppard, R. 1993. *Ireland's Wetland Wealth*. Dublin: Irish Wildbird Conservancy.

—— *Chapter 12* ——

The Green Checkerboard

I f the call of the corncrake was the special sound of summer nights in the mid–twentieth-century Ireland of farm carts and milk churns, the special fragrance of summer was that of coumarin, chemical essence of sweet vernal grass. It was a prime wild grass of hayfields and also a sweet stem to chew, both for the casual country dweller and the abundant Irish hare. Fifty years later, this sweet grass is in dramatic retreat over most of the island, hay-making has virtually ceased, the hare is in decline in northern counties, and the smell most memorably associated with summer meadows is the stench of cattle slurry, sprayed to fertilize a second growth in the patchwork of silage stubbles.

The Emerald Isle is, if anything, greener than at any time in its history, as the climatic conditions that give the island its exceptional growth of grass (almost year-round in the southwest) are matched to a farming regime that aims to mow it for silage two or three times a year. Reseeding of the old, permanent pastures has steadily replaced their seminatural range of meadow grasses with a sward dominated by rye grass, which thrives on added nitrogen in fertilizer granules or slurry.

The history of lowland grasslands in Ireland spans at least 12,500 years, beginning with the steppelike vegetation, rich in sedges, docks (*Rumex* spp.),

226

and mugwort *(Artemisia vulgaris)* that helped to feed large herds of reindeer and the giant Irish deer. This was a high point of species-rich grassland, because after the Nahanagan cold snap 10,600 years ago, a final postglacial warming brought a rapid encroachment of trees and eventually dense forest cover. For almost 5,000 years, small green meadows beside lakes and rivers were all that remained of the former expanse of grass. Then, with Neolithic farming clearances about 5,000 years ago, open grassland began to develop, in a shifting, ever more complex mosaic of cattle pasture, arable plots, scrubland, and trees. In the landscape of the recent historic period, documentary records of rapid forest clearances sometimes seem to be contradicted by the pollen record. But in the northeast Irish lowlands, pollen studies graphically confirm the later rise in grasslands at the same time as the start of hedged enclosure. The last 300 years have seen a return to an area of grassland unmatched on the island since the end of the Ice Age, but of late it is grossly diminished in diversity.

Some 250 species—about one quarter of the Irish flora—make up the native weave of the grasslands, most of them localized to some degree in the patchwork of plant communities. There are indicator species, such as the toad rush *(Juncus bufonius)* and marsh fox-tail *(Alopecurus geniculatus)*, on the wet soils pocked with the hoof prints of cattle, or the red fescue, quaking grass *(Briza media)*, and bird's foot trefoil of very dry land. But drainage, reseeding, and fertilizing has been homogenizing farm grassland for almost half a century, so that a field in Wexford or Limerick or Sligo is all the same green. On old grassland surviving on shallow soil in the Burren, lime-rich, untouched by manure, and waving with oat grasses, there may be as many as forty-five species of plant in a few square yards; in the intensive swards of the lowlands dominated by rye grass and clover, one might search a field to find ten.

Oat grasses *(Helictotrichow* spp.) replaced the original woodland all over Europe on suitable soils, those that were neither too wet nor too dry. These were also the soils most suitable for agricultural improvement, so little natural oat grass survived. In Ireland its refuges are rough roadside verges and neglected graveyards, but also in the meadows of the Shannon callows, where it has adapted to what Stephen Heery in *Shannon Floodlands* called "the extreme end of dampness," sometimes even mingling with reeds.

In the "improvement" of grassland, the small rosette species of herb have suffered particularly. The cowslip, for example, with its richly scented clusters of yellow flowers, was common at the middle of the century in the wide pastures of Leinster; now it is seldom seen there, though it flourishes still on limestone farther west, and in Northern Ireland it is a protected species, not to be picked. In wet grasslands, among the first plants to disappear were the marsh helleborine and the Irish broadleaved marsh orchid *(Dactylorhiza majalis)*, crowded out as drainage and chemical nitrogen promoted the faster growth of grass.

An increasing intensification of grassland farming not only changed the nature of the sward but its management. The switch to silage from hay-making spread rapidly in the late 1900s and was accelerated particularly in the small-farm counties of the west by the introduction of new technology that baled the mown grass in tight-wrapped plastic sheeting. Hay-making has always been a problematic undertaking, vulnerable to wet summers and very dependent on the availability of neighbors and kinfolk, rotating their labor cooperatively under the traditional *meitheal* ("work party") system. The mowing and baling of silage by contractors, carried out swiftly in almost any weather, leaves a winter's fodder lined up in take-away packages and has simplified the farmer's life at a stroke.

For insects and ground-nesting birdlife, however, the new regime has brought profound change. Hay meadows are mown once, in June or July—even August—when grasses and herbs have flowered and seeded. Silage is cut a month or more earlier, beginning in May, and there may be two or three cuts in a season, four to six weeks apart. For many birds—the corncrake a notorious example—this is a schedule for disaster: they cannot complete their nesting cycle before the first cut, and the second can destroy their second attempt as well. If mowing does not actually macerate the eggs or chicks, it exposes them to predators, just as it exposes the invertebrates that share the meadow's food and shelter.

Even where mowing for silage is not the problem, pastures heavily stocked with cattle leave less space for nesting waders such as lapwing or snipe; trampling is a major cause of breeding failure. And as the changes in farming push waders into unaccustomed habitats, or into nesting more closely together, casualties from predators increase. Studies in Northern Ireland, where curlews are in marked decline, found 80 percent of the eggs taken by foxes, gulls, and crows. Lapwing nesting close to water have more mammal predation to contend with, not least from feral American mink, introduced by Irish fur ranchers in the 1950s, which have now colonized most of Ireland's waterways.

On the island as a whole, the decline in breeding lapwing—some 40 percent between 1970 and 1990—has been masked by the familiar and continuing winter presence of huge (if probably diminishing) immigrant flocks from Britain and continental Europe. Abundance of the lapwing has always reflected changes in the landscape. "Of late years," wrote the Dublin naturalist John Watters in 1853 in *The Natural History of the Birds of Ireland,* "it has somewhat diminished in numbers, in consequence of the great advance of drainage and reclaiming of waste lands in Ireland." In place of the "waste lands" came a mosaic of grassy meadows and plowed fields, to which the lapwing adapted. Today, in Britain, its preference is for nesting in fields of spring-sown wheat or even on bare earth among seedling vegetables, both of which give a clear view of predators. Along the River Shannon, alarming collapses in breeding on the callows have been balanced by the discovery of lapwings nesting on the bare peat of abandoned cut-away bog. But both in Ireland and Britain, the lapwing is typical of waders in retreat as breeding birds that improvise their nests in a dwindling range of habitats.

The corncrake's cultural significance has helped to dramatize its decline on grassland, but long-term changes in Irish farming have brought other bird species to just as perilous a scarcity. Among resident birds, the gray partridge *(Perdix perdix)* has been declining since 1850, and a sharp contraction of range toward the end of the twentieth century saw many populations disappear altogether. In 1995 the total population of wild partridge was estimated at fewer than eighty pairs, most of them in County Wexford. The partridge, like its diminutive and now extremely scarce migrant relative the quail *(Coturnix coturnix),* flourished in the Ireland of many little fields of tillage—potatoes, turnips, flax, and grain—with generous, weedy margins full of insect life. Its numbers fell when the countryside emptied after the Famine, rose when wheat tillage was promoted in the 1930s, and fell again with the decline of mixed farming and the great postwar increase of grassland. Once the primary quarry of hunters throughout the island, the partridge has dipped perilously close to extinction for reasons mainly to do with habitat. Conservation experiments have focused on a small population of partridge discovered on cutaway bogland in County Offaly (the same tract beside the Shannon that has drawn the lapwing). The birds were attracted by vegetation naturally recolonizing the abandoned peatland, and this is being augmented with wild grasses for nesting cover and with clover and grains to provide insect food for the chicks.

The quail, meanwhile, is now a scarce summer visitor to its traditional strong-hold in the wheatfields of County Kildare.

With the demise of the working farm horse at the middle of the twentieth century, along with farm-house flocks of hens and geese, the 1-acre (.5-ha) patches of fodder oats and barley, so long a part of the mixed small-farm landscape of the west, lost their purpose and disappeared. The spreading tide of rye grass, mowed before flowering, replaced the free-seeding herbage of hayfields. In the grain-growing region of the southeast, the plowing and re-sowing of fields in the fall put an end to winter stubble, with its fallen grain, and herbicides eliminated the weeds and grasses of field margins. Within a few decades, many of the island's seed-eating farmland birds were in decline. By 2000 the corn bunting *(Miliaria calandra)* was considered extinct as an Irish breeding bird, and there was a marked contraction in range of the yellow-hammer *(Emberiza citrinella)*, a bunting of brilliant color and cherished way-side song that had once accompanied human settlement even on the most re-mote and exposed of Atlantic islands.

The loss of diversity in Ireland's grassland would seem to have inevitable consequences for the distribution and welfare of the Irish hare. "The hare is a creature that yields great delight and recreation to every gentleman that useth to hunt her," wrote the ebullient Arthur Stringer of Antrim in *The Experi-enced Huntsman*—not least, he added, because "it is a game easily found; for there is scarce any place or part of a country, but it hath hares." On the big es-tates, indeed, the shooting and netting of hares in the nineteenth century pro-duced some striking game-bag totals: almost 500 hares were killed at the Crom Estate, County Fermanagh, in November 1864; 355 in December 1884 at Finnebrogue. But by the year 2000, on the intensive lowland farmland of Ulster and elsewhere, the density of hares was very different.

The Irish hare *(Lepus timidus hibernicus)* is a subspecies of the arctic hare, which, in races of mountain hare, has a range extending across northern Eu-rope and Scandinavia. It took up residence in Ireland perhaps 8,000 years ago and long before the Romans brought the southern European brown hare *(Lepus europaeus)* to Britain for coursing greyhounds. On that island, over time, the brown hare pushed the native hare into the Scottish uplands. In Ire-land *Lepus timidus* made itself at home from mountain summit to sea level, and although there were nineteenth-century introductions of the brown hare for hunting on estates, their populations have remained small and local.

The Irish hare is rather larger than the Scottish and is ruddier in color. The Scottish hares, like those of the Alps and northern Scandinavia, turn mostly white in winter, but only the hares of the highest Irish peaks are likely to show any white at all. With regard to their temperament when hunted (as they still are, in open season, by the island's twenty-five beagling clubs), Stringer's eighteenth-century opinion may stand: "For I never see anything like craft, subtilty or policy in a hare, but what I may rather call natural innocence."

The wide range of habitats used by the hare is reflected in its diet. In a setting of relatively protected vegetation (that is to say, within the fences of a national park), a hare on bogland grazes a variety of moor-grasses and sedges and significant amounts of heather, whereas a hare in "unimproved" fields eats mostly grasses of the finer species: bents and fescues. But in many Irish uplands, there is little but the toughest sedges and a wiry mat grass, and rarely is there heather, either to eat or to shelter the hare. As to the grass of reseeded farmland, this is less and less suited to the hare's diverse appetite.

In a survey in the 1990s of 58 square miles (150 km²) of Northern Ireland, Karina Dingerkus found hares in 90 percent of the 1-kilometer squares sampled in the uplands, but in only half the squares in the improved rye-grass farmland of the lowlands. One in three landowners spoke of decline over the past ten to twenty years or even of local extinction. In the hills, most sightings were of one or two hares per square, flushed from the cover of tall *Juncus* rushes or hedgerows in farmland with sizable areas of seminatural grassland. An accompanying study of diet (from droppings) suggested that the Irish hare prefers grasses, regardless of habitat or season, and that it eats a wide range of plants: on average, twenty-four species, mostly the finer bents (*Agrostis* spp.) and fescues, with the sweet vernal grass, a particular favorite of upland hares. At the County Antrim study site, the animals showed a positive preference for grazing semi-improved pasture and avoided the improved grassland areas.

It is the belief of James Fairley, formerly Professor of Zoology at the University College, Galway, however, that "high ground and moorland are not, in fact, optimal habitat for the Irish hare." In his *Irish Beast Book*, he cited hare counts on blanket bog around a peatland experimental station in north County Mayo that gave an average of only one hare per square kilometer. "On areas that had been turned into agricultural grassland by surface seeding, liming and fertilizing, numbers shot up dramatically to around 40 or 50 per square kilometer." This certainly seems to demonstrate a preference for grass of any kind, even rye grass, but a feeling of safety may also be important. In the northern survey, hares were quick to run from a distant intruder on the open rye grass fields that offered no cover. The experimental Mayo grassland was in a remote and protected zone. Similarly, high densities of hares—about one hundred to the square mile (40 to the km²)—can be seen at the North Slob at Wexford, which is specially sown with early shooting rye grasses for the benefit of wintering geese. But this is also the one Statutory Hare Reserve in the republic, offering year-round protection from hunting, trapping, and coursing.

The dominance of rye grass in Irish pastoral farming, already somewhat modified by the addition of clover and other grass species, may need to be rethought in light of a European study, called Biodepth, originally born of concern with the current mass extinction of species by human activity. Scientists in seven countries, including Ireland, set up grassland test plots in which native plants were grown in a range of species, from monoculture to a spectrum typical of local diversity (a "natural" corner of an Irish field could have about eight or nine species). They found that diversity brings higher productivity. Fewer species in the plot meant a less substantial leaf mass, a greater incidence of insect attack, and eventually a greater loss of nitrogen from the soil.

The great changes in Europe's farming objectives and subsidies, and in Ireland's own economy, seem sure to reshape much of the rural landscape, reducing the overall area of grassland for dairy cows, beef, and sheep to a core of big commercial farms and halting the pressures for intensively managed pastures and meadows on marginal land. In the northwestern half of the island in particular, small-farm livestock will dwindle and forest plots increase. The march of bracken and rushes across redundant fields will set bounds of its own to any slow resurgence of timothy, cocksfoot *(Dactylis glomerata),* and Yorkshire fog.

THE LACEWORK HABITAT: HEDGEROWS AND WALLS

To tilt down from the clouds and land in Ireland on a showery day is to be impressed, first of all, by the intense shades of green revealed as the cloud whisks away, and then by the intricate pattern of boundaries below. This is farmland as human history: a patchwork of family ownership in which distances between neighbors are short and the scale of their holdings generally small.

In most Irish counties, 124 acres (50 ha) of grass provides a substantial cattle farm, and a field of more than 7.4 acres (3 ha) is considered big. On poorer land, both farms and fields are scaled down, and the weave of boundaries is tightened. The total length of hedges, banks, and walls is put at some 515,679 miles (830,000 km), accounting for more of Ireland's land surface than is covered by deciduous forest or protected in national parks and nature reserves. Indeed, these leafy selvages are even more valuable ecologically than the remaining shreds of old broadleaf woodland. They are a linear refuge for diversity; a network of microclimates and microhabitats; a corridor, endlessly looped, for species survival.

The nature of the boundaries ranges between two broad ecological extremes. At one end are the tall, leafy hedgerows of the sheltered eastern counties; at the other, the lichen-crusted stone walls and close-cropped grassy field banks of the hills and the windy western seaboard. The density of hedgerows and tree lines can be remarkable. In one area of the midlands there were 14 miles on 250 acres (22 km in 1 km²), and the intricacy of walled enclosure is most extreme in the special conditions of the Aran Islands (see Chapter 8). Both forms are dramatically different from the continental European farming systems that organize land in open fields with unobtrusive divisions.

The antiquity of enclosure in Ireland is confirmed by the Neolithic field walls discovered under the peatland of North Mayo, and their parallel pattern on the hillsides is oddly similar to that resulting from the reform of farm structure some 5,000 years later. The raths or hill forts of early, pre-Norman farming were often surrounded by small fields and enclosures, and the dimensions of stone walls, trench-and-bank ditches, and wattle fences topped with boughs of blackthorn (barbed wire is still called "thornwire" in some parts of Ireland) were given in the law texts of the seventh and eighth centuries. There was no mention of a hedge for the earth-and-stone bank, but natural vegetation (blackthorn, hawthorn, willow, and gorse) would soon have supplied one, much as it has on many unplanted field banks (or ditches) of modern Ireland. Remnants of ancient banks often cast their shadows as crop marks in aerial photographs of modern field systems, and the chance certainly exists that ancient Irish hedgerows have survived.

The possible antiquity of hedgerows became an ecological issue in Britain in response to their widespread destruction by farmers after 1950. It was argued, against the protests of conservationists, that nearly all hedgerows are a modern creation, the product of the Enclosure Acts of the eighteenth and nineteenth centuries. Over much of Britain this was manifestly untrue: early maps and other documents featured hedges and hedgerows, natural or planted, as far back as Anglo Saxon times. Many ancient hedges were destroyed in the collectivization of agriculture that created the medieval open-field system, but many others survived and were managed by coppicing (the regular harvesting of new branches). The great enclosure of land between 1750 and 1850 established some 200,000 miles (320,000 km) of hedges, mostly hawthorn, but this may have affected no more than one-fifth of England.

The intensive study of hedges that began in the 1960s demonstrated the widespread survival in Britain of hedgerows much older than 1700. It also produced Hooper's Rule—a rule-of-thumb for dating hedges from the number of tree and shrub species they contain. Max Hooper, examining hundreds of hedges with ages known from written records (and as old as 1,100 years), found that the number of species in a 30-yard (27-m) stretch roughly equaled the age in centuries. This was certainly adequate to distinguish hedges of earlier times from those planted in the Enclosure Act period. Hedges acquire species over time. Those forming naturally on banks would have had more species to start with, and the planted hedges of the pre-Enclosure period were often of mixed species to supply a range of needs: food, fuel, and timber, as well as shelter.

In Ireland, with a different agrarian history, the correlation of hedgerow age and diversity is accepted, but Hooper's Rule is seen as having only limited potential. The first enclosure from the medieval open-field system came about with the arrival of the colonial landlords in the seventeenth century, when almost the whole of Ireland was carved into large estates. For another century, great stretches of farmed land were open or "champain" landscape, on which farmers made temporary fences with bushes in summer. The enclosure movement arrived from England in the late 1700s and took most rapid effect in the more "scientific" and "improving" farm estates of Leinster, where big new stone-and-sod banks were raised around rectangular fields and planted with hawthorn quicks, or cuttings. (The gentry, hunting foxes on horseback, learned to leap the new banks, with their double trenches, by alighting, momentarily, on the top.)

Oliver Rackham, the British landscape historian, resists the assumption that almost all the present hedged countryside of Ireland results from the enclosure movement: "Hedge systems in Ireland are just as diverse as in England," he urged in *The History of the Countryside*, "It is unreasonable to suppose that all the different kinds should have arisen within 200 years. 'Quickset hedges' are quite often mentioned in Civil Survey perambulations of 1654." Rackham called Hooper's Rule "an empirical relation which holds over most of England." He found it surprising how little influence management or soil have on what composes a hedge.

Irish botanists, however, show disillusion with Hooper's Rule, finding that some areas are simply much richer in hedge-forming species than others, and that lime-rich soils, for example, have many species that were unlikely to be recruited to the hedges of acidic uplands, no matter how old. In lime-rich lowlands, the natural hedges mix hawthorn with hazel, spindle *(Euonymus europaeus)*, and guelder-rose *(Viburnum opulus)*. But the thin hedges of the uplands are commonly dominated by hawthorn, gorse, or blackthorn, or, near the Atlantic seaboard, by the introduced frost-tender shrub *Fuchsia magellanica*. At the wetter levels of hedge banks, the native scrub willows display an endless confusion of hybrids, and rooted in drier soil high on the banks are the even more variable brambles, or blackberry bushes *(Rubus fruticosus)*, of which more than eighty forms or microspecies have been recorded for Ireland.

Striking differences between brambles in leaf shapes, flower color, prickliness, size, and form of fruit can be obvious in one brief session of wayside blackberry picking, but sorting out their microspecies can fill a botanical lifetime. The English bramble specialist Alan Newton once sampled Ireland's briars on circuits within the island, stopping five times a day to examine all the brambles at hand: he ended with a total of fifty-seven varieties ("out of the currently accepted total of 73 for Ireland"), among them several ancient, distinctive, and unnamed microspecies. The fascination of such local endemics lies in the clues they offer to vegetation history and biogeography.

The hedgerow as a linear orchard is a pleasing Arcadian image, and it is attractive to suppose that those of more provident centuries would have been stocked with at least the basic wild crab apple and wild plum or bullace *(Prunus domestica)*. Many certainly were, but their surviving distribution is distinctly patchy. The crab apple, with its small, sour, yellowish fruits, is still plentiful in hedges around the shores of Lough Neagh but relatively rare in general, even though recorded from all forty vice-counties (botanical/geographical divisions) of the island. In early Celtic Christian times, the *aball* was included among the "seven nobles of the wood," but already there was a generally recognized distinction between the sour wild apple and sweeter, cultivated kinds. Crab apples were certainly planted in hedges to provide grafting stock for orchard fruit, but most "wild" apples in today's hedgerows can be traced to apple cores tossed away at the roadside.

The wild plum, from which the damson *(Prunus institia)* was cultivated, was probably introduced and planted quite extensively. It is still common in the roadside hedges of north County Dublin, an area with a long tradition of horticulture, and quite frequent in hedges in the plantation counties of Ulster. A thicket of wild plum found near an old monastery in County Clare suggests one ingredient of the abbot's wine.

Gorse, with its densely spiny stems and brilliant panicles of golden-yellow flowers, lights up much of the Irish countryside from March to May, notably in the hedgerows of the drumlin counties and the small-farm foothills of the mountains. It is remarkably well-equipped to survive on dry, shallow soil and in parching winds. As a legume, it supports root bacteria that can fix nitrogen from the air; its spines, which are modified leaves, defend it both against drought and casual grazing; and its explosive seed pods promote colonial growth that crowds out competition.

Unlike the dwarf species, which flowers in late summer on thin, peaty soils in the south of the island, the tall bushes of gorse were an important resource to country people from medieval times until well into the twentieth century. In 1958 a national folklore survey found that furze (the most popular common name) had uses ranging from fencing and fodder to harrowing fields, cleaning chimneys, and heating bakers' ovens.

Among hedging shrubs, fuchsia has few such ancillary uses (the sap, perhaps, as a purple dye), but its explosive vigor on Ireland's mild Atlantic coasts— and, of course, its beauty in flower—have carried it, mostly by human agency, along many hundreds of miles of roadside. The full name of the predominant Irish hybrid, *Fuchsia magellanica* cv. Riccartonii, speaks for its South American parents and the cultivar hybridized at Riccarton, near Edinburgh in Scotland, about 1830. It reaches its greatest luxuriance in Kerry, in hedges sometimes up to 13 feet (4 m) high, but continues north to Connemara and Donegal and even round to Antrim. There is a second form *(Fuchsia gracilis)*, with more slender buds and less intense coloring, quite outshone by the vivid scarlet-and-purple lanterns of Riccartonii. Seedlings of either are exceptionally rare, but a twig thrown down in a moist place, or pushed into a hedge bank, will root itself and often flower within a year, and heavy, wind-blown branches layer to form thickets (Praeger mentions "a single bush" on Valencia Island in Kerry with a circumference of 98 yards, or 90 m).

Decorative or not, Ireland's hedgerows are being steadily whittled away. Many have been bulldozed in making bigger fields, this despite arguments that, on such a windy island, hedges can boost yields substantially for both livestock and grain by keeping soil warmer and promoting earlier growth of grass and crops. An even more dramatic removal, not only of hedges but also of roadside banks and dry stone walls, began with the current boom in house building and road construction—a loss of corridor habitats for wildlife unlikely to be made up by suburban gardens furnished with exotic shrubs and cement-block walls.

Many road engineers have planted new embankments and cuttings with dense bands of small native trees such as birch and alder. At the same time, the nationwide leap in car ownership and road traffic often forces regimes of mechanical hedge-cutting, even on the most minor roads, that dispossess nesting birds in spring or destroys their winter berries in autumn. This regular flailing, together with cutting by power and telephone utilities, checks the growth of sapling trees, such as ash and beech, that formerly turned many hedgerows into treelines. Many older roadside trees have also been felled or mutilated unnecessarily for fear of the impact of winter storms.

The hedgerow ash tree or sycamore swathed in ivy is characteristic of the Irish wayside, its dark silhouette at sundown an instantly evocative image. Some visitors find it repellent, even grotesque, suggesting an uncared-for landscape. Worthy Irish tree lovers have expressed alarm at the possibly baleful effect of unchecked ivy growth, and they urge campaigns of stem-hacking.

Yet the profusion of climbing ivy seems distinctive of many hedgerow ecosystems and no more grotesque than the banners of Spanish moss (*Tillandsia* spp.) that drape the oaks and cypresses of the southern United States. As a bromeliad, Spanish moss takes its nourishment from the air; ivy is nourished from roots in the ground and the tiny, adhesive roots it uses for climbing do not penetrate the bark or take nutrients or moisture from the tree. In *The History of the Countryside*, Oliver Rackham wrote of finding, on a lake island in County Offaly in the Irish midlands, "an extraordinary wood of great ancient oaks, hung with ancient ivies (one ivy trunk is thicker than a fat man). . . ." So trees and ivy can clearly coexist for centuries.

Once ivy has ascended the tree trunk, its stems no longer cling but, growing firm and woody, spread out into a bush that bears nectar-rich flowers in autumn and black berries the following spring. However, this mass of growth (an ivy tod, to use an old English word), billowing into the crown of branches, also offers evergreen resistance to the gusts of winter storms and can hasten the fall of a dying or weakened tree.

At a quite different scale, ivy twines across the old, lichened field walls of Connacht, adding both great beauty and extra shelter for insects, robins, and wrens. As Charles Nelson observed in *The Burren*, "the walls of the many roads seem stitched together by plaits of ivy woven between the stones. . . ." He counts the plant among the finest and most conspicuous of the Burren's wildflowers, and having an Atlantic subspecies *(Hedera hibernica)*, most readily discernible by the sweeter smell of its sap.

Nelson is also the rapt observer of honeysuckle *(Lonicera periclymenum)*, the fragrant climber of Irish hedgerows and dry stone walls alike. He noted it as

a superb keeper of time. The [flower] buds are vertical to begin with, and between 6 o'clock and 7 o'clock in the evening the anthers, still enclosed in the buds, burst open to release pollen; some while later the bud itself opens, the lower lip separating to let the anthers poke out. The flower now changes its position to horizontal, when the upper lip splits slightly to release the style which curves downwards. The perfume exuded by the trumpet is most powerful at this time of the evening, attracting hoverflies which collect the pollen, and hawk moths that sip the nectar. . . .

The sight of a stoat skipping along the top of a dry stone wall, by no means an everyday occurrence, is a characteristic encounter for Irish naturalists in relatively unfrequented countryside. Wall cavities make ideal dens for stoats, as do the holes in hedge banks formerly occupied by the brown rat *(Rattus norvegicus)*. In bogland, stone walls provide dry refuges for the field mouse (more properly called the wood mouse, as it is known in Britain), and the elaborate burrows this little mammal makes in hedge banks are often used by the far smaller pygmy shrew in its ceaseless hunt for insects, and by the queen bumble bees *(Bombus* spp.) for nest building in the spring.

The field mouse is Ireland's commonest small mammal, found even at high altitude on Ireland's highest mountain, Carrauntoohil in County Kerry. Fairley, who trapped it there, described the charm of an animal rarely encountered by day:

> There is something exceedingly attractive about *Apodemus* at rest, poised, alert and quivering. There is a distinct impression of sharply-tuned senses imparted by the large eyes, the ears, twitching at the slightest sound, and the long whiskers . . . the characteristic means of progression is by leaping, when the large hind feet are employed to advantage.

Fairley is also attentive to mammal parasites, noting one mouse with eighty-three fleas in its fur and another with four beetles, eight fleas, and fifty-one mites of four different species—a roster suggesting the potential biodiversity within one small hole in a hedge bank.

The wider Irish hedgerows also accommodate most of the active setts, or burrow complexes, of the island's biggest mammal, the badger (see Chapter 16). Some of the setts are shared with foxes, which may even breed in an active main sett. Such neighborliness, however, does not extend to the hedgehog, a competitor for earthworms. Where badgers are flourishing, hedgehogs tend to disappear, leaving behind, perhaps, spiny coats chewed to the skin and tufted with badger fur (the usefulness of hedgehog skins in combing wool for spinning may have accounted for the animal's introduction to medieval Ireland).

The birdlife of Irish hedgerows attracts most public concern for their conservation, as they shelter the nests of Ireland's four most widespread species: wren *(Troglodytes troglodytes)*, robin *(Erithacus rubecula)*, blackbird *(Turdus merula)*, and chaffinch *(Fringilla coelebs)*, in that order. There has been constant pressure on local authorities to confine mechanical cutting of roadside hedges to the winter months, and the revised Wildlife Act of 2001 set March 1 as the general date to stop. At the same time, none of the species on the Irish lists of "birds of conservation concern" is threatened by loss of hedgerow nesting habitat. Indeed, a survey in Northern Ireland at the century's end found substantial increases in the breeding populations of typical hedgerow birds such as chaffinch, wren, and blackbird, and almost a trebling over five years in numbers of the dunnock, or hedge sparrow *(Prunella modularis)*—this probably related to a run of mild winters.

In the farmland of Ireland's midwest, with its thick, tall, and overgrown hedgerows, wrens have been mapped as holding one fifth of breeding territories, twice the share held by robins, blackbirds, dunnocks, or chaffinches. Because the dense, leafy habitat and mildness suits all of these birds, it may be that the sheer adaptability of the wren gives it such advantage. It was, after all, the only one of the twenty-six species and subspecies of Troglodytidae in North America to reach Europe across the Bering land bridge in the Pleistocene era, and in Ireland it nests even among the rocks of exposed Atlantic islands. In the hedgerows, its minuscule size and powerful legs and feet give it remarkable maneuverability in its ceaseless search for insects. While the robin watches from a lookout point and flies down to seize its food and the dunnock is adapted to explore the ground, the wren, in E. A. Armstrong's description in his monograph on the species, "can squeeze through small apertures, hop and run, move backwards out of an enclosed space, cling like a tit, hover momentarily like a humming-bird, jerk up a tree-trunk like a tree-creeper, and even pose head downwards on a vertical surface, nuthatch fashion. . . ."

The wren figures famously in Irish folklore, notably in a Christmas ritual wherein one is killed (or merely captured) and taken from house to house by a masked group that chants at the door:

> *The wran, the wran, the king of all birds,*
> > *St. Stephen's Day was caught in the furze [gorse bush],*
> *Although he is little his family's great,*
> > *Put your hand in your pocket and give us a treat.*

The origins of the custom, shared historically by many European peoples, is traced to early and magical relationships with nature. In pagan cultures an animal might be chosen for worship but then ritually killed once a year and promenaded from house to house so that people could share in the virtues flowing out at the death of a god. The convenient wren, made king, was considered extremely unlucky to kill, except for this one occasion. Where the custom persists in Ireland today (as in parts of Connacht), its ancient significance is rarely appreciated.

ARISTOCRATIC BATS AND OTHERS

Among the evocative icons of the Irish countryside are its historic ruins: medieval castles and tower houses, jagged silhouettes in a soft dusk; shells of Georgian mansions, gutted by fire in the Irish War of Independence (1919–1922) and now often muffled in ivy. Scattered among them are the big houses that came through intact and occupied, their estates pared down but still grandly furnished with trees.

These buildings have achieved an almost organic function in the countryside. Many, indeed, built of quarried limestone, can be imagined as an extra stratum of cliffs and caves. Old bridges and railway tunnels echo their geological origins even more readily. All this is important to Ireland's bats, and to the very survival of a species now in sharp decline in Europe as a whole.

Of the island's nine species of bat, a special treasure is the lesser horseshoe bat *(Rhinolophus hipposideros)*, a small, delicate animal that flutters like a butterfly around hedges and through woodland, hawking after insects or gleaning them from foliage or stones. Beetles and spiders are as much its prey as craneflies, moths, and mosquitoes, and the splendid warbling of its echolocation signals thrills a growing number of bat watchers listening with electronic detectors.

The bat's horseshoe-shaped noseleaf marks it as the only member of the Rhinolophidae in Ireland: the other eight species are Vespertilionidae, most of which emit their signals through open mouths. The lesser horseshoe is also, archetypically, a species that hangs upside down by its toes. Unlike, say, the tiny pipistrelles, which often pack together horizontally in crevices and cavities, it hangs without touching its neighbor in a well-ventilated roost. The spacious, warm attic of an old rural mansion is ideal for a summer maternity roost and may hold upward of two hundred bats. At basement level, the cellars, ice houses, and underground passages common in such buildings also give the stable temperatures and high humidity the bats need for hibernation. Thus, the lesser horseshoe has been called "the bat of the aristocracy," frequenting stately homes, or their ruins, all year round.

Its dramatic decline outside Ireland, like that of many other European species, can be blamed on increasing use of pesticides (including toxic chemicals used to treat and preserve roof timbers) and on the heedless reroofing of old buildings. Within Ireland, conservation operations directed by bat expert Kate McAney for the Vincent Wildlife Trust have repaired and protected existing sites, sometimes with grills to prevent human entry.

After the pipistrelles (*Pipistrellus* spp.), the bat most often recorded in Ireland is the brown long-eared bat *(Plecotus auritus)*, whose range includes even many offshore islands. Its strikingly long ears, about three quarters the length of head and body combined, equip it to listen for the wing beats of prey such as moths, rather than hunting by echolocation. Like the lesser horseshoe, it seeks out large, airy roof spaces, but its colonies are small—most with fewer than fifty bats—and these roost in clusters under the roof ridge rather than hanging freely.

If the lesser horseshoe is the bat of the aristocracy, then long-eared could be called the bat of the clergy, as old rural churches are its most distinctive and important roosting sites. As a consequence of Ireland's colonial history, the oldest church buildings still in use in the Republic are those of the Church of Ireland (part of the Protestant Anglican communion), and their congregations outside the cities are mostly very small. A survey of some forty of these churches in the counties around Dublin, with an average age of 178 years, found the brown long-eared bat in more than half of them, including one in a densely populated Dublin suburb. Pipistrelle roosts were few (per-haps, the survey suggested, because the churches were usually too cold for them), but in the attic of one built in 1712, in the village of Slane, County Meath, pipistrelle guano reached a height of 2 feet (60 cm).

On an island of so much cattle pasture, it might be expected that some special affinity would show up in the diet of bats. Leisler's bat *(Nyctalus leis-leri)* is Ireland's largest species and also its most significant in conservation terms. Although at the northernmost end of its range, the island now holds the largest surviving populations of the bat, including a breeding colony in west Cork of up to one thousand individuals, the biggest known roost in the world. In Britain and continental Europe, Leisler's is widespread but rare; in Ireland it is abundant. Emerging early at summer sunsets from roosts in buildings, it flies high and fast among hawking swallows, and dives after pas-toral insect prey, notably a common yellow dung fly *(Scatophaga stercoraria)* and dung beetles (family Scarabaeidae).

A quite different hunting strategy is used by Daubenton's bat *(Myotis daubentoni)*, which feeds over water, often by snatching insects from the sur-face with its feet. It roosts regularly in small crevices under Ireland's thou-sands of road bridges, most of which were built by stonemasons in the 1800s. As lorry traffic intensifies, in increasingly heavy vehicles, local authorities have sought to reinforce the bridges by spraying liquid cement under the arches, often killing the bats and destroying their roosting sites.

There are two scarce but quite widely distributed species. Natterer's bat

(Myotis nattereri), noted for sometimes using its tail to flick insects into its mouth, is most often found roosting, in small numbers, in crevices of stone archways and the roofs of caves and tunnels. The whiskered bat *(Myotis mystacinus)* is even scarcer, and the only Irish bat found roosting regularly with other species, often in houses.

The smallest of the European bats, the pipistrelles are also the commonest, from the Mediterranean to Scandinavia. One cave in Romania has a winter roost of 100,000, but cave roosting is rare in Ireland and Britain. Here the pipistrelles are predominantly a house-dwelling, even suburban, species, clustering into small cavities of roofs and walls and feeding in flight in many different habitats: gardens, hedgerows, riverside trees. Roosts can be large for such domestic settings, some with up to six hundred bats, and their chatterings and scratching noises can make them disquieting neighbors. Public attitudes to bats have been improved by the enthusiasm of a new wave of researchers (notably young women zoologists), the promotion of conservation by the Vincent Wildlife Trust, and the setting up of amateur bat groups in the cities.

Until 1996 the number of Irish known bat species stood at seven, but recent studies have discovered another two. They are both in the pipistrelle family, and only one of them is actually new to Ireland. Until recently, the "common" pipistrelle on the island was thought, in fact, to be the common pipistrelle *(Pipistrellus pipistrellus)*. But the increasing use of bat detectors to analyze hunting calls alerted researchers to the presence of a pipistrelle with a higher voice. *Pipistrellus pygmaeus* (now known, predictably, as the soprano pipistrelle) calls at 55 kilohertz, compared with the 45 kilohertz of the common pipistrelle. It is also slightly smaller, has a darker face, catches smaller insect prey, and roosts separately.

The soprano now seems, in fact, to be the more common of the pipistrelles in northern Europe and may be Ireland's most abundant hedgerow bat. The entirely new pipistrelle is Nathusius's pipistrelle *(Pipistrellus nathusii)*, a woodland species slightly bigger than the common pipistrelle. It is widespread in continental Europe, where it migrates between winter and summer roosts. It has been known as an occasional winter visitor to Britain, and there are records from North Sea oil platforms, but a maternity roost of 150 bats, discovered in a nineteenth-century stable block in County Antrim in 1997, is the most westerly colony yet. It suggests an expansion of the animal's range and, in Ireland's mild climate, a possible change of lifestyle to more sedentary habits.

The island's checkerboard of grassland, with its weave of hedgerows, offers most of the "forty shades of green" in the Irish ballad. But missing across wide tracts of the island are the colors of the broadleaf woodland that once stretched almost shore to shore. How much of this was the fault of English colonialism? The next chapter sorts out some of the myth in the story of Ireland's deforestation.

SELECTED REFERENCES

Aalen, F. H. A., K. Whelan, and M. Stout. 1997. *Atlas of the Rural Irish Landscape*. Cork: Cork University Press.

Fairley, J. 1984. *An Irish Beast Book*. Belfast: Blackstaff.

———. 2001. *A Basket of Weasels*. Belfast: Author.

Hayden, T., and R. Harrington. 2000. *Exploring Irish Mammals*. Dublin: Town House.

Heery, S. 1993. *The Shannon Floodlands: A Natural History*. Kinvara, Galway: Tir Eolas.

Kelly, F. 1997. *Early Irish Farming*. Dublin Institute for Advanced Studies.

Webb, D. A., J. Parnell, and D. Doogue. 1996. *An Irish Flora*. Dundalk: Dundalgan Press.

Nelson, E. C. 1991. *The Burren*. Kilkenny: Boethius Press.

Royal fern of the bogs

Otter

Badger

Enjoying the cool shade of Hill Wood, Moore Abbey, near Monastervan, County Kildare

Lackeen tower house in County Tipperary

population is not at all surprising. Their diet is as varied as a Roman feast: it is headed by birds, including ducks, pigeons, waders and their eggs, and small mammals (wood mice, hares, and squirrels, for example), and their nutritional gaps are filled with earthworms, insects, frogs, lizards, and snails. In autumn, martens share the blackberries with their predator, the red fox, and the hazel nuts and crab apples with the mice.

The little bronze lizard poised above its shadow on a rock is Ireland's only native reptile, the common lizard, which gives birth to its young alive. A good place to look for them, according to Gordon D'Arcy's *Natural History of the Burren,* is in the inner enclosures of the hill forts, which in summer are stony crucibles of warmth and light. The lizards' need to soak up heat externally finds them basking in sand dunes and on dry-stone walls but also among the heather stems of dry hummocks in the bogs. They are, indeed, quite common in Ireland—more so, perhaps, around the coasts—but so unobtrusive and well-camouflaged that many people are surprised to know they exist. Others confuse them with the amphibian smooth newt, which is common in ponds and in the spring-fed wells (often referred to as "holy" in the folk traditions of the Burren). The newt figures, sometimes interchangeably with the lizard, in a magic tale in which the "mankeeper" (a name of complex Scots-Irish origin)

Common or viviparous lizard *(Lacerta vivipara).*

may leap down one's throat, give birth in one's stomach, and have to be induced to leave the same way by the eating of a meal of corned beef.

The evident suitability of the Burren as lizard habitat has encouraged intermittent and anonymous introductions of exotic species. In 1958 a number of the large green Mediterranean lizard *(Lacerta viridis)* were released (by unknown hands) into the rocks and seem to have survived a few years before dying out. In the 1970s the slow worm *(Anguis fragilis)*—actually a legless, burrowing lizard native to Britain and Europe—was first sighted and identified in the northeastern corner of the Burren, and the species is now well-established at several locations as a handsome addition to the region's fauna. This introduction is popularly credited to "New Age" travelers for whom the sparsely populated Burren, with its mystical wilderness atmosphere, became, for a time, a borrowed Eden.

Attempts to establish alien species have always been a sensitive topic in Irish natural history, especially among scientists and naturalists trying to sort out the island's erratic inheritance of flora and fauna. In his years as editor of *The Irish Naturalist,* Praeger roundly denounced any "forgers of nature's signature." A particular target was the impressively named Captain E. Bagwell-Purefoy, a British Army officer who in the early 1900s imported a large number of buckthorn bushes *(Rhamnus catharticus)* into County Tipperary as the particular food-plants needed by 250 caterpillars (similarly imported from Britain) of the sulphur-yellow brimstone butterfly *(Gonepteryx rhamni)*. Not only were shrub and butterfly naturally rare in the region, but the Irish brimstone is of a distinctive race, sufficiently different in its shades of yellow to rank as a subspecies. Whether Captain Purefoy's imported colony survived long enough to breed with the nearest Irish brimstones is unknown.

Buckthorn joins blackthorn and hawthorn in the sometimes spiny shrubbery of the Burren, and the brimstone is just one of the exhilarating range of butterflies on the wing between April and October. At least thirty species have been recorded in the region, twenty-six of them as indigenous residents and the others migrants from Europe. This is only a handful short of Ireland's total butterfly list of thirty-four, of which twenty-nine are resident. Some are found in small, local concentrations elsewhere on the island, but they are abundant in the Burren because their food plants flourish in its well-drained, limy habitats. The brimstone is one example. Others are the common blue *(Polyommatus icarus)* and the small blue *(Cupido minimus)*, whose caterpillars

— *Chapter 13* —

Killarney and the Woodland Heritage

O utside a county market town in the very center of Ireland stands a tree with magnificent presence, a giant gesticulation of seamed and weathered limbs. The "King Oak" in the Charleville estate at Tullamore in County Offaly may be four centuries old: it has not been ring-counted. Four of its massive lower branches spread out to rest their elbows on the ground and one of them reaches out horizontally for some 75 feet (23 m). This is clearly not a tree that had to compete with close neighbors or fight for sunlight. Its grandeur epitomizes the notion of ancient oak, and the thunderbolt that gashed its trunk from crown to base a few decades ago seems something of a cosmic conferral of nobility for endurance.

It evokes a vision of oakwoods from films about Robin Hood: spacious and lofty, with plenty of room for archery. There were such primal woods in the Irish lowlands, and their supposed destruction by English colonists to smelt iron, make barrels, and build warships for the British fleet was a popular charge in the rise of Irish nationalism. All these exploitations did, indeed, go on, and were sometimes dramatically destructive, but they happened at a great historical remove from the clearance of primeval wild wood. This resulted routinely from the early spread of agriculture, as modern revision of Ireland's woodland history makes clear. The supposedly great age of surviving ancient oaks has also taken a knock, from ring counts carried out for the momentous Irish dendrochronology (see Chapter 5). In thirty years of study, only a handful of living oaks were found to have started growth in the seventeenth century.

The debunking of myths about woodland history has been the special mission of Oliver Rackham, a historical ecologist based in Cambridge University. Both at home in England and in Ireland he chides the modern notion of a tree as artifact, a piece of timber with a finite life. He explores a long history of a woodsmanship that, through sympathetic coppicing and pollarding (cutting back), harnessed a tree's stubborn life and capacity for renewal. Matching this history to what is known of the intensive farm settlement of Ireland's Iron Age and Early Christian periods, in *The History of the Countryside* he wrote,

> We are asked to believe that in Ireland, of all countries, large tracts of wild wood somehow remained intact and unused until 400 years ago, and then were effortlessly destroyed. What was there about Irish trees that prevented them from growing again after felling? Why did depletion lead to destruction, as it did not in England? How were the hundreds of thousands of men found to dig up hundreds of millions of trees, and what did they live on while doing it?

Valerie Hall, using analysis of peatland fossil pollens dated by volcanic ash layers (see Chapter 5), is another effective explorer of Irish landscape history, revising in particular the picture of woodland wealth and its consequent destruction in the sixteenth century, as portrayed in contemporary documents. Sampling at named sites in three northern counties—Down, Antrim, and Derry—she found no pollen evidence, either of the highly praised and supposedly extensive oakwoods or of any rapid, irreversible felling 400 years ago.

In reality, relatively little original forest awaited the Elizabethan colonial settlers who took over most of Ireland in the wake of the Tudor military campaigns. Even before the Normans, in the seventh and eighth centuries, the main woods were already confined to poor, marginal land and the acid soils of the hills. In most of the countryside, as Fergus Kelly concluded in his *Early Irish Farming*, "[T]he general picture we get is of woods and copses, very often privately owned, whose resources are limited and need careful protection by the law."

Kelly based his book on the rich detail of early Irish law texts. Documentary sources also guided Eileen McCracken in her earlier study, *The Irish Woods since Tudor Times*. She judged that by 1600 "about one-eighth of Ireland was forested" and that English-controlled exploitation reduced this to about 2 percent by 1800. But forests marked on maps did not always correspond to trees on the ground. Rackham, delving deeper into the meticulous Civil Survey of 1655—the equivalent of England's Domesday Book—arrived at a generous estimate of 3 percent of Ireland under woods at that time, with the main impact of commercial exploitation still to come. By the time of the next island-wide survey, in 1835, perhaps one-tenth of the 1655 woodland remained in existence; the rest had disappeared under farmland. Recent pollen studies, from peatland and lake muds, have confirmed the dramatic colonial clearances of woods in some areas, but in others have seemed to contradict written records of forest "slaughter" in the seventeenth century (see Chapter 6). Pollen profiles from the last millennium chart a steady decline of deciduous woods and a loss of diversity in the few fragments of native forest that survive.

Most of the remaining forests are quite small. In the whole of the north of Ireland, only a few acres of native woodland remain. Other isolated pockets survive along the Atlantic littoral, from Glengariff in west Cork to the national park at Glenveagh in County Donegal, but in the heavily farmed interior such vestiges are rare. On a peninsula in Lough Ree, County Roscommon, are what Rackham ranked as "the best-preserved ancient woods that I have seen in Ireland." The mixed coppices of St. John's Wood include coppiced "stools" of oak up to 11 feet (3.5 m) across, along with Ireland's endemic whitebeam and an exceptional number of crab apple trees. (Stools are the stumps of felled trees from which new branches have sprung up, to be harvested in turn.) The wood's name recalls the medieval seat of the Knights Hospitallers, crusaders of the order of St. John of Jerusalem, whose large thirteenth-century castle stands nearby.

One fragment of a rare native woodland survives in the area known as the Gearagh, east of Cork City in the flat-bottomed valley of the River Lee. Here, up to the mid-1950s, a floodplain forest of 1,500 acres (600 ha) grew thickly on ungrazed alluvial islands webbed by a dense network of clear streams—one of the few such braided anastomosing river systems in northwest Europe. All except 250 acres (100 ha) were felled and flooded for the reservoir of a hydroelectric scheme, and blackened stumps of the old forest now stick up from the water. The remaining woodland of oak, ash, hazel, and alder is still ranked as internationally important, and its ground flora and dragonfly fauna are exceptionally rich.

By far the largest native woods, however, covering some 3,000 acres (1,200 ha), carpet the mountain valleys and lakeshores of Killarney in County Kerry. They are utterly acceptable as ancient oakwoods, but hardly of the kind conjured up by the King Oak of Charleville. In some of their remoter reaches—as one climbs, perhaps, along a stream in one of the dark mountain valleys—their atmosphere is dramatically different than that of Robin Hood's Sherwood Forest. Dense mosses and liverworts cushion the rocky ground and girdle the tree trunks; a garden of epiphytic ferns stretches out along the lowest boughs. The slow, silvery drizzle of Killarney's extreme oceanic climate keeps the air constantly humid. One thinks not of Robin Hood but of mountain mist-forest, and goblins—perhaps of J. R. R. Tolkien's *Lord of the Rings*.

Even such ancient-seeming oakwoods, their shadows darkened by the evergreen companionship of holly, probably bear little resemblance to their primeval woodland ancestors. Fossil pollens held in small wet hollows within the woods speak of a dense woodland in which oaks were mixed with pines. This was the picture until about 2,000 years ago, when the pines declined, possibly with the help of Iron Age people. This opened the woods to a new diversity, with a greater abundance of birch, hazel, and even the exotic *Arbutus* (discussed later). Their decline, in turn, and the ascendancy of the oaks, was brought about by centuries of human disturbance and the pressure from grazing livestock and introduced deer.

Mixed with pine or not, Killarney's primeval oaks were undoubtedly among the first in Ireland, as oak arrived in the south, along with elm, about 9,000 years ago. We cannot be certain of the species, because fossil pollen is not sufficiently discriminating, but the oaks of Killarney were almost certainly *Quercus petraea*, the sessile oak of acid, well-drained mountain soils and silica-bearing bedrock. *Quercus robur* is the pedunculate oak preferring fertile lowlands, tolerating soggy soil and even flooding but also enjoying the limestone of elevated drumlins and eskers. The King Oak of Charleville is pedunculate, and its prevalence in plantings on the landlords' estates of the central plain and eastern counties raised questions about its Irishness. Enough pedunculate oaks survive in fragments of old woodlands to speak for the species' credentials as a native, but something in the image of sessile oak as a tough, mountainy survivor may have clinched its official adoption as Ireland's national tree.

The descriptive names of the two species are of limited help in telling them apart in the field. *Quercus robur* has acorns on long stalks, or peduncles, and *Quercus petraea* has acorns that are mounted tight to the twig (sessile means stalkless). But oaks spend many years with few, if any, acorns. There are also classic profiles of the two species, in which the pedunculate oak thrusts its branches outward from a large, stout trunk and the sessile is slim and fan-shaped, thrusting its branches upward. But environment can influence this, too. Even experts must often get down to details of leaf shape, aware that while the two species do have a partial barrier to hybridization, every degree of intermediate form exists between them.

In the native forest on Killarney's sandstone, however, where the canopy sets its own smooth contours above the rocky ground, the oaks are nearly pure sessile, as they were in the great mountain wilderness of Kerry that survived, almost undisturbed, to the late sixteenth century. To the Elizabethan armies of the 1580s it was to be cursed not only as an obstacle demanding long detours but as a shelter for the "woodkernes," the displaced or defeated Irish. Dramatic and impracticable proposals for clear-felling such obstinate forests, haunt of wolves as well as rebels, already had a long history. "In 1399," related Eileen McCracken in *The Irish Woods since Tudor Times*, "when McMorough, king of Ireland, lay in the woods west of Kilkenny with 3,000 men, Richard [the Second] ordered the mobilisation of 2,500 natives to cut down the wood and burn the trees." Such plans usually petered out, but enough were carried through, especially in Ireland's eastern counties, to help fuel the long story of colonial abuse.

At Killarney the assault on the woodlands followed quickly on military victory and the award of land to English settlers. Remoteness protected the trees from the first eager plunder for the timber trade (notably, says McCracken, to make casks for the wine of France and Spain), but they were soon being stripped of bark for the tanneries, and clear-felled to make charcoal for the ironworks that lined the rivers of the southwest. W. A. Watts, using written accounts, is left in no doubt of the "immense destruction" that the ironworks caused between 1580 and 1700, especially in the remoter valleys. Kenneth Nicholls, in *Gaelic Ireland c. 1250–1650,* confirms that "unmanaged, unsustainable exploitation" of the Irish woods remained the norm of the settlers.

He gives as one example the destruction of the woods around Kenmare Bay, west of Killarney, where oaks that had been judged as fit for naval ships were burned for ironworks charcoal, an exploitation that continued in the area for sixty years.

The end of iron working, however, still left large native woods in the Killarney Valley, particularly on the great estates owned by the Herbert and Browne families. They were a resource for coopers, turners, boat builders, and hoop makers who in summer lived in huts in the forest, their fires reflected at night in the dark lakes below the mountains. But any active management and replanting of the woods waited for the economic boom and demand for timber that accompanied Europe's Napoleonic Wars of the early 1800s. At that time there was large-scale felling and coppicing of standing oaks and some extensive replanting.

In Tomies Wood, for example, across the glitter of Lough Leane from today's Killarney Town, forty-nine thousand oaks were planted in 1805, and enough survived to give the present regular spacing of trees, all looking somewhat alike. Here and there, however, the original wood breaks through in big, coppiced stools of oak and holly. Along with oaks, the plantings introduced Scots pine, larch, and spruce, a foretaste of the commercial conifer forests established by state foresters in the twentieth century. In the modern Killarney National Park, extended around Muckross House, the estate woods include many alien and exotic trees. But the main expanse of oakwoods, out on the sandstone hills, is still satisfactorily mossy, ferny, and indigenous.

Here and there among the oaks, but growing much more vigorously in the open, among the cliffs of the mountain slopes and at the rocky edges of lake islands, is one of the rarest and most beautiful, and also the most puzzling, of Ireland's native trees. The strawberry tree *(Arbutus unedo)* looks its best in September and October, when its clusters of creamy white flowers hang beside the ripening red and yellow fruit. These are not juicy but do have a textured skin somewhat like the strawberry. The tree had names in Irish long before the English arrived, and in their documents of the late sixteenth century terms such as "crankanny" and "wollaghan" are attempts to render *crann caithne* or (for the berries) *ubhla caithne*. The strawberry tree figures, indeed, in *Bretha Comaithchesa*, the eighth-century legal document famous for its economic ranking of twenty-eight of Ireland's trees and shrubs. The seven "nobles of the wood" were led, predictably, by the oak, the seven "commoners" by the alder.

Caithne, or the strawberry tree, checks in even lower down, between the white-beam and the aspen *(Populus tremula)* in "the lower divisions of the wood."

The strawberry tree really is, however, somewhat of a pretender among these Celtic arboreal ranks. Its main allegiance is to southern Europe, where it grows as a shrub among the *maquis* and thrives on the burnings that maintain that spiny, drought-resistant Mediterranean scrub. It grows at scattered locations up the Atlantic Coast of France, as far as Brittany, and then its distribution jumps to the southwest of Ireland, most abundantly at Killarney. Beyond that, the rocky shore of Lough Gill, in County Sligo, is an isolated northern outpost. The strawberry tree is particularly sensitive to frost and could not possibly have survived in the vicinity of the last Irish glaciers (it is only marginally hardy at Kew Gardens in England). Long-range dispersal by birds seems most likely. But distribution happened long enough ago to give it Irish naturalization. Fossil pollen from Muckross in Killarney has offered a date 4,000 years ago, and a pollen profile from a small island in Lough Inchiquin to the south—an island still dominated by it—shows a continuous record for 3,000 years. The tree was a lot more abundant in the past, probably as part of the spread of secondary forest, and could become so again, as global warming takes hold.

Another of Killarney's ecological surprises is confined to 62 acres (25 ha) of limestone pavement on Muckross Peninsula, separating two lakes. On an outcrop totally blanketed with moss grows the only yew wood in Ireland. At more than 3,000 years old, Reenadinna Wood is also the oldest known stand of yew in Europe.

In the Holocene history of Irish trees, the yew had arrived within the first millennium of warming, but it seems to have waited for expansion until disease struck the elms about 5,000 years ago. On the limestone of the Burren, yew burgeoned at that time and then declined catastrophically as farmers arrived with grazing livestock. Yew leaves contain an alkaloid that can be highly poisonous to cattle, but in Killarney it is the tree that has suffered most from herbivores. The forest there has been heavily grazed for two centuries by deer and domestic animals, including goats. Today many trees have been ringed and killed, scored by the antlers of the introduced sika deer *(Cervus nippon)*, and the rare yew seedlings in Killarney's mainland woods are heavily browsed.

In the oddly muffled and dimly lit wood of Reenadinna, the yews spread their roots out over the rock surface and clench them deeply into fissures under the moss. There are other small, natural-seeming stands of yew on limestone outcrops elsewhere, and individual trees are scattered over rocky terrain, some of it siliceous, in the west and north of Ireland. But there are fewer of these truly wild yews than of planted ones, and many people would not recognize a wild tree as being the yew they know in the landscaped gardens of big country houses and, even more familiarly, in graveyards (not only in Ireland, but Britain, North America, and the Antipodes).

The common yew of the hills, and of Killarney, has a rounded, unremarkable profile of radiating branches. The Irish yew is a chance mutation in which the branches grow stiffly upright and close together, shaping a form as darkly sculptural as an Italian cypress. It dates from about 1740, when one of the tenants of Lord Enniskillen found two oddly slender, seedling yews growing among the limestone rocks of Cuilcagh Mountain in County Fermanagh. He planted one in his own cottage garden, which died, and presented the other to his landlord to plant in his demesne at Florencecourt. This one grew vigorously and was multiplied through cuttings for friends. It became, in time, the mother tree of *Taxus baccata* cv. Fastigiata, a garden variety cloned commercially in the early nineteenth century and planted in millions across the world. The original tree still survives but in a bedraggled state, its branches now drooping and crusted with moss and lichens.

In the often subtropical ambiance of Killarney, the many exotic trees and shrubs planted by the estates' past owners merge seamlessly with native vegetation and have largely kept to their place. But one shrub has become an invasive menace, sprawling through the woodland clearings and smothering the seedlings of native trees and shrubs by its dense, oppressive shade. Although rhododendron found a place on the interglacial heaths of 500,000 years ago, its reintroduction to Ireland in modern times has been ecologically disastrous.

Rhododendron ponticum, whose native habitat is the Pontic Mountains of northern Turkey, was introduced into Britain in 1763 to beautify woodlands with its masses of rosy-purple flowers, to make windbreaks and provide extra shelter for game. It was brought to Ireland, probably later that century, and planted widely on estates in the south, west, and north. It thrives on acid or humus soils where rainfall is high and will even flourish on the open bog. Ecologists in Ireland and Britain have been brought near to despair by its virulent success. With trees to lean against, rhododendron can reach 26 feet (8 m) high, and its mature bushes form dense and almost impenetrable thickets. Under the whorls of stiff and leathery leaves, light reaching the ground may be as little as 2 percent of daylight, shading out even lichens and mosses growing on the lower tree branches. Few Irish organisms, unfortunately, have a taste for the shrub. It contains a toxin unpalatable to herbivores and probably most insects: even its leaf litter piles up uneaten on the forest floor. It offers little to birds, whose breeding diversity is sharply reduced in infested areas. Burning only encourages the shrub, and herbicides are costly.

At Killarney the explosive spread of rhododendron was recognized in the 1930s, and by the middle of the century the shrub had spread prodigiously, forming a dense understory, crowding out the native holly and smothering seedling oaks. By 1969 almost half the native forest had been colonized, and the shrub was invading even the swamp forest of alder on many of the lakeshores. Major operations began at that time to hack the rhododendron back. They have continued since, in Killarney and in other oakwood remnants in the west, a campaign assisted each summer by teams of conservation volunteers, but while any rhododendron bushes remain as a seed source nearby, invasion by this lovely but lethal shrub soon resumes.

Historically, its way into the woods was opened up by an unnatural absence of undergrowth. After centuries of heavy grazing by native red deer, cattle, sheep, and goats, the importation of sika deer in the nineteenth century tipped the balance even farther against regeneration. Their browsing extended even to the generally poisonous yews, many of which were ring-barked and killed by sika antlers. In the oakwoods, their browse line shows up in the holly understory. "Below a level of 1.2m no leafy shoot survives," observed the botanist Daniel Kelly in 1981 in the *Journal of Ecology*—this in woodland supposedly protected by the state. His assessment was followed by a major cull of sika deer, but their place was quickly taken by trespassing sheep. As the century ended, the pressure on unprotected sections of the woods was still too severe to allow the main tree species to regenerate, and even behind extensive new fences the hoped-for seedlings were slow to appear. Much of Killarney's decay seems irreversible, as gaps in the woodland soak up the Kerry rain and turn to bog, tussocked with purple moor-grass and too waterlogged for acorns to sprout.

The birdlife of Killarney and the fragmentary oakwoods elsewhere in Ireland lack a good many species that would be found, at a modest biogeographical distance, in sessile oak woodlands in western Scotland or Wales. There are no woodpeckers, for example (though bones of the great spotted woodpecker [*Dendrocopus major*], found in caves in Clare, show that it was once part of the Irish avifauna). Nor are there breeding marsh or willow tits (both *Parus* species) or the nuthatch *(Sitta europaea)*. The rarity of other woodland birds with a predominantly western distribution in Britain has puzzled ornithologists, especially as several of them are annual passage migrants in Ireland.

Some reasons may have to do with scale: Irish oakwoods are now simply too small to support more diversity. Scavenged of fuel wood and severely grazed and browsed, they have lacked the standing dead wood so necessary to woodpeckers, the nuthatch, the marsh and willow tits, and the pied flycatcher *(Muscicapa hypoleuca)*. Where Killarney's sika deer have browsed away the brambles and ivy, even Ireland's ubiquitous wren feels unwelcome.

So it can be argued that human activity led to the extinction of many woodland species in Ireland and are what keep the woods unattractive to passing migrants. But with the exception of the great spotted woodpecker, there is no fossil record to support this, and some ornithologists find it more plausible that, in postglacial colonization, specialists such as woodpeckers found their niches filled by more generalist species of bird.

The reduced competition has led to some species expanding their niche. Ireland has only four tits, for example, compared with Britain's seven, and the smallest of these is the coal tit, which caches seeds to see it through lean periods in winter. In Britain, both male and female tits forage in the same part of the tree and have bills of the same size. In Ireland the female's bill is deeper, improving her ability to deal with hard-shelled seeds. This capacity in turn extends her foraging range lower on the trees, or on the ground, which reduces competition between the coal tits themselves.

This tit is more common in coniferous woods, where an even smaller bird, the goldcrest *(Regulus regulus)* keeps it restless company. The goldcrest is of the same family as the kinglet *(Regulus satrapa)* of America's forests, but even more diminutive; its shrill "zee-zee-zee" is a remote and tantalizing whisper in the treetops.

In the oakwoods of Killarney the blue tit *(Parus caeruleus)* and the great tit *(Parus major)* more often catch the eye and ear. Both are drawn to breed among oaks by the abundance of spring caterpillars feeding on their leaves, and they time their first broods accordingly. The hatch of caterpillars is, in turn, timed for their own food: they must be ready to eat the oak leaves just as they break from their buds and before the trees pump too much tannin and other phenols into them as a defense against herbivores. Caterpillars that eat the leaves early grow faster and survive better; tannins hold them back. Young blue tits, too, grow better on a tannin-free diet.

Blue tits reach their greatest density in oakwoods, but the one bird in Killarney especially adapted to trees and dead wood is seen only with luck. The treecreeper *(Certhia familiaris)* works its way up a tree trunk like a small brown mouse, propping itself against the bark on an extra-stiff tail and using a thin, curved bill to probe for insects. It nests behind loose bark and roosts in crevices, where, with fluffed-up plumage, it can look much like a pinecone wedged in a crack—a cryptic device lingering, perhaps, from Ireland's vanished native pine forests.

The treecreeper is a sedentary bird and vulnerable to hard winters. At Killarney severe winter cold is rarely a threat—but heavy rainfall can be. The treecreeper was one of a dozen species breeding in the Reenadinna yew wood that were studied for changes in population in relation to winter weather. Small birds such as treecreeper, wren, goldcrest, and chaffinch were all affected by long periods of heavy rain, perhaps because it chilled them, perhaps because they had to spend too much time sheltering rather than feeding. The rain may also have made their invertebrate food more scarce. Findings like these have uncomfortable implications for the little birds of Ireland's woods and hedgerows, as global warming conjures even wetter and windier winters from the Atlantic.

KILLARNEY'S WILD RED DEER

In 1847, at the height of the Great Famine, a band of peasants living among the bleak mountains of Erris in County Mayo managed to corner and kill the last of their local herd of red deer. In that same harsh period the few red deer still living in the glens of north Donegal and Connemara were hunted to extinction. Only in Killarney, on the great estates protected by the Herbert and Browne families, did the native deer survive. Today, on the mountain slopes and pastures of the Killarney National Park, they form the largest pure-blooded reservoir of Ireland's most impressive wild mammal.

The red deer is the animal dramatically familiar from British Victorian paintings with titles such as "Stag at Bay" and "The Monarch of the Glen," portraits of majestic male deer with many-pointed antlers and wearing the dark mane of the rutting season. They were painted in Scotland just as the last stags of Connacht were being brought down, and the *"scoticus"* in the species name unites the red deer of the two islands while marking them off from those of the rest of Europe. But the genetic integrity of the Irish race, under insidious assault for some 750 years, is still of great concern to Irish zoologists, and the wild herds of the Killarney mountains represent the pure core of its conservation.

Ecologically, the deer belong to the transition zone between forest and open steppe. The dating of bones recovered from Irish caves includes them in the island's fauna both before the advance of the last ice sheet and again in the late-glacial period (at about 11,800 years ago), but the date and manner of their arrival in the postglacial period is still uncertain. The earliest red deer samples date from some 4,000 years ago, and antler artifacts are extremely rare even among Neolithic remains. It seems possible that the deer was introduced to Ireland, as it was to some Mediterranean islands. It ranked among the gods of Celtic mythology, however, and appeared in carvings on the high crosses of early Christianity, being hunted by men with spears and dogs. By the middle ages, it was widespread and fairly abundant.

The introduction by the Normans of deer parks common in medieval England, and often called deer "forests" even where trees are few, created a parallel world of hunting on private property. The importation of alien red deer, as well as the much smaller fallow deer began as early as 1244, and introductions and amalgamations continued at intervals through the centuries. In the 1800s the landlords of big estates were still introducing new species for ornament, including color variations of red deer from England and Europe. Momentously, in 1860, Lord Powerscourt in Wicklow made the first introduction to Ireland or Britain of the Japanese sika deer, closely related to the red deer and, despite a smaller size, reproductively compatible.

Hybridization between red and sika began in the Powerscourt Park and later elsewhere after deer escaped from the estate during the civil disturbances of the early 1920s. They multiplied in the mountains and conifer forests, and today hundreds of red × sika hybrids, widely different in color and form, travel the open slopes of the Wicklow hills. The sikas and sikalike hybrids keep to the trees and have rapidly expanded their population. Red deer are known to be good colonists, and both in Ireland and Scotland it is the hybrids that seem to be leading the sika advance through the commercial spruce forests.

A few years after the introduction of sika to Powerscourt, two hinds and a stag were passed on to Killarney. Within a century they had outnumbered the red deer, reaching perhaps 1,500 animals, and their browsing and antler-rubbing were causing serious damage in the woods. Heavy culling in the 1980s reduced their numbers by half, but the sika herd does have its own ecological virtue. Internationally it is considered an endangered species, and the Kerry sika are some of the purest in the world.

Cross-breeding with the reds had seemed inevitable, but remarkably had not been recorded even by the end of the 1900s. Hybridization seems to need the pressure of confinement, but, as Killarney's herds of red deer increase in size, and reds and sikas graze more often side by side, the consequence may be the same. Once begun, hybridization seems to spread unstoppably. The rise in commercial deer farming in Ireland has also increased the risk that a red × sika hybrid could arrive by chance at Killarney, launching a damaging compromise of ancient genes.

It was with this in mind that, at the end of the 1970s, the nucleus of a pure herd of red deer was established in a new national park in the Connemara mountains (the national park at Glenveagh in Donegal has a herd of some five hundred red deer, descended from nineteenth-century stock introduced from Scotland and England). Also, in a bold experiment, a nucleus of Killarney deer was introduced to Inishvickillane, a small, 250-acres (100-ha) island, walled with high cliffs, among the Blasket Islands off the coast of Kerry. By the turn of the century, some forty adults and calves were sharing the mineral-rich grass of this unlikely sanctuary, sheltering among the rocks from ferocious ocean storms.

At this point, too, the Killarney herds had been built up over three decades from a low point of about 170 animals to approximately 650, probably the most the park could sustain in company with so many sika deer and thousands of trespassing sheep from surrounding farms. The red deer range from the lakeshores and woodland to contours at 2,000 feet (600 m) on some of the highest mountains in Ireland. They lead totally wild lives, without any supplementary feeding, traveling on long, slender legs across the bogs and rushing winter streams. Winter is wet and windy, rather than severely cold, and many of the mountain deer migrate to the valleys, feeding openly on the pastures or browsing in the woods on holly, ivy, and fungi. But some of the older stags spend their whole lives on the heights, depending on heather for winter food

and sheltering in crevices. In the lowland herd, grazing on calcium-rich vegetation, stags grow their first antlers earlier and bigger, and may in maturity
produce magnificently branching horns with as many as twenty points. The
mountain stags rarely pass fourteen points, and their discarded antlers (shed
annually in March) are quickly chewed up by both stags and hinds in an effi-
cient recycling of minerals.

A notably vivid and thorough picture of the year-round lives of the deer
has been provided in the first book to be written in Ireland about a single species of wild mammal. *The Wild Red Deer of Killarney* (1998) is the work of an
amateur naturalist, Sean Ryan of Cork, an accountant who spent thirty years
of weekends and holidays stalking the deer of the mountains. His long-lens
photographs are like Edwin Henry Landseer portraits plus eyelashes, and his
fieldnotes have amassed some unprecedented observations, such as this bizarre
behavior of some anxious red deer mothers:

> In twirling—and there may be several bouts—the hind turns clockwise,
> rapidly and tightly within her own length. In effect, she pivots on her hind
> legs, using her forelegs to spin round. . . . Once, a hind with calf, who had
> spun six times in her first bout, then went deliberately to the top of a hill-
> ock, and twirled again, 13 times. This time she revolved so rapidly, and in
> such tight clockwise circles, that she leaned inwards, towards her right,
> and at an angle of about 30 degrees from the vertical. On stopping, she
> appeared quite normal, neither confused nor dizzy, and cantered away,
> followed by her calf.

Twirling seems designed to distract a human intruder, rather as a ground-
nesting bird may try to lead a human away from her chicks by feigning a broken wing. The activities of red deer stags in the rutting season can be even
more arresting, and the reward for Ryan's hours of stalking are powerful and
precise descriptions of excited wallowing (in black, peaty bog pools), threshing
(of the nearest shrub), spraying (of urine), roaring, fighting, and copulating.

The first roars of the rutting season echo through the mountains at the end
of September, as the powerful master stags of Killarney begin to round up
their harems, herding them along in a mobile territory and defending them
against competitors. Fights become common by mid-October, as roaring con-

tests and awesome goring of the ground give way to actual locking of antlers, pushing, and heaving—all this, very often, on steep rocky slopes where one or other stag seems bound to get hurt. A scarring injury does happen often, and, occasionally, death. Ryan gives due warning in his book that wild red stags cannot be trusted and are especially dangerous in the fall. The best time for seeing a herd of the deer is from mid-July onward, as family groups crowd together on level and open expanses of bog, sometimes together with stags in groups. At other times, visitors to the park may be quite unaware of deer, grazing with their heads down among tussocks of moor-grass, long ears swiveling to every sound.

RED SQUIRRELS

The wildlife filmmaker Stephen Mills wrote in *Nature in Its Place: The Habitats of Ireland* of Killarney's red squirrels "brandished by gusty winds like little mops in the tree-tops." Such exuberance seems fitting in one of Ireland's more secure populations of red squirrel, a native species that is nearing extinction in many of the woodlands of Britain.

The red squirrel made its first home in Ireland's postglacial upland forests of Scots pine. When changes in climate and widespread felling reduced these to a scattering of copses (perhaps, indeed, to a total extinction within the first millennium A.D.), the squirrels adapted to the fast-diminishing patches of hazel and oak. The records of the export levy on their skins testify to their medieval abundance, but by the end of the eighteenth century the squirrel was considered extinct in Ireland.

Between 1815 and 1876 it was reintroduced from England to at least ten sites in the northern half of the island and by the early twentieth century was back in every county. In 1911, however, another squirrel reintroduction launched a serious ecological misadventure. Eight American gray squirrels, introduced at Castleforbes in County Longford as a wedding present from the Duke of Buckingham to one of the daughters of the house, multiplied in a rapid colonization. By the 1990s grays were established in at least twenty counties in the eastern half of the island and moving outward at about 2 miles (3 km) a year. The broad River Shannon still acts as a barrier, but a relentless spread to all counties is accepted as inevitable.

The red squirrel, only half the weight of the gray, is naturally adapted to living in the canopy of conifers; the bigger, less agile gray squirrel prefers broadleaf trees and spends more time on the ground. Even when they meet and coexist in the same woodland they generally get on together. In Britain, however, where the gray was introduced in the late 1800s, it has ousted the red to extinction over most of England and Wales. In Ireland, too, the red has disappeared from parts of its former range now occupied by grays.

The reasons for this competitive exclusion are still not certain, but one big factor relates to diet, especially in woods that include a substantial number of oaks. Both species have a broad range of foods. Both, for example, chew on woodland mushrooms in autumn and supplement their staple tree seeds with berries, tree buds and flowers, bark sap, and insects. But while both will eat acorns, the grays are much better able to digest their toxic tannins even when the fruit is unripe. From late summer onward they can cache them in the ground, and compete directly with the reds for the current crop of hazelnuts.

Killarney is one of the few large woodlands where the reds can take their time in choosing what to eat, including seeds held through the winter in the cones of Scots pines, naturalized from old estate plantings. In Northern Ireland, a Forest Service "squirrel strategy" is keeping oaks out of selected red squirrel forests, and planting a mix of conifer species, including Scots pine, to guarantee a succession of seeds. Conifer forestry, much criticized for its lack of broadleaves, could offer the reds their needed sanctuary, but it seems that only tracts of about 5,000 acres (2,000 ha) or more in area are really likely to leave them secure. Meanwhile, red squirrels may still be seen in many urban parks, such as the National Botanic Gardens and St. Anne's Park in Dublin, while grays raid the bird tables in the capital's suburban gardens.

The wild promise of Kerry's natural landscape has attracted invertebrate specialists for more than 150 years. Along with the Victorian fern hunters came collectors of butterflies, beetles, and moths. The beating of birch trees to shake out the exceedingly rare white prominent moth *(Leucodonta bicoloria)* is a classic image of the time, and doubtful claims of rarities from the nineteenth century continue to haunt the butterfly list of Killarney National Park. Among the twenty-three butterflies recorded as currently native there are the scarce purple hairstreak *(Quercusia quercus),* which flits high in the canopy of old oak woodland, the brilliant yellow brimstone, and the marsh fritillary. Among dragonflies, too, Killarney has a reputation for rarities, notably the downy emerald *(Cordulia aenea)* and its fellow hawker, the northern emerald *(Somatochlora arctica)*.

THE BEAUTIFUL SLUG

One of the most handsome invertebrates of Killarney, however, is the large and long-lived Kerry slug *(Geomalacus maculosus)*, first discovered beside Caragh Lake in 1842. It was later collected in parts of northern Spain and Portugal, which seem to be the only other haunts of the species in Europe. This makes it part of Ireland's Lusitanian fauna, one of the more colorful threads to the story of postglacial colonization discussed in Chapter 4, and its rarity has earned it protection under the Bern Convention.

The Kerry slug comes in two color forms, each appropriate to its habitat. On boulders and bluffs of old red sandstone on Killarney's open moorland, the slug has a charcoal body sprinkled with white spots, a coloration that tones well with the lichened rock (though it stays for much of time within deep moss protruding from the cracks). Within the woods, on tree trunks draped in lichens and moss, the animal is ginger to dark brown, with spots of yellow gold—a match for its refuges under the bark of rotten logs. In the warmer climate of Iberia it avoids the day, disappearing into a crevice at sunrise with all the pliancy of a worm and emerging after dark to graze on mosses, lichens, and fungi. But in Kerry and Cork it is often active during the day, especially in rainy weather and at most times of the year. It can stretch itself amazingly (a large specimen of 1.5 to 2 inches, or 4 to 5 cm, can reach up to 5 inches, or 12 cm), but a touch from a finger tip will curl it into a ball.

Because the open Irish hillsides are the result of forest clearance, the supposition is that this slug is an animal of old woodland, specializing in cropping the lichen and moss of ancient trees, and that only the particular mildness and humidity of Ireland's southwest has allowed it to roam out from the woodland microclimate. Alternatively, it could be an open-country species that adapted to tree trunks in the postglacial forest cover. Iberia offers no certain guidance, as the slug's haunts there have been subject to the same deforestation processes. For ecologists trying to frame a conservation policy for this attractive and highly significant slug, the puzzle is tantalizing, but it seems secure enough meanwhile, especially at the three protected sites of Killarney National Park, Glengariff Forest, and Uragh Wood Nature Reserve.

SELECTED REFERENCES

Carruthers, T. 1998. *Kerry: A Natural History.* Cork: Collins.

Duffy, P. J., D. Edwards, and E. Fitzpatrick. 2001. *Gaelic Ireland c. 1250–1650.* Dublin: Four Courts Press.

Hayden, T., and R. Harrington. 2000. *Exploring Irish Mammals.* Dublin: Town House.

McCracken, E. 1971. *The Irish Woods since Tudor Times.* Newton Abbot, Devon: David and Charles.

McCusker, P. 1990. *Butterflies of Killarney National Park.* Dublin: Stationery Office.

Mills, S. 1987. *Nature in Its Place: The Habitats of Ireland.* London: Bodley Head.

Quirke, B., ed. 2001. *Killarney National Park.* Cork: Collins.

Rackham, O. 1986. *History of the Countryside.* London: Dent.

Ryan, S. 1998. *The Wild Red Deer of Killarney.* Dingle: Mount Eagle.

— *Chapter 14* —

In a Moist Shade

Much of the natural world is scaled down in Ireland; only its ocean setting is at all enormous. The island itself spans so few degrees of latitude, so few thousands of feet of altitude, that the scope for diversity of species seems quite limited. Its mountains, by global standards, are mere hills: even the highest peak, Carrauntuohill in County Kerry, is a relatively modest 3,410 feet (1,039 m). Yet they have the *feel* of mountains—an intimate intensity that seems to make up for what they may lack in height and grandeur. In vegetation, too, it is the closer look that counts: an exploration of folds in the landscape, of moist, shady places, of crevices. In this minor world of microhabitats and minirefuges, Ireland can boast exceptional luxuriance and abundance, as well as the rare and mysterious.

The key to the richness of Irish ferns, mosses, liverworts, and lichens is the steady moisture of an extreme Atlantic climate. Ireland as a whole is actually drier than Scotland or Wales in rainfall volume, but the rain is measured out in frequent showers (a "skift" of rain is one expressive country word). In western areas of Ireland it rains, on average, about 240 days in the year, but many of the daily totals are of less than .04 inches (1 mm). As summer progresses, pressure rises over the Atlantic but falls generally over Europe, drawing a cloudy and moisture-saturated flow of air eastward from the ocean. Prolonged humid spells—with a relative humidity of 90 percent—are quite common, and average values for the year exceed 80 percent. Over and above this sustained and exceptional humidity, the western mountains provide heavy annual totals of rain, often exceeding 79 inches (2,000 mm) a year. On the summit of Mangerton Mountain, at the edge of Killarney National Park, the annual average is a peat-hammering 125 inches (3,184 mm).

In such conditions, the exuberance and luxuriant growth of ferns and mosses (in the Killarney oakwoods, in particular) has invited comparisons with the mossy mist forests of tropical Central America and Asia. The diversity of liverworts along the western coast stands with that of British Columbia, Japan, and the Himalayas. Overall, Ireland's cryptogamic flora (reproducing by spores or gametes, not by seeds) is of exceptional beauty and worldwide importance.

Although the forms in which the plants grow—great hummocks of moss in the oakwoods, ferns sprouting from branches 40 feet (12 m) above the ground—can conjure more exotic analogies, Ireland's ferns and mosses fit into a continuum of Atlantic habitats that stretches south, beyond the coast of Europe, to the Macaronesian archipelago off the western coast of Africa: Madeira, the Canaries, and the Azores. On these islands, in the cool, misty laurel and heather forests of steep mountain slopes, are the richest fern habitats of the northern temperate zone. Several of their species flourish also in humid niches along Ireland's western coast.

The so-called Killarney fern *(Trichomanes speciosum)* is the largest of Ireland's native filmy ferns, its finely cut, translucent fronds sometimes reaching more than 14 inches (35 cm) in the spray-filled mountain gorges of Kerry. It used to be at least frequent in the southwest of Ireland, but during the Victorian popular craze for ferns, plants were uprooted for sale to gardeners and tourists: the fern was devastated around Killarney and actually wiped out in County Wicklow, where the species was first found in Ireland. Today its remaining locations at Killarney are kept secret by local botanists, and it survives elsewhere at a few scattered sites in Donegal and Fermanagh.

Two other small but exquisite Macaronesian filmy ferns might almost be mistaken for large mosses. *Hymenophyllum tunbrigense* and *Hymenophyllum wilsonii* cover rocks in the Killarney woods in soft, feathery shawls. They twine through the saturated hummocks of the mosses and even climb into the trees as epiphytes. (Epiphytes are plants that grow on other plants, mainly trees, for support but not for sustenance). They are typically plants of deep shade in wet and humid habitats. But they can also survive in moisture-laden winds on exposed, north-facing mountain slopes above the sea, as they do on Achill Island and in northern Connemara. At Killarney, *Hymenophyllum wilsonii* grows on shaded, mossy rocks at sea level but flourishes even more strongly at heights above 2,600 feet (800 m).

On Achill Island these filmy ferns are joined by a third Macaronesian species, the hay-scented buckler fern *(Dryopteris aemula)*, which has its headquarters in the Azores. Its fronds unroll only slowly through the summer and then stand green through the winter, so that the plant needs an exceptionally long growing season in permanently moist and cool conditions. It seldom grows far from sea level on extreme western coasts, but Christopher Page, of the Royal Botanic Garden in Edinburgh, an expert in the ecology of the ferns of the Atlantic fringe, has been impressed to find buckler fern at exceptional heights on mountain slopes in Kerry and in the Mournes. He thinks it may indeed be a direct Atlantic Coast survivor from the evergreen forest flora before the Quaternary.

Page includes among the ferns of the Atlantic continuum two striking species of radically different habitats. The royal fern *(Osmunda regalis)*, the tallest in Ireland, often rises to head height or more and forms massive clumps of rhizomes, some extremely old. Its foliage dies down completely in winter, but May sees the rise and unfurling of its boldly wrought golden-green croziers, and by August the fertile fronds terminate in elaborate, cinnamon-brown plumes of sporangia. The royal fern is most abundant in wet, peaty conditions on the cutaway bog, field ditches, and streamsides along the west coast, and it rarely grows above 164 feet (50 m)—except in Kerry's Macgillicuddy's Reeks, where it climbs to 984 feet (300 m).

Sea spleenwort *(Asplenium marinum)*, on the other hand, is so vulnerable to frost that it is prepared to tolerate doses of salt that would kill most other plants. It occurs all around the Irish coast, but is most abundant in the west, where it tucks itself away in crevices, arches, and caves at the foot of cliffs at the most exposed and wave-lashed headlands. Here in dark, still niches of undercut rock, shaded for summer coolness and watered at the roots by freshwater seepages, its waxy evergreen fronds shrug off an almost continuous mantle of sea spray: indeed, it depends on the spray to bathe the rocks with the stored winter warmth of the ocean. The fern also finds a home in deep crevices in limestone pavement on the coast of the Burren, where it grows to a size and vigor more appropriate to spleenworts of the tropics. "It is certainly a wonderful sight in the middle of January," wrote Charles Nelson in *The Burren*, "the majestic, dark green fronds of *Asplenium marinum*, dripping wet with rain and spray and just reaching the pavement surface, highlighted by the brief rays of the low sun against the water-black limestone."

The Burren's principal showpiece fern, the maidenhair fern, has been discussed in Chapter 8, but for sheer luxuriance on the limestone, if not botanical glamour, it is the rusty-back fern *(Ceterach officinarum)* that catches the eye. This is a light-demanding fern, flourishing in dense colonies on the open rock pavement and on the tops of the Burren's limestone walls. It responds to the desertlike regime of the rock wilderness, curling up in summer droughts around its brown, felty underside (in Irish, it is *raithneach rua*, the red fern). When rain returns, it drinks it in greedily and spreads its fronds again.

Farther north, among the cliffs of Ben Bulben in County Sligo, alpine and southern ferns grow side by side in lush, limestone valleys. Here Christopher Page found all three native species of *Polystichum* growing together: the mostly alpine holly fern *(Polystichum lonchitis)*, the mainly northern hard shield-fern *(Polystichum aculeatum)*, and the generally southern soft shield fern *(Polystichum setiferum)*. In this hospitable glen they hybridize, throwing up forms unique in the world.

Overgrazing, however, has seriously depleted the diversity of ferns on the hills. The mountain male-fern *(Dryopteris oreades)*, for example, generally rare in Ireland, was abundant on the granite cliffs of Eagle Mountain in the Mournes in the late 1800s; now it seems extinct. The rich fern flora of Ulster, so intriguing to the teenaged Robert Lloyd Praeger, is today largely confined to the more inaccessible crevices of the Mournes and to the deeper glens of the Antrim coast. In the west of Ireland, too, the ravines in the hillsides, with their waterfalls and brief, rock-walled gorges, have become the last refuge of mountain species. Below the outward-jutting oaks, rowans, and hollies, on cliffs and ledges inaccessible to sheep, there are hard and lady ferns *(Blechnum spicant; Athyrium filix-femina)*, and the bucklers *(Dryopteris* spp.) and polypody ferns *(Polypodium vulgare)* root into mossy branches in a brave imitation of a mist forest. Even in many of the old Atlantic oakwoods, inaccessible ravines like these become linear and aerial fern gardens while, beside them, the incessant grazing allows just one species, the globally ubiquitous bracken, to spread by deep rhizomes and engulf the woodland floor.

For most Irish country dwellers the word "fern" is synonymous with bracken. Its advance began with Neolithic forest clearance on the hillsides and continued under later regimes of moorland burning and woodland grazing. When cattle were kept on the hills, their trampling helped to control the fern, as did the centuries of harvesting for human and animal bedding. But by the late twentieth century, the rural depopulation of Ireland and the rise in sheep numbers encouraged the spread of bracken to blanket whole hillsides and dominate the unprotected western woods. In the oakwoods of Killarney, fencing out sheep and deer has brought a surge of new vegetation: bracken has lost ground, seedling trees have appeared, and many of the former forest ferns have returned.

MOSSES AND LIVERWORTS

The profusion and luxuriant growth of Ireland's mosses and liverworts has been a source of considerable pride for the island's botanists. David Webb, founding author of the standard *Irish Flora*, documented "the enormous sheets of *Hylocomium splendens* and cushions of *Breutelia chrysocoma* in the Burren, or the rich tufts of *Pleurozia purpurea* and *Herberta hutchinsiae* on the Kerry mountains." But the actual total of bryophyte species is not correspondingly generous, and Webb, writing in the 1940s, thought especially surprising the absence of a distinctive subtropical flora among the mosses and liverworts to match that of the ferns. This suggested to him that the last glaciation was more disastrous for bryophytes than for vascular plants, and that the early postglacial stages offered them less favorable conditions.

Half a century later, however, opinions can be markedly different. Ireland's liverworts, in particular, are judged as some of the most richly diverse in the northern hemisphere. And the southern species among them are so mystifying in their origin that they have suggested an ice age scenario quite contrary to Webb's.

The minute, lightweight spores of cryptogamic plants such as liverworts and mosses should make it easier for them to spread over long distances. Yet many of the liverworts (hepatics) that flourish in western Ireland belong to groups that never produce spores, and in any case are hundreds or even thousands of miles from the nearest known colony. This suggests to David Long, of the Royal Botanic Gardens in Edinburgh, as noted in *Glasra* in 1990 that "the ancient cryptogamic flora has been much more successful at surviving the effects of glaciation than the vascular flora, and that the unglaciated oceanic extremities of Britain or Ireland provided sufficient micro-refugia for much of the present-day flora to have survived."

Achill Island is particularly rich in the so-called disjunct groups of liverworts that are scattered between tropical mountains on both sides of the Atlantic and parts of Atlantic Europe. (Disjunct populations are so far separated from each other that they are unable to exchange genes.) Of Achill's seven species of this sort, the most famous and intriguing is *Adelanthus lindenbergianus*, first discovered there in 1903. It occurs also in Connemara and Donegal, but otherwise, in the northern hemisphere, only in Mexico and East Africa. While spore and gemma production has been noted in the species elsewhere, it has not been seen in the Irish plants.

In Connemara, rare leafy liverworts grow in a variety of shaded habitats: on dripping cliffs of crystalline mica schist high in the Twelve Bens; tucked under wiry canopies of heather, where this has escaped the sheep; and on boulders and tree trunks in the sessile oakwoods of Derryclare and in the Kylemore Valley in the north of the region. Kylemore's exceptionally shaded and rainy wood, tangled with rhododendron, is one of the finest sites for southern Atlantic liverworts in Ireland, and the *Lejeunea* species grow there in particular profusion. Across the valley, the regrowth of heather inside the fencing of the Connemara National Park may have arrived just in time to save the leafy liverworts so dependent on their shade, *Adelanthus lindenbergianus* among them. This type of heath is shared only with the mountains of western Scotland and so becomes a habitat of global significance and value.

Virtually all the liverworts of bogland are of the leafy kind, and in the wetter bogs grow so luxuriantly that they can smother the sphagnum mosses. One of the biggest and most distinctive species, particularly on blanket bogs, is *Pleurozia purpurea*, which forms tufts like a bunch of copper-colored worms (it has a disjunct distribution, scattered between northwest Europe, China, the Himalayas, Alaska, Guadeloupe, and Hawaii). Its rows of closely overlapping leaves are shaped to store water—also, possibly, to trap minute animals for nourishment, as related liverwort species do in the tropics. This modification of leaves, forming upside-down helmet-shaped pitchers, is even more marked in the group of *Frullania* species, found not only on bogs but as pioneer plants on rock and on the bark of trees, as in the oakwoods of Killarney.

The dense carpets of bryophytes on the boulders and trees of these Kerry woods have brought comparisons with the moss forests of Malaya. In Derrycunnihy Wood there are about forty species of mosses and fifty of liverworts, some climbing to the tree tops as epiphytes to meet the lichens on the uppermost twigs. On oak trunks at head height, the feathery, golden-green moss *Isothecium myosuroides* forms a thick, almost continous sheath, and it spreads out along the horizontal branches to make a thick carpet in which polypody ferns and the filmy wisps of *Hymenophyllum wilsonii* take root. The smaller, more upright branches carry a reddish growth of a liverwort, *Frullania tamarisci*, which at the upper twigs shares the light and wind with pale green tufts of the moss *Ulota crispa*. After a storm, the forest floor is strewn with wads of epiphytes blown from their perches; they go on growing where they fall. The darker the wood, with tall trees and little undergrowth, the higher the epiphytes grow, but in lighter areas, their communities move downward. In this they are behaving like the orchids of a tropical rain forest.

The darkest wood of Killarney is Reenadinna, where the low canopy of yew trees casts a deep and year-long shade. Much of the wood is bare of any understory of smaller trees or shrubs, and the limestone rock is carpeted as lushly as a hotel foyer with just a few species of moss. Where the sward grows directly on the rock it is an almost pure growth of *Thamnium alopecurum*, a beautiful dark-green moss with feathery fronds like tiny palm trees. The wood's epiphytes are mainly close-leaved liverworts, creeping across a smooth and flaking bark that does not really invite colonization. *Frullania* species, with their self-sufficiency in water, dominate the branches of the yews.

There could scarcely be more of a contrast to the gloom of Reenadinna than the open, short sward of coastal dunes and machair closely grazed by sheep, yet this is the Irish habitat that holds exceptional populations of petalwort *(Petalophyllum ralfsii)*, a liverwort ranked in European concern with plant rarities such as the Killarney fern and marsh saxifrage. Many of the former stations in Ireland and Britain have been lost, notably in creation of golf courses on sand dunes, but field surveys by Neil Lockhart of Dúchas (the Irish heritage service) have found stretches of machair on the Galway and Mayo coasts, with some of the largest colonies known anywhere in Europe, their millions of plants growing in the tightly grazed, mossy turf like microscopic lettuce.

THE LICHEN BROCADE

The weave of Irish lichens is often astonishingly rich and luxuriant, especially on the rocks and walls of the west, and the color and texture added to the landscape more than makes up for any want of rarity. Ireland's mountains are not alpine, and the usually acidic nature of their high ground is not lichen-friendly, but the flora of exposed coastal habitats is full of interest and aesthetic pleasure, even for those without much special knowledge of lichens. The high humidity of the west nourishes the growth of the symbiotic partnership of alga and fungus that constitutes the lichen. It has no roots or protective cuticle, and in sunshine can become dry and crisp in less than an hour. Moisture is what activates its metabolism, often within minutes, so that the alga can photosynthesize carbohydrates both for itself and its fungal partner.

Connemara, in particular, has attracted lichenologists for more than a century. Its mix of rock types and habitats supports almost seven hundred species, or some two-thirds of Ireland's total, and there are great areas of relatively undisturbed rock surface. Crusty lichens may grow as slowly as ⅟25 of an inch (1 mm) a year and persist for centuries on rock or tree bark.

Peat bog covers most of the Connemara lowlands beneath the quartzite mountains and has its own, characteristic *Cladonia* lichen flora, but it is broken by massive blocks, rugged outcrops, and broad sheets of ice-scoured granite scattered with erratic boulders. The blocks and boulders are colonized by leafy lichens, among them massed blue-gray lobes of *Hyopgymnia physodes* and rosettes of the silver-gray *crotal (Parmelia saxatilis),* historically used as a wool dye (coloring socks a warm reddish-brown). Other species more commonly belong on the bark of trees, such as the long, hairlike branches of *Bryoria fuscescens* or the warty crusts of *Mycoblastus sanguinarius.* These may be relict species surviving from the postglacial pine forest, which became extinct about 2,000 years ago, and the erratics both of Connemara and the Burren could be important reservoirs of Boreal lichens in a now mostly treeless landscape.

Boulder lichens are often vividly colored, as in the bright yellow-orange crusts of *Caloplaca, Candelariella,* and *Xanthoria* species. These identify particular rocks as regular perches for birds, their droppings rich in nitrogen in the form of uric acid. Birds clearly regulate the populations and growth of particular species, and bright lichen colors become a mark of good landing places. This coevolution of birds and lichens has intrigued lichenologist Howard Fox, who has explored its consequences through a wide range of habitats in Ireland. He has found, for example, that trees chosen as roosts by flocks of rooks *(Corvus frugilegus)* and starlings *(Sternus vulgaris)* are quite distinct in their lichen flora and that those forming a canopy of perching twigs are different from those lower down.

An "orange zone" of *Caloplaca* and *Xanthoria* is typical of exposed, smooth surfaces on the steeper rocky shores of the west. Below it is the black, tarlike skin of *Verrucaria maura* and above it the "gray zone," dominated by the tangled, yellowish tufts of sea-ivory *(Ramalina siliquosa)*. *Ramalina* species can produce spectacular, leafy growths, some of them centuries old, on the walls of island ruins and on coastal dry-stone walls. Study of some Connemara walls has found as many as sixty lichen species in a 11-yard (10-m) stretch, with a special richness in walls about 100 yards (100 m) from the shore. In the limestone Burren, away from the sea, old walls are often bright with various species of *Caloplaca* and the white *Aspicilia calcarea*, both of them numbered among the special lichens of bird perches.

Investigation of the lichens of the Burren hills, with their extension in the limestone of the Aran Islands, was slow. As late as the 1960s, no more than thirty species were known, but two decades of fieldwork directed from University College, Galway, produced another three hundred or so, about one-third of these on silica-rich rock such as the hundreds of granite erratics and outcroppings of chert and sandstone. As many as fifty species were found on a single bird-frequented boulder (*Xanthoria parietina* once again heads the list, its orange aureole pinned to the rock like some glowing imperial decoration). The lichens of the limestone pavement are generally less eye-catching, but their role in pitting the surface with eventual solution hollows, in which the rock is dissolved by tiny pools of rain water, forms part of the long weathering of the stone.

Extensive and sometimes near-impenetrable stands of hazel, ash trees, and blackthorn scrub in the Burren have a luxuriant growth of lichens, among them leafy species that vouch for the humid, unpolluted conditions of Ireland's oceanic woodlands. Most striking of these is the tree lungwort *(Lobaria pulmonaria)*, growing in wreaths of embossed, leathery fronds up to 7 inches (18 cm) long. These are also found prolifically on the branches of oaks at Killarney and in vestigial Atlantic oakwoods protected as nature reserves, such as Derryclare Wood, deep in a valley of the Twelve Bens in Connemara, and Old Head Wood on the coast near Louisburgh in County Mayo. Derryclare extends northward many of the Lusitanian lichen species found in southwest Ireland and particularly excites lichenologists with its luxuriant tree-bark swards of the blue-green *Sticta canariensis*. Like many of Ireland's rarer bryophytes, this has a strangely scattered, disjunct distribution, occurring in the Canary Islands, rarely in western Ireland and Britain, and at a handful of wet, rocky sites in Norway. Even more intriguingly, it has been colonized at Derryclare by a parasitic fungus, *Hemigrapha astericus*, otherwise known only from Australia and South America: the spores of microscopic fungi have been recovered even from the stratosphere.

The modest height of Ireland's moisture-saturated mountains rules out many of the lichens that depend on the conditions of bleaker, colder altitudes. The acidity of summits and ridges, often of quartzite and coarse sandstones, does not encourage great richness of species, and heavy grazing by sheep has restricted those that grow on soil. Diversity increases on the lower ground, where there are many more outcrops of iron-rich rocks. Some compensation for lichen hunters is the abundance of strongly western oceanic species in the mountains (including one rarity, *Gyalidea hyalinescens*, that also flourishes in the Azores).

The richest site for montane species is Mount Brandon, on the Dingle Peninsula in County Kerry, Ireland's second-highest peak at 3,131 feet (954 m). A chain of stream-fed tarns in the cliff-bound, glaciated valley that bites deep into the mountain from the east is the outstanding lichen habitat. Large leafy species grow in spectacular abundance on the sides of boulders splashed by waves. They are part of a remarkable community of lichens maintained, possibly since late-glacial times, by some concentration of alkalinity in the tarns.

Threats to Ireland's abundance of lichens are many and insidious. Most direct is the simple eradication of habitat in the wave of suburban and road development now sweeping along the coasts of Connemara and other rocky western littorals, covering thousands of acres of bedrock and sweeping away old networks of dry-stone walls. Overgrazing and steady encroachment on peatland are degrading and reducing the fabric on which many more of the island's species depend. The quality of air, notoriously crucial to lichen survival, seems certain to suffer in the boom of economic development that has overtaken the country. In 1988 important studies that related sulphur dioxide pollution to low lichen distribution were carried out by An Foras Forbartha, then a state environmental agency, and the School of Botany, Trinity College, Dublin. Some 2,200 trees in Dublin were surveyed with the help of senior school-children to create a baseline map of air quality zones outside the lichen "desert" of the city center. The studies found Dublin's lichens vulnerable to even lower than usual sulphur dioxide levels, perhaps because Ireland's high rainfall both delivers more acid and keeps lichens growing more actively to absorb it. In Galway at the same time, the presence in urban areas of sensitive *Caloplaca* and *Parmelia* lichens spoke for the high quality of air in a rainy but windswept city; but Galway has since been almost doubled in size by intense building development.

So far, 1,050 lichens have been recorded in Ireland, some 30 percent of the European total. While some parts of the island have been well-studied, the full distribution of lichens, as of the bryophytes, needs much more research to complete the picture of Ireland's floral diversity.

SELECTED REFERENCES

Doogue, D., D. Nash, J. Parnell, S. Reynolds, and P. W. Jackson, eds. 1998. *Flora of County Dublin*. Dublin: Dublin Naturalists' Field Club.

Hackney, P., ed. 1992. *Stewart & Corry's Flora of the North-East of Ireland*. Belfast: Queen's University Belfast.

Nelson, E. C. 1991. *The Burren*. Kilkenny: Boethius Press.

Page, C. N. 1988. *Ferns: Their Habitats in the British and Irish Landscape*. London: Collins.

Phillips, R. 1980. *Grasses, Ferns, Mosses, and Lichens of Great Britain and Ireland*. London: Pan.

Pilcher, J., and V. Hall. 2001. *Flora Hibernica*. Cork: Collins.

Watson, E. V. 1981. *British Mosses and Liverworts*. Cambridge University Press.

—— Chapter 15 ——

FishThat Came in from the Cold

"T he gillaroo," wrote Justice Kingsmill Moore in *A Man May Fish*, as respected a judge of trout as he was of human misdemeanor,

> is the panther of the water, the loveliest of our fish. On a background of satinwood are scattered spots and blotches of red, orange, umber and burnt sienna, so thickly as to touch and interfuse. [Gerard] Manley Hopkins must, I think, have had in mind the gillaroo when he wrote "Rose moles all in stipple upon trout that swim."

Dressed for a day's angling in tweed knickerbockers, thick woolen socks, a soft-brimmed hat to soak up rain, Kingsmill Moore cast his fly for gillaroo *(Salmo trutta stomachius)* on Lough Melvin, a lake amid fields in a glacial valley on the borders of Sligo and Leitrim. He fished them up to 3½ pounds (kilos were not in his culture) in rough weather along rocky shores. Sometimes he caught, instead, a sonaghen *(Salmo trutta nigripinnis)* or even a ferox *(Salmo trutta ferox),* though this hook-jawed, predatory trout had to be trolled for and that was never his sport.

The three Melvin tribes of trout (four, if we include the ordinary brown trout, *Salmo trutta*) are sufficiently distinct in livery, form, and lifestyle as to count as separate species. Their coexistence in one comparatively small lake speaks for the rich ecological history of Ireland's native freshwater fish.

Until the Normans introduced the first coarse fish species (probably the pike, *Esox lucius*, in the twelfth century) all of the fish in Ireland's lakes and rivers had arrived by sea from continental Europe. There were about a dozen species. Together with the eel, the sea and river lampreys *(Petromyzon marinus* and *Zampetra fluriatilis)*, two herringlike shads *(Alosa fallas* and *Alosa alosa)*, and at least one stickleback *(Gasterosteus aculeatus)* came the fish of the salmonid family that would mean so much to humans, for food and then for sport.

At the end of the last glaciation, some 13,000 years ago, the only open streams and rivers in Ireland were those draining into the Celtic Basin from the southern margins of the island. These may already have supported salmon and trout that survive as distinctive Irish races today. But as the thaw spread north, these "Celtic" salmon and trout were soon joined by other races—and species—of salmonids, released from ice-bound parts of Europe, where they may have been isolated for some 20,000 years. These boreal or subarctic races of salmon and trout began to colonize the Irish rivers, together with other salmonids: the arctic charr *(Salvelinus alpinus)* and whitefish or pollan *(Coregonus autumnalis)*. The trout were to become uniquely successful, as nowhere else in their range, reaching from the Arctic Ocean to Eurasia and North Africa, was there a habitat so free of competitors.

All the colonizing species were cold-water fish conditioned to the habit of anadromy—feeding in the sea but migrating into freshwater to spawn. However, as the island landmass rose, relieved from the weight of ice, many fish found themselves isolated in small mountain lakes or in streams above impassable waterfalls. Below such rocky barriers, trout shared the lakes and rivers with salmon, whose need for migration to sea was overriding. In the high streams and tiny mountain lakes, groups of the more adaptable trout survived. Their forms became stunted by the lack of nourishment in dark and peaty acidic water, but they lead distinctively long lives, and in narrow moorland streams or the lakes of silent corries, quite undisturbed by anglers, they have preserved genetic strains that are otherwise extinct.

For Victorian angler-naturalists fishing the richly productive trout waters of Ireland, the distinction of species from the different markings, forms, and habits of the fish sometimes improved their pleasure. The most notable of the trout strains are still indeed those that look markedly different from each other, but the fish's ability to mold itself to the environment can sometimes mask its true heritage.

In modern genetic fingerprinting, through electrophoresis, a particular protein in the eye of salmonids has been found in two significantly different forms. One of these alleles, labeled LdH-5 (100), is widespread among all salmonid species and is taken to be the ancestral form, whereas the other, labeled LdH-5 (90), is found only in brown trout and is thought to be a subsequent mutation.

From the Caspian and Black Seas of southern Europe, through southern France, to the Atlantic Coast of Spain, there are undisturbed brown trout populations that have the ancestral 100 allele alone. But the farther north one moves into the formerly glaciated parts of western Europe, the greater becomes the dominance of the later allele 90. Indeed, only three pockets of trout with the allele 100 alone have been found so far in northern Europe: in a high mountain lake in Iceland, a group of mountain lakes in northern Scotland, and a population of trout in the Erne-Macnean system of northwestern Ireland, at a point farthest from the sea. Elsewhere in the island, high frequencies of the ancestral gene occur only in long-lived trout populations above the impassable waterfalls and in the ferox trout of Lough Melvin. "Thus," mused fisheries scientist Edward Fahy, "like the human Celtic race, the ancestral trout was driven by successive invasions . . . to occupy the fastnesses of the inaccessible western mountain ranges."

Lough Melvin is neither high up nor inaccessible, yet it seems to be one of the few lakes in northwest Europe to have retained a pristine salmonid community since the end of the Ice Age. Its waters are home to a distinct stock of Atlantic salmon, to arctic charr, and to the three kinds of brown trout that have attracted scientific discussion for more than 100 years. As distinct types, they were described by the British Museum in 1866: the red-spotted trout, or gillaroo; the black-finned trout, or sonaghen; and the ferox, or black-lough. But until the crucial electrophoretic studies of the 1970s, led by Andrew Ferguson of Queen's University, Belfast, a basic question still remained: were the fish simply ecophenotypes of the same stock, produced, for example, by different feeding habits acquired by chance early in life, or were they genetically distinct? The studies showed not only that they differed genetically, but that they journeyed to separate spawning areas in the rivers flowing into the lake. The fish even segregate their feeding: the sonaghen eats mid-water zooplankton, especially water fleas; the gillaroo browses on bottom-living snails; and the adult ferox chases fish, including Melvin's arctic charr. DNA fingerprinting has confirmed the genetic differences between the three, and Ferguson has urged their recognition as separate species to highlight the need to conserve them: the "sympatric" Melvin salmonid community must be unique in the world.

Many of the genetically unique Irish trout populations have already been lost through habitat destruction, pollution, and introductions of hatchery stocks. Another modern development has devastated the sea-trout stocks of Connemara and its adjacent rivers. Sea trout *(Salmo trutta trutta)* are simply brown trout, living mostly in short and poorly nourished coastal rivers, or upland lakes, that have kept the habit of running to sea to feed in richer conditions. Those of Connemara were especially attractive to anglers, who fished the strong, lively "white trout" (turned silver and blue for their migration) as they returned to their spawning grounds up scenic, rocky rivers.

A dramatic collapse in their stocks began in the 1980s in sheltered bays and sea loughs where salmon farmers had moored their cages. The trout were returning prematurely to their natal estuaries, heavily infested with debilitating sea lice *(Lepeophtheirus salmonis)*, and their marine survival fell rapidly toward zero. Crowded salmon cages and poor infestation control were clearly the major source of these parasites, but a direct, forensic link with the lice on the sea trout could not be shown, and the salmon-farming industry denied the connection.

Successive investigations reinforced what was obvious to many, and lice management improved on some farms in response to licensing controls. By 2000 the state's Environmental Protection Agency asserted the continuing critical state of the Connemara sea trout stocks and the key role of the farm lice. Controls have been aimed at a near elimination of the parasites through use of pesticide baths or the use of wrasse *(Labridae)* as cleaner-fish. But the Irish government may have to follow moves in Norway and Scotland to relocate the fish cages well away from sea-trout rivers.

In warmer climates than Ireland's, even brown trout lose their urge to run to sea: Northern Spain and Portugal seem to mark the boundary of anadromy in *Salmo trutta*. Indeed, although the natural range of brown trout may include the colder, low-nutrient alpine lakes of Europe, the continent has no counterpart to the great lowland trout lakes of western Ireland, with rich limestone feeding and plump specimen trout. Their crucial advantage is mainly their oceanic climate of cool summers and strong Atlantic winds.

The waves that rock the anglers' boats on Loughs Corrib, Conn, or Mask are vital to the welfare of the trout, which in these lakes qualify for specimen status at 10 pounds (4.5 k). Their turbulence keeps the water well-mixed and full of oxygen from top to bottom: the lakes do not stratify in summer for long, and water temperatures rarely reach 68°F (20°C), the point at which salmonids begin to be uncomfortable. In continental Europe, the summer temperatures in most lakes are higher, and the water stratifies into two layers: warm above and cold below. The lower layer loses its oxygen as the summer progresses and is no longer available to the trout as a cool retreat. In such warm, weedy lakes, the salmonids give way to the cyprinids, notably the carp of monastic fish ponds, which need water warmer than 64°F (18°C) even to spawn.

Ireland's prime game fish remains *Salmo salar*—even if, as the species fights for its future in the island's rivers, the angler's bag from the spring-run fish has been limited (in 2002) to one fish per day.

The upward-leaping salmon, gleaming arc in the tumult of a waterfall, is so simple and familiar an image that it stamps a single, homogenous brand on what, worldwide, is an extremely various fish. Some salmon are short and squat, others long and streamlined. Some travel 2,000 miles (3,200 km) upstream to spawn (in the Yukon); others lay eggs in estuary gravel, washed over by saltwater. Some spend one winter at sea and others several. Some have lost the migratory habit and remain in freshwater lakes throughout their lives.

Such differences might be expected within the five species of Pacific salmon, of the genus *Oncorhyncus*, that use the rivers and lakes of North America. But even the Atlantic salmon *(Salmo salar)* shows great variation in behavior, not merely between the two sides of an ocean but even between neighboring islands and rivers. Salmon run into Scotland's River Tweed for eleven months of the year. Anglers on the Liffey, Dublin's river, can fish from January 1, and the capture of the first big fish of the year, perhaps a 12-pounder on New Year's Day, commands a ritual front-page photograph in the newspapers. In summer comes a second run of fish half that size, the grilse or juvenile salmon that have spent only one winter at sea. Yet in the short rivers of western Ireland the summer run of grilse is the only passage of the year, and one that may have to wait for heavy rain and a rushing spate of water to make it possible.

These younger, smaller fish have returned from a brief and modest migration, north around the coast of Scotland and on toward the Faroes: a journey of some 600 miles (1,000 km). Migrations of "spring salmon" from other, more substantial Irish rivers travel northwest across the Atlantic to eddy systems off West Greenland, where the fish may spend several winters feeding on small shoal fishes, such as capelin *(Mallotus villasus)*, before returning to Ireland. Fat stored in their muscles sustains them from their spring entry to the river until they spawn at the end of the year. Grilse, too, cease feeding once they enter fresh water, and the fact that curiosity or aggression prompts a salmon to snatch at a fishermen's fly or lure has been one of the cherished enigmas of angling.

Another far more spectacular mystery has been the nature of the mechanism that brings migrating salmon back precisely to their birthplace and their own genetically distinct population. Ireland's overall stock of wild salmon has been steadily reduced by drift-netting at sea, which catches up to 80 percent

Salmon parr (young salmon under two years old) distinguished by the "fingermarks" on their flanks.

of fish returning to the rivers. Now, increasing risks of interbreeding with escaped farm salmon also threatens to disrupt genetic purity and behavior that guides the wild fish on migration and preserves each river's stock from the impact of local disasters.

The importance of the olfactory sense to the salmon's navigation ability is long-established; even blinded grilse have found their way home normally. But what is it that makes up the particular scent of a river? One hypothesis suggests that young salmon (smolt) become imprinted with the characteristic odors of their home water, composed perhaps from aquatic plants or soil. Another suggests that salmon can detect metabolic attractants, or pheromones, released by their relatives—perhaps by the parr, the infant fish, living upstream from the estuary. Neither proposal explains the role of olfaction, if any, in the salmon's navigation between the deep ocean and the coast, but they suggest that layers of varying temperature in the sea may also contain specific odor information, along with electromagnetic and other cues. Fish biologist H. Nordeng in his 1977 article in *Oikos* suggested that the pheromone mechanism extends into the ocean: the migrating smolt, arriving at local feeding areas in the far Atlantic, trigger an inherited response in the mature salmon, already feeding there, which sends them home along the smolt route to spawn in their rivers.

Such a tightly woven pattern of behavior might even extend to a genetically programmed pattern of variation within a particular stock or race. As Noel Wilkins proposed in *Salmon Stocks: A Genetic Perspective,* the loyalty of a tribe of salmon to its river is self-reinforcing because there is virtually no gene flow from one tribe to another. Its precise homing may be a story of adaptation to the physical, chemical, and biological conditions of a particular river and a genetic distinctiveness produced by natural selection over time. As this diversity increases within the species, its prospects for survival are enhanced. This is why the interbreeding of wild salmon with fish escaping from farmed-salmon cages gives such cause for genetic concern.

The salmon has an aggressive competitor in the brown trout, which can dominate the calmer habitat of pools and restrict the salmon to the fast-flowing, energetically expensive riffle or broken water in mid-river. The salmon's superb streamlining and large pectoral fins that use water flow to hold the fish in position are adaptations for these water conditions.

Competition from Ireland's alien freshwater fish, however, can prove to be a more direct threat. Pike eat substantial numbers of salmon smolts, and the roach *(Rutilus rutilus),* used by many anglers as live bait for pike, has spread vigorously in recent decades, occupying salmonid nursery areas in many catchments and competing for food and living space.

Ireland's capacity to produce wild salmon has been substantially reduced historically by the damming of four major rivers for hydroelectricity. Their populations are enhanced by hatchery smolts, but with unproven results and perhaps at some risk to the genetic integrity of the native fish. Drastic drainage program in the later 1900s, designed to speed the flow of water from farmland, destroyed the natural contours of long stretches in many of Ireland's 154 salmon rivers, stripping banks and leveling riverbeds. But in the spawning catchments of the prime salmonid fisheries of the west, much has now been done to restore the natural form of lowland streams and their sequences of riffle, glide, and pool. The banks have been replanted with willows and hollows scooped beneath them as hiding places for fish. To follow these waterways now is to see birds and insects that had vanished from the barren gutters of the drainage schemes.

Such restorative moves come at a time of critical decline in the salmon stocks of the northeast Atlantic and an international consensus increasingly opposed to all drift-net fishing for salmon at sea. Ireland is alone in the North Atlantic in not moving directly toward ending coastal drift-netting, but instead aiming at reducing the commercial take through catch limits while improving wild salmon prices for local fishing communities. On some angling rivers, the raising of salmon smolt from native stock has helped maintain sporting catches, and millions of ova, or eggs, from the state's hatchery on the Burrishoole River near Newport in County Mayo have been helping to revive the salmon stocks of Germany's Rhine River. One giant fish of Irish origin, taken in the Rhine in 1998, weighed 21 pounds (9.6 kilos).

In such difficult times for Ireland's aquatic environment, what has been the fate of the other salmonids that arrived with the salmon and trout? Both pollan and arctic charr have been threatened. The herringlike whitefish known today in Northern Ireland as pollan probably entered the island's lake-strewn interior by way of the Shannon. It succeeded so well in Lough Neagh in the north of Ireland, in particular, that it came to be netted for food as it approached the lake shores in summer—a fishing tradition that persists today. The exact taxonomic status of the Irish pollan was uncertain until the 1970s, when electrophoretic analysis by Ferguson and his team showed it to be the same fish as the arctic cisco *(Coregonus autumnalis)* and thus a glacial relict found nowhere else in western Europe. In its regular habitats in arctic rivers, from the White Sea to Alaska and Canada's North West Territories, the cisco runs to sea and is fished extensively by native people. In Ireland, after many millennia of landlocked isolation, it survives not only in Lough Neagh but in Lower Lough Erne and Loughs Ree and Derg of the Shannon system. To Ferguson, it is not only unique among Ireland's fish, but one of the most important species in the whole Irish fauna.

The quality of water in Irish lakes, potentially some of the purest in Europe, has come under great pressure from artificial eutrophication, as the runoff from intensive agriculture has added to the outflow of sewage from country towns. This enrichment had boosted algal growth and has, at times, made severe demands on the lakes' dissolved oxygen. In the 1970s Lough Neagh was considered to be one of the most eutrophic lakes in the world, and it was remarkable that the pollan survived the worst summers: only mixing by the wind could have stirred enough oxygen into its bottom waters.

Elsewhere, the arctic charr has been the first species to be lost as oxygen levels decline, disappearing progressively from one lake after another. While the surviving charr of Scotland are found mainly in deep and nutrient-poor lakes in glaciated basins, Ireland has had large numbers living in shallow lakes rich in productivity. In the midlands, however, the fish were last reported from Lough Ennell during the 1920s. By the 1980s even the great western lakes were affected, as runoff of farm fertilizer and cattle slurry raised pollution past critical thresholds. Lough Conn in County Mayo had a large population of charr in the 1970s and 1980s; by 1990 it had suddenly crashed. Among the lake's 500,000 trout, the average fish had grown dramatically plumper, a sign of worrying changes in the lake's ecology. But charr do still flourish in several western lakes (the largest in Lough Mask in County Mayo), and the promotion of controlled angling for them may be their best guarantee of conservation. The general recent improvement of water quality in Irish lakes and rivers has, meanwhile, been slow but steady.

The Atlantic migrations of the freshwater European eel continue to tease science with circumstantial evidence. It has long been accepted that, once in their lifetime, all European eels leave their rivers and lakes and swim southwest to the Sargasso Sea to breed—a journey that, from one Irish study, may be delayed to the great age of fifty-seven. But no breeding eel, male or female, has ever been seen in the Sargasso: merely millions of the tiny larvae, *leptocephali*, that appear in February. These are carried back toward Europe with the North Atlantic Drift and metamorphose in autumn, at the edge of the continental shelf, into transparent young "glass eels." Their migration continues into estuarial waters where, as schools of dark elvers about 2.5 inches (6 to 7 cm) long, they slowly adapt to freshwater life and swim into rivers in spring.

There is no evidence that young eels return to the rivers or even the country of their parental origin. But those that do can be very determined about it. At the head of Killary Harbour, for example, on Ireland's west coast, the immigrating elvers meet the Aasleagh Falls in the entrance to the River Erriff. Three small side streams enter the river between the falls and the tide, yet the elvers virtually ignore these. They swim straight for the falls and climb the mosses where the water trickles down: a brief but dangerous journey made even in bright sunlight.

The European gourmet market for smoked eel has made the fish a profitable resource. In Lough Neagh, a long-established fishery uses draft nets and

Aasleagh Falls on the Erriff salmon fishery.

longlines with a thousand or more hooks to catch 880 to 1,100 U.S. tons (800 to 1,000 metric tonnes) a year. In smaller lakes and rivers, substantial catches are made with fyke (funnel-trap) nets, introduced from the Netherlands. Compared with Lough Neagh, Ireland's southern lakes have seemed under-stocked with eels, a situation now substantially redressed by the release of captured elvers.

While feeding, chiefly on invertebrates, and growing slowly to perhaps 20 inches (50 cm), the eels are known as "yellow eels" (though in fact they are dark brown). At some point beyond their sixth year, but usually between the tenth and fifteenth, they stop eating and turn silver in readiness for migration. This begins in autumn and peaks on dark nights of storm and flood, the fish perhaps alerted by the Earth vibrations, microseisms, that run ahead of Atlantic depressions. In their passage through Toomebridge, where the River Bann leaves Lough Neagh, the fat silver eels are caught by the thousands. But a recent severe decline in the number of eels returning to European rivers may warn of ocean changes set in motion by global warming.

FOREIGN FISH AND NATIVE SHAD

Of Ireland's twenty species of freshwater fish, the majority were introduced after the twelfth century. The modern Irish name for the pike, *gaill iasc*, identifies it bluntly as a "foreign fish," and most of those who fish for it are themselves from France and Germany and substantial contributors to angling tourism. Pike is a swift and powerful predator on brown trout, and control of its numbers in the prime angling lakes, beginning fifty years ago, has been Ireland's longest-running venture in conservation management. Intensive culling in the Connacht lakes of Corrib, Mask, and Carra is intended to halt the decline of trout and restore the lakes as salmonid-dominated ecosystems.

Most of the modern introductions have been made in the cause of sport angling, sometimes despite state prohibition of import and transfers. Tench *(Tinca tinca)* and carp *(Cyprinus carpio)* are at the northern limit of their range and present no ecological problems, but roach and dace *(Leuciscus leuciscus)* have affected salmonid production and brought declines in brown trout. Notoriously, both fish entered Cork's Blackwater River in 1889 when a visiting pike angler, using them as live bait, let two tinfuls be swept away in a flood. Despite legislation banning fish transfers within Ireland, roach have been introduced to many lakes and rivers, rapidly achieving dominance and some-times displacing trout stocks almost to the point of extinction.

Among the rarest species of fish breeding in Irish freshwaters are two spe-cies of shad—large, herringlike fishes that joined the Irish aquatic fauna with retreat of the ice. They live today in shallow coastal waters of western Europe, from Norway to Spain, and enter large rivers to spawn. The twaite shad and the allis shad have both suffered from river pollution and habitat destruction and are accorded special protection in the Habitats Directive of the European Union.

In Ireland the allis shad has only a handful of records, but the smaller twaite shad breeds above the reach of tides in several of the island's southern rivers, notably the Munster Blackwater, and the Barrow, Suir, and Nore. In the Barrow, indeed, its arrival in May brings anglers to St. Mullins especially to spin for shad in a catch-and-release competition. The twaite also has some interesting nonmigratory populations in large European lakes such as Como, Lugano, and Maggiore. In Ireland, its landlocked home is Lough Leane at Killarney, where it is known as the "goureen." It seems to have been isolated here for thousands of years, spawning on gravel bars, and its size (rather smaller than its marine cousins: rarely longer than 10 inches—25 cm—compared with perhaps 16 inches—40 cm) has helped to earn it subspecific recognition as *Alosa fallax killarnensis*. Stretches of the twaite shad's Irish breeding rivers are now included in the island's Special Areas of Conserva-tion, a protection also now being extended to the sea lamprey *(Petromyzon marinus)* with similar migratory habits.

PEARL OF PURE WATER: MARGARITIFERA

In 1094, Gilbert, Bishop of Limerick, presented a pearl to Anselm, Arch-bishop of Canterbury, which was "graciously received"—the earliest acknowl-edgment on record of the treasure to be dredged from Irish rivers. Over the next 800 years the fishing of pearls was to offer intermittent plunder, mainly to poor peasants living along the banks. Today the freshwater pearl mussel *(Margaritifera margaritifera)* has become ecologically precious in itself, and pearl hunting, now illegal in Ireland as elsewhere in Europe, is only one of the threats to its survival.

The life cycles of some species seem more than usually designed to demonstrate the workings of chance, and the exceptional lifespan of the fresh-water pearl mussel—up to 100 years and more—might almost be a recognition of its luck in existing at all. Such longevity, once thought to make the mussel the oldest living invertebrate, actually reflects its eccentric choice of habitat. Rather than develop on limy substrates that generally favor mollusks, the mussel thrives in fast-flowing rivers rising in mountains of sandstone or granite, often rich in silica but markedly lacking the calcium needed for shell-building. The species slowly matures as a heavy, robust bivalve, clam-shaped, sooty-black, and up to 5.5 inches (14 cm) long: small wonder that the same poor peasants valued the shells, with their tough, nacreous lining of mother-of-pearl, as uncommonly beautiful and serviceable spoons.

The mussels spend most of their lives two-thirds buried in gravel and sand, often in pools shaded by leaning alders or willows, their siphons extended to filter the water of food particles. Although not entirely sedentary (grooves in sandy riverbeds suggest the mussels can move a couple of yards relatively briskly), they lead what has every appearance of being a settled existence. Their early days, by contrast, are a positive whirl of uncertainty.

Most freshwater mollusks deposit their eggs securely in gelatinous masses or fixed to aquatic plants. From field studies on the reproductive biology of the freshwater pearl mussel carried out in Scotland and in University College, Galway, we know that, unusually for an animal living in fast-flowing water, the freshwater pearl mussel sets its young adrift in the current. In midsummer the male releases sperm into the water, which is carried to the eggs through the female's inhalant siphon. They are brooded in the female gills and then released as a cloud of larvae called *glochidia*—an average of 9.8 million from each animal. They are microscopically small but equipped with a pair of shells and a muscle to snap them shut.

As the *glochidia* are swept away in the river, their survival depends on being passively breathed into the gills of a salmon or trout, whereupon they clamp to a filament of the soft, red tissue and begin to absorb nourishment from it. In their twenty-four hours of viability (much less if, in the short, spate rivers of Connacht, they are flushed into the sea) 99.9996 percent of the *glochidia* will fail to find a host. Of the forty in ten million that do, all but two will be lost during the fourteen days it takes to grow to an independent size. At this point, the infant mussels drop off their hosts and bury themselves in the gravel of the riverbed, where they stay for five years, filter-feeding and gaining the necessary bulk and weight to withstand the tug of water at the surface of the gravel and the impact of rolling stones.

Why should the mussel launch its young in such a reckless manner? The attachment to salmonids could act to spread the mussel populations upstream, or it may have helped the species colonize new rivers as salmon and trout explored postglacial coasts. No Ice Age fossils of the freshwater pearl mussel have been found in Ireland, and its parasitic phase is specific to the native salmonids—it cannot use nonnative fish, such as rainbow trout *(Salmo gairdnerii)* for attachment. Indeed, Russian scientists have suggested that the relationship of pearl mussels and salmon is symbiotic—the fish provide nourishment at a critical phase in infancy (a parasitism that seems to do the salmon no harm at all), while the adult mussels help to maintain water quality for the salmon. In one Russian river, the mussels have been shown to filter 90 percent of the volume in low-water years.

In Europe the freshwater mussel achieved a wide postglacial distribution, from Ireland to the Urals, from Scandinavia to the Pyrenees. In Ireland, it has been partnered by a unique form that lives in hard, lime-rich water and is now on the verge of extinction. *Margaritifera margaritifera durrovensis* has a single community in the River Nore in the southeast of the island, but this has been dwindling rapidly in water of deteriorating quality, with no sign of younger mussels.

The early appreciation of pearls by people is an intriguing facet of human aesthetics. For the mussel, creation of a pearl is defensive. It surrounds a foreign body such as the embedded larva of a trematode, a parasitic flatworm, with the same organic layers that compose its shell: fine membranes with calcium between them. A perfect pearl, the size of a pea, has no particular color of its own but shares the opalescence of the nacreous lining of the shell—a shimmer magically enhanced by the perfect pearl's sphericity.

Up to about 1970, pearl fishing was still the main cause for decline of freshwater mussels in the British Isles. British pearl fisheries were famous in Roman times and later supported thriving commercial operations. Scotland still has Europe's largest colonies of the mussel, but, although protected, pearl fishing continues there and in Ireland, pursued by a knowledgeable and ruthless caste of adventurers.

In 1947 the Scottish ecologist Frank Fraser Darling, in *Natural History in the Highlands and Islands*, described methods traditionally used by "the tinkers" (an itinerant underclass of former tinsmiths, common to Scotland and Ireland):

> They stand in the bed of the river, slowly quartering it, bent double, a glass box in one hand in order to see the bottom clearly, and in the other a long stick notched at the end. They press this stick over the mussel, and pull it from the bed of the river and put it in a sack slung over the shoulder. The load of bivalves is brought ashore, where they are opened with a knife. Certainly not more than one mussel in a hundred carries a pearl and many of the pearls are too small to be of value. The wastefulness of the fishery is apparent.

As early as 1911, a leading Irish authority on mollusks, Arthur Stelfox, was concerned about the growing scarcity of *Margaritifera margaritifera* in many rivers, "perhaps owing to the depredations of the pearl-searchers and their wanton destruction."

The mussel is still widespread in Ireland, especially in hill rivers of Donegal, south Mayo, Galway, Kerry, and Cork, but the health of its populations is a different matter. Few rivers have young mussels, and in some, the last successful recruitment of young dates back to the 1960s. Pearl fishing may have been the first cause of decline, but the main cause since the 1970s has been deteriorating water quality, so that the largest populations of mussels are in remote areas with the least change to river channels and little intensive agriculture, forestry, or industry. There are certainly many other freshwater creatures whose disappearance would sound alarms, but the beauty of *Margaritifera* and the complexity of its life cycle make it a fitting guarantor of pure water.

An alien freshwater bivalve, the zebra mussel *(Dreissena polymorpha)*, which arrived in the River Shannon in the 1990s, is, however, reckoned a serious threat to Ireland's freshwater diversity. Native to rivers in Eastern Europe and already notorious for its smothering colonizations in North America's Great Lakes, this mussel has been spreading rapidly in Ireland's central waterways, carried on the hulls of leisure craft. Unlike the freshwater pearl mussel, it attaches itself to hard surfaces by tough byssal threads and multiplies to densities of more than 80,000 per square yard (100,000 per m^2).

The filter feeding of the zebra mussel can clear water wonderfully to human eyes, but its consumption of phytoplankton and zooplankton takes food from native aquatic species and makes it more difficult for larval fish to hide from their predators. If it were introduced into Lough Melvin (and it is now abundant in nearby Lower Lough Erne, a few miles away), Ferguson fears it would almost certainly have a disastrous effect on planktivores such as sonaghen and charr.

AMPHIBIAN TRAVELS: ENIGMAS OF FROG AND TOAD

In different places in the West of Ireland, at the margins of the wild areas, you are shown places to which, and no further, St. Patrick is said to have gone, and when he saw the desolate country beyond he said, "I'll bless yees to the West, but the de'il a foot I'll put among you." This is told of Iveragh, West Kerry, which accounts for the Natterjack toad being found there.

—GEORGE KINAHAN, *NOTES ON IRISH FOLK-LORE* (1881)

One toad, one frog, and one newt: Ireland has just three amphibians. Aside from reptiles, no sector of wildlife is so poorly represented; even neighboring Britain can offer half a dozen native amphibian species. The impoverishment seems all the more striking because both the European common frog and the rare natterjack toad have had a persistent question mark over their origins in Ireland. While the smooth newt is typically explained as having probably arrived by whatever land connection may have existed postglacially, frog and toad have been widely regarded as anthropogenic introductions.

This follows on the categoric statements of their absence by early churchmen seeking moral parables in nature. As David Cabot noted in his *Ireland: A Natural History:*

Noxious animals and their evil associations were an obsession of early Christian commentators who placed the frog in the same category as toads, snakes and lizards because of a superficial similarity. Thus when St. Patrick, in one generous swing of the crozier, drove all the pernicious creatures away, the frog left the country—or so Christian mythology claims.

There is a possibility that the Normans brought frogs in for food in the tenth or eleventh centuries, and also that they all emerged from a clump of frog spawn imported from England and tipped into a ditch at Trinity College,

Dublin, by a Dr. Gwithers in about 1696. If the animal was already present postglacially, it is hard to understand why it should not have been widespread historically and commonly known. Its absence, as a cold-tolerant and vigorously colonizing species, would be equally perplexing. Yet as one research team, led by Chris Gleed-Owen, commented in the *Bulletin of the Irish Biogeographical Society* in 1999: "It seems to have been almost unanimously accepted that the common frog was introduced—despite the lack of any scientific backing, and the weight of biogeographic and palaeontological evidence in favour of natural colonisation."

The earliest evidence of a seemingly prehistoric presence was subfossil frog limbs found with arctic lemming bones in deep clay in one of many caves in a limestone cliff at Keshcorran Hill, County Sligo, in 1899. Their position and condition satisfied Robert Scharff, keeper of the National Museum, that the frog belonged to Ireland's late-glacial fauna, but doubt persisted. A century later, a sample of these bones was among the museum fossils chosen for radiocarbon dating in the Irish Quaternary Fauna Project. It proved to be a mere 500 years old, rather than the 10,000 that Scharff had supposed. A similar disappointment met researchers in the year 2000, when ten similar museum fossils, collected from bone caves in Sligo, Clare, and Cork, all yielded dates of less than 400 years.

Until the Quaternary Project in the 1990s, almost no work had been carried out on Irish subfossil amphibians. If the frog has had to wait for its immigrant credentials, there has been even less palaeontological evidence for the antiquity of the natterjack toad, the curious and charming creature that survives, in a few hundreds, at small and scattered sand-dune habitats in Kerry. The first subfossil bones of the toad, recovered from a megalithic cemetery at Carrowmore, County Sligo, in the 1970s, were later shown to be those of frogs.

The natterjack is worth spending time with, both for its own engaging qualities and specific ecological needs and as an animal whose travels have warranted exploration through the most advanced genetic analysis. At the turn of the twentieth century, the credentials of the natterjack as a native Irish species seemed considerably more secure.

The toad's common name in English probably derives from old colloquial words: "natter," for idle or grumbling chatter, and "jack" for male. *Cnádán*, the Irish name for both frog and toad, carries similar overtones of querulous complaint. Certainly, the call of the male around the breeding ponds in spring is among the most resonant noises of any European animal, a single, monotonic note that can carry for more than a mile. Other local names in Ireland round out the picture: the "running toad" from its habit, dictated by short hind legs, of running like a mouse rather than hopping; the "golden back" from the yellow dorsal stripe; and the "black frog" from its dark back and its small, jet-black tadpoles. The natterjack is primarily adapted to warm and arid environments, and although it has ranged widely through western and northcentral Europe, it is confined to specialized habitats, mostly on riverine flood plains, coastal sand dunes, inland heaths, or in gravel or sand pits. All these evoke the raw and treeless landscape that was general as the last ice retreated. As a pioneer burrowing species, the natterjack is thought to have spread rapidly northward from a single glacial refuge in the Iberian peninsula.

At its first record in Ireland in 1805 by the botanist James Mackay, the toad was extremely common around Castlemaine Harbour, a sandy estuary at the head of Dingle Bay in County Kerry. The early distributional records spread to nearby sandy peninsulas and lakes, and then to sand dunes at Castle-gregory on the northern coast of the Dingle peninsula, beyond a mountain barrier. Later in the nineteenth century, the suggestion gained ground that natterjacks could have been imported in sand ballast dumped out from boats trading with Europe. There was, after all, no reference to them by the naturalist Charles Smith, in his 1756 description of the area, and some modern zoologists have found it "incredible that . . . one with such an eye for natural curiosities would have failed to mention them." Thus their immigration in a cargo or in ballast has remained a persuasive alternative hypothesis. Robert Lloyd Praeger, however, was dismissive in his 1950 *Natural History of Ireland:* "Could misdirected ingenuity go further than to suggest the importation or shipwreck of a cargo of toads on that lonely and harbourless coast?" The toad, he held, was an indigenous Kerry Lusitanian, like the lovely large-flowered butterwort and the spotted Kerry slug.

For an animal with such particular needs, the natterjack's range across Europe is remarkable. Its central abundance is in Portugal and Spain, but its populations extend through seventeen countries, including northern England and southern Sweden (where they breed in the rock pools of tiny offshore islands), and stretching east to Poland and Belarus. The British herpetologist Trevor Beebee, an authority on the natterjack, believes that its postglacial history in Europe is one of a rapid but relatively short-lived expansion from

Iberia, followed by a much more patchy distribution, as human activity fragmented and isolated its habitats.

All this made the natterjack an outstanding candidate for a study that
bears on some basic evolutionary questions. What, apart from ecological conditions, sets the limits to the range of a species? Why does natural selection
not work at the boundaries to achieve an infinite extension of them? Could it
be that genetic diversity falls off as range extends, and that this reduces a species' ability to adapt? The natterjack's history offered a pattern of dispersal and
diversity that should be frozen in time in the DNA of its scattered and isolated populations. Beebee, with geneticist Graham Rowe, set out to explore it.

They sampled the DNA of tadpoles and toadlets collected from pools
across Europe and subjected it to microsatellite analysis (microsatellite loci
are relatively short, repetitive nucleotide sequences that show the relatedness
between individuals). As they suspected, genetic diversity declined with distance from the home base in southern Spain. Even in east Poland, where the
natterjack's populations are large and numerous, diversity was low. Just as intriguing was the picture they obtained of the interrelatedness of populations.
The toads in Kerry, for example, have strong affinity with those in northwest
England, and both groups are widely separated genetically from those in the
southeast of England. When this divergence is calibrated against real-time
events such as sea-level changes and the Late-glacial "cold snap," the data fit
the proposition that the Kerry toads are descended from animals that crossed
a land bridge from England some 10,000 years ago. A calibration based on
the sand ballast theory, on the other hand, produces dates for the separation of
British and Continental natterjacks that are far too recent.

Such findings raise, in turn, more questions. Why, having arrived across
the Irish Sea, should the animals have established themselves ultimately at
one local stretch of sandy shore and nowhere else? Why are there no recorded
subfossils of the toads when southwest Britain can produce bones spanning
10,000 years in regions where the natterjack is extinct today? How could such
a distinctive little creature have gone unnoticed in Kerry for so long, either by
naturalists or in local knowledge and folklore? And why, if the natterjack
made it across the bridge, does Ireland lack the common toad *(Bufo bufo)* that
has an even wider British and European distribution?

The new native credentials for Ireland's toads should, however, help the case for conservation. The natterjack is both rare and declining at the north-west of its range (the British population fell by three quarters in the second half of the twentieth century). In Ireland in the early 1990s it bred at fewer than fifty sites, most of them tiny and residual. A resurvey in 1997 found the toads scarce in many places and apparently quite absent from some former strongholds, such as the Inch Peninsula, the great sweep of dunes at the head of Dingle Bay.

Incursions by the sea have played some part in loss of habitat, but most habitat destruction has resulted from human activity. In particular, the creation of golf courses in pristine coastal dunes in the 1980s found Ireland singularly unprepared with planning control and ecological research, as was discussed in a previous chapter. In 1991 Beebee reported to the Council of Europe on threats to the Kerry natterjacks. Among his concerns was that a new golf course, created in prime natterjack habitat at Castlegregory, had replaced the shallow, temporary pools in the dunes, used for breeding, with a series of deeper, permanent ponds. These, he warned, would develop plant life and attract aquatic insects, especially the rapacious larvae of large water beetles and dragonflies, which could wipe out whole populations of natterjack tadpoles. But the new ponds at Castlegregory continued to attract a flourishing population of toads, and, after almost a decade, U.S. herpetologist John Kelly Korky still found hundreds of adults in the dunes. Spawn had also been translocated to ponds created in sand dunes at the Raven Nature Reserve in County Wexford, where breeding has been successful. Now, according to Korky, it may be time to establish some inland natterjack populations, safe from any threat from rising sea level.

The island's widespread frog population, meanwhile, is listed in the Irish *Red Data Book* as "internationally important," given the serious decline of the species in the rest of Europe and worldwide concern over reports of mass mortalities among other kinds of frog. The main threat in Ireland appears, again, to be loss of habitat. About half the island's ponds disappeared as a result of agricultural and other development in the twentieth century, probably causing local extinctions, and fragmentation of habitat is leaving "islands" of frogs in large tracts of inhospitable farmland. As in Britain, the colonizing of ponds in new suburban gardens may prove to be a lifeline as the farmed countryside grows more hostile.

294   Ireland

SELECTED REFERENCES

Beebee, T. J. C. 1983. *The Natterjack Toad* (Bufo calamita). Oxford University Press.

Fahy, E. 1995. *The Brown Trout in Ireland.* London: Immel.

Kingsmill Moore, T. C. 1960. *A Man May Fish.* Gerrards Cross, Buckinghamishire: Colin Smythe,

Moriarty, C. 1978. *Eels—A Natural and Unnatural History.* Newton Abbot: David and Charles.

Phillips, R., and M. Rix. 1985. *Freshwater Fish of Britain, Ireland, and Europe.* London: Pan.

Whelan, K. 1989. *The Angler in Ireland.* Dublin: Country House.

Whilde, A. 1993. *Threatened Mammals, Birds, Amphibians, and Fish in Ireland, Irish Red Data Book 2: Vertebrates.* Belfast: Her Majesty's Stationery Office.

Chapter 16

Tribe of the Mustelids

Some [badgers] are born to serve by nature. Lying
on their backs, they pile on their bellies soil that has
been dug by others. Then clutching it with their four
feet, and holding a piece of wood across their mouths,
they are dragged out of their holes with their burdens
by others who pull backwards while holding on
here and there to the wood with their teeth.
Anyone that sees them is astonished.
—GIRALDUS CAMBRENSIS, A.D. 1185

Ireland's largest carnivore is a creature of the night, a rustling presence in the
dark beneath the hedgerow. The island's population of the badger is among
the highest in Europe, and it digs its elaborate setts, or dens, alongside the
richest of pastures. Yet people rarely meet the animal, and country dwellers
who malign it have rarely ever seen one, except as a gray ghost caught in
headlights or as a corpse on the morning highway. An even deeper shadow
now seems to fall across the badger's future on an island that gives cattle first
claim on the lowlands.

Since the 1970s, in Ireland as in England, the badger has stood accused of
infecting dairy herds with tuberculosis, and many thousands have been snared
and killed in official and unofficial clearance measures. But it is a stubborn

species that has survived long centuries of hunting and persecution across Europe, and it remains a protected species in Ireland. The uncertainties about its role in disease have focused research into badger populations and behavior, revealing a pattern of social organization apparently unique among carnivores.

The Eurasian badger *(Meles meles)* is a larger and more powerful animal than the American badger *(Taxidea taxus)*. Its body, up to a yard long, has a hunched, wedge-shaped poise: a small, neat head, held low, and a surprising bulk of soft, long-haired fur and muscle. The legs are short but immensely strong, the front claws long and tough for digging. Bold black and white stripes flow over the head, perhaps as an emblem most easily seen in the dimness of an underground tunnel, averting collisions and misunderstandings. There are few vocal sounds associated with badgers, even though they live in groups: they lack even an alarm call. As the behavioral ecologist Hans Kruuk has written in *The Social Badger,* the species is "as inarticulate as any solitary mustelid." The tentative and undeveloped state of its social life, when compared with highly socialized animals such as hyena, coyote, or wolf, emerged in his long-term study of the badger clans of Scotland.

The Eurasian badger occurs throughout the Palaearctic, from Ireland to Japan, but is probably most abundant in northwest Europe, in Ireland, Britain, and Scandinavia. At current estimate, Ireland has, perhaps, some 250,000 adult badgers, a population coincidentally equal to Britain's, despite the great difference in the area of the islands. But eviction or extermination has left most of lowland rural Britain without them, so the mean density of the animals in Ireland is not exceptionally high.

The question of how the badger reached Ireland invites familiar options and objections, discussed in the earlier chapter on postglacial colonization. Archaeology finds the badger present by the Neolithic, when it would have been valued for meat, fur, and medicine. Deliberate introduction is appealing as an explanation—but one that cools on reflection. Badgers can be formidable fighters, and more than a match for dogs urged into their setts. The repeated capture, sea transport, and introduction necessary to establish a wild population appears a heroic enterprise. Given the special empathies with animals, magical and otherwise, imputed to Stone Age culture, a traffic in tamed cubs may seem more possible.

From this distance in time, what intrigues is the lifestyle of badgers 5,000 years ago, before the development of agriculture brought changes in the animals' food supply. The typical picture of a badger clan in Ireland today is of a group of about six adults, living in a main sett of interconnecting burrows, excavated deep in a dry bank and with half a dozen or more entrances; sometimes many more. Each group has several smaller, outlying setts within a strongly defended territory of perhaps 500 acres (200 ha). Earthworms are the staple diet, seized on the surface of short-grazed pasture on a moist or dewy night, and their supply is probably crucial in deciding the size of the clan's living space.

At one prosperous extreme, as in the hilly farmland of southwest England, badger groups may number up to twenty or more adults, territories are 125 acres (50 ha) in size, and a single sett, developed over hundreds of years, may have forty or fifty entrances (forty-four holes was the record in the Irish national badger survey conducted by Chris Smal in 1995). But in wilder, less productive country, the animals revert to a more solitary and wide-ranging lifestyle. The badgers studied by Kruuk on coastal moorland of the Scottish Highlands were smaller than their farmland counterparts and roamed between small setts or even slept under boulders. They eat sheep carrion and rowan berries *(Sorbus aucuparia),* and even the earthworms in their diet are of a smaller species, the red worm *(Lumbricus rubellus)* rather than the long and glistening common earthworm *(Lumbricus terrestris)* so abundant in the lowland cattle pastures.

In Ireland, also, most of the badgers of the rockier, boggier counties of the west lead leaner, more solitary lives and travel farther to forage for more scattered and varied kinds of food. "If there had been elephants in Europe," wrote Kruuk, "it is likely that, sooner or later, they would have turned up as an item in a badger's diet." In dry Mediterranean habitats, without earthworms, badgers are solitary animals, subsisting on insects and fruits. In Ireland and Britain, too, badgers dig out the nests of bumblebees and wasps, especially when a dry summer sends earthworms into estivation beneath the surface, but worms are its consistent and specialist prey, and this was probably as true in the Neolithic as today.

Surviving the clearance of forests and reordering of boundaries that went with the colonial plantations of Ireland, the badger is discovered in the early eighteenth century ensconced in "strong earths" and already accustomed to disturbance and persecution. In his huntsman's manual of 1714, Ulster's Arthur Stringer noted that "sometimes huntsmen do hunt a fox into an earth where a badger is, and then do commonly, if possible, dig him out." He gives hints and tips on this procedure ("encourage the terrier, and . . . keep him lying at the face of the badger") but otherwise disdains the animal as "a very melancholy fat creature [that] sleeps incessantly, and naturally (when in season) very leacherous [sic]."

The gamekeepers of country estates systematically persecuted the badger as part of their general paranoia about predators, even though the animal's only serious victims among vertebrates are the hedgehog and the rabbit, neither of much concern to sporting gentry. By the mid-nineteenth century, the badger was considered practically extinct in England and a declining species in Ireland. The Famine, however, and the mass emigration that followed, depopulated many areas of countryside, and large tracts of land that had grown potatoes and oats were left to revert to permanent pasture. World War I and political and agrarian unrest in Ireland made trapping an irrelevance and helped promote a resurgence of the wildlife known as "vermin." By the late 1930s the naturalist Charles Moffat wrote of the badger: "As with the fox, the increase has been so great as to lead to a somewhat insistent cry for the destruction of the animal, even in districts where it had previously been regarded as worthy of strict protection." The newly formed Connemara Vermin Destruction Association, in a region that would seem the least likely to nourish a plague of badgers, claimed to have destroyed 120 in its first year.

World War II—or "the emergency," as it was known in the neutral Irish Free State—again reordered priorities, and the badger population settled to its normal, self-effacing equilibrium in the countryside. "When not exceptionally abundant," as Moffat recorded, "the badger seems . . . to have been looked on as very nearly harmless." It was not until the 1970s, when tubercular badgers were identified in dairy-farming counties of both England and Ireland, that opinion changed and persecution was renewed.

The bacterium that causes bovine tuberculosis *(Mycobacterium bovis)* is similar to the human germ *(Mycobacterium tuberculosis)* and was a significant killer of people until the arrival of antibiotics and the pasteurization of milk and its products. The continuing presence of tuberculosis in dairy cows has been of huge economic importance, and Ireland has spent several billion dollars in trying to eradicate the disease. Regular veterinary testing of cattle and the slaughter of up to forty thousand cows a year has done little more than contain the problem. The spread of infection from herd to herd, and its introduction with purchased animals, account for the great majority of breakdowns in cattle health, but the badger has been implicated as perhaps the main wildlife reservoir of the disease and the animal most likely to bring tuberculosis into contact with cows.

Badgers have an unusual ability to survive extremes of the disease and to go on living long into the stage of failed immunity. On national data so far, some 11 percent of Irish badgers carry tuberculosis, and thousands have been snared and shot in experimental clearances of disease-prone areas. In Britain, too, experiments in culling have continued, even though, in a new century, the low residual incidence of tuberculosis seem as intractable as ever and its mode of transmission from badger to cattle remains a mystery.

For almost three decades, scientific and public controversy erupted with each new report that appeared to scapegoat the badger but offered no progress in the reduction of tuberculosis (this notably in Britain, a society in which feeling for animal welfare is more acute and that has a wider base of independently funded scientists). But another result of the uncertainties about the role of badgers was an almost unprecedented scrutiny of the animal's social system and behavior in the field.

Kruuk's portrayal of group life and diet has supplied a framework for much of the study in both islands. The clan comprises a number of individuals, most of them related, which collectively inhabit an area and tacitly agree on its defense. They establish latrines at its boundaries and act aggressively toward trespassing neighbors. Yet they never go around in a pack, but remain, in Kruuk's words "a tight community of solitary animals." Their communication seems limited to the maintenance of a clan smell: a huge gland under the tail, filled with a creamy paste, is used in a mutual squat-marking in which the animals swap bacteria. The gland is quite unknown in other mustelids and could, according to Kruuk, have evolved as an adaptation for communicating in the group territorial system.

The large main sett is the center of the territory: the place where they meet, often spend the day together, and where the cubs play. The badgers have dug it as a communal effort, but acting alone and at different times. They sleep in a huddle for warmth, yet forage alone. Only the dominant female in the clan gives birth to cubs, and she cares for them on her own; the low-ranking animals in the group often spend the day at a distance, lying up below ground in subsidiary setts.

Finding the right kind of soil, well-drained and friable, in which to dig the main sett is obviously crucial to badger settlement. In fact its importance as a vital resource has been suggested as an alternative to food supply in determining the size of the group and its territory, the density of clans, and even as "an ultimate reason for the evolution of sociality and territoriality in badgers." This argument, presented to a special conference on the badger in the Royal Irish Academy by the British biologist T. J. Roper, has had a worrying significance for wildlife conservation. With so little woodland available, most badger setts are dug into the wide Irish hedgebanks. Many of these, with their trees and wild herbage, have already been bulldozed to make larger fields, and main setts have frequently been dug out or blocked up by farmers. The new indictment of the badger is also used to vindicate the organized digging and baiting of the badger with dogs: a traditional, though illegal and clandestine, gambling "sport," both in Ireland and Britain.

The path of tuberculosis infection from cattle to badgers seems straightforward: earthworms and dor beetles *(Geotrupes stercorarius)* are major decomposers of cow dung, where foraging badgers seek them out for food. The reverse path is still stubbornly obscure: in controlled experiments, cattle fenced in with tubercular badgers proved remarkably slow to catch the disease. Decades of costly research have still to provide proofs of cross-infection to the level demanded by conservationists. Most badgers actively avoid cattle in fields at night, and better fencing could keep cows away from the badgers' boundary latrines and the pathways where they urinate, but most farmers do not see preventive costs as worthwhile. The nosing of sick or dying badgers by curious cattle is one widely credited route of infection; the use of barns and drinking troughs by badgers is another. Differences in cattle densities, or in distance from herd to herd, could also be crucial to any cycle of disease, as the wide variation in the prevalence of tuberculosis between cattle in Northern Ireland and those in the Republic has suggested (the incidence of the disease in the North has generally been lower).

An alternative to culling infected badgers, or even a general extermination of badgers in dairy cattle areas, would be to give them an oral vaccine, like the one that has eliminated rabies in foxes over wide areas of continental Europe. A tuberculosis vaccine developed in Britain was tried out for five years among badgers in West Cork in the 1990s; they were happy to consume it in biomarked, chocolate-coated baits. It was hoped that the sick animals among them would quickly die and that vaccinated animals would multiply to replace them. At the end of the trial, however, the population's tuberculosis levels showed no improvement, perhaps because the vaccine could not survive the badgers' robust digestive juices. The level of tuberculosis breakdowns in local cattle, too, was finally just as high. But research on the vaccine continues vigorously, now that a way to identify the dosed animals has been proven in the field.

The intensive research into badger ecology has thrown up some unexpected insights into the animal's role in woodland ecosystems. The stinkhorn *(Phallus impudicus)* is common in Irish woods throughout the year: this somewhat obscene-looking fungus first forms off-white, rubbery "witches' eggs" about the size of a large hen's egg and then erects a penislike fruiting organ. Its conical cap is coated with a sticky, olive-colored gel that gives off a sweet stench like rotting carrion. It can be smelled 33 feet (10 m) away and attracts several species of blowfly. They feed on the spore-laden mucilage, which is laxative, and defecate rapidly in the vicinity, depositing large numbers of spores.

Stinkhorns seem to concentrate remarkably around main badger setts, a fact noted by the research team from University College, Cork, led by Patrick Sleeman, that has been studying badger behavior. The concentration can be explained by the frequent death of badgers below ground, especially as cubs. One carcass may produce almost sixty thousand blowflies, so dispersal of stinkhorns fits into a three-way benefit that feeds the flies, removes potentially dangerous carrion, and spreads the fungus.

Compared with the stocky wrestler's body of the badger (Linnaeus first thought it might be a bear), that of its fellow mustelid, the Eurasian otter, matches strength to streamlining. The tail is long and powerful and the length of the male, at about 54 inches (130 cm) from nose to tail point, makes it Ireland's largest land mammal next to the deer. There is, however, a great variation in actual size and weight, from a 35-pound (16-k) male with a large, broad head to a 11-pound (5-k) female with neater and more sensitive features.

As its name suggests, the Eurasian otter has an enormous geographical range, from Ireland to Indonesia, Lapland to northern Africa. But there are now some striking gaps in its distribution, with a virtual extinction in Japan and much of Europe. Only at the Atlantic fringe, and notably in Ireland, does it remain a common animal of rivers, lakes, and undisturbed coasts. It still fishes near the heart of most estuary cities and towns—even, by night, in Dublin itself. The presence of otters is easily monitored from their distinctive droppings (or "spraint") at prominent territorial sites at the waterside, and a survey of one large catchment in Munster recorded otter signs at 97 percent of 195 sites, the highest positive result in any distribution survey.

Actual sightings, however, are uncommon away from the quieter shores of the west and south. Here otters often fish by day (usually at low tide when the bottom-dwelling fish are least active and most easily reached) and bring their larger prey out on to rocks or up into sand dunes. I once watched a hungry raven harassing such an otter at the crest of a dune until it took its flounder back to the sea. The splayed, rather catlike tracks that otters leave in the sand between the waves and coastal lakes and ponds point to a frequent pursuit of eels and trout, but also to the otter's need to take regular baths in freshwater. This washes the salt from its pelt, maintaining the air-trapping fluffiness of its underfur, so essential for insulation in cold Atlantic water.

To watch an otter fishing in a cold, rough sea in winter is to marvel at its robust lifestyle, yet Kruuk, who has made intensive study of otters along the shores of Shetland, north of Scotland, stresses the inherent vulnerability of their "life at the edge of the precipice." They are strung out thinly along a linear habitat (perhaps two otters per mile) and travel long distances daily, back and forth, to get their food. They cannot dive too deeply, or too far out, yet their fishing is so costly in energy that they have to catch a lot in a very short time.

On the other hand, otters at the coast swim generally in unpolluted water, they rarely need to cross a busy road, and the shoreline is often rich in undisturbed sites for their holts. These are excavations rather like badger setts, with multiple entrances, dug in peat, sand dunes, or soft banks, their chambers often lined with seaweed or plastic flotsam. Male otters lead largely solitary lives, often sheltering in a holt at night, and some of the holts are used by females to produce two or three cubs who may then keep her company for months.

Inland otters tend to lead far more nocturnal lives and locate their holts more discreetly, often among the roots of large deciduous riverbank trees such as ash and sycamore. Some have entrances underwater; others are burrows originally dug by badgers, foxes, or rabbits, and some maternity holts may be quite distant from the waterside. The remedial drainage of many Irish rivers, widening and deepening them, raking back their banks and stripping them of trees, has cost otters not only their refuges but also much of their food supply as fish populations fall. Farm pollution that produces fish kills and a steady decline of water quality must also raise mortality levels, but mostly in an unseen way. Of some six hundred dead otters that figured in a national survey, more than half were killed on the roads, most of them the more mobile and adventurous adult males.

Deaths incidental to human activity almost certainly eclipse those still caused by "sport" hunting with packs of otter hounds (ostensibly now with mink as the quarry). Before licenses for such hunting were finally withdrawn in 1989, kills were typically small—perhaps nine or ten otters a year between the four Munster packs—and some wildlife scientists actually regretted losing such regular samples for the monitoring of local populations.

The otter's unchallenged niche along Ireland's waterways and coasts has been under invasion by a much smaller, but quite as determined, mustelid predator. The American mink was first introduced into Ireland for fur farming in the 1950s, and through escapes and releases very quickly showed its affinity for Irish rivers and lakes. By 2000 its final advances to the coast of Connacht had given the mink a feral presence in virtually every corner of the island. Its progress has often been marked by dynamic population explosions into virgin areas, with wholesale predation on waterbirds and wildfowl, breeding colonies of terns and gulls, and nesting waders. These local assaults on fish farms, game birds, and poultry have often been dramatic and shocking; however, the density of mink will settle down over three or four years to the carrying capacity of the territory.

The highest populations are along the shores of slow-flowing midland rivers, where reed beds give cover to the animals at perhaps two per mile of reedy shoreline. Abundant freshwater crayfish *(Austropotamobius pallipes)* join eels, smaller fish, frogs, and birds in the diet. It was feared at first that the mink might displace the otter in river habitats, but while they do share some foods, the otter's superior speed and underwater vision make it the better hunter of bigger, faster fish. On some rivers and lakes, the droppings of mink and otter can be found almost side by side, such is their apparent coexistence.

At the west coast, the mink again picks a different niche among resources, taking crabs and fish from rock pools and finding more of its food on land (rabbits among sand dunes, for example). However, the mink can swim across at least 2 miles (3 km) of sea, giving it access to island colonies of seabirds. At the coast, also, it seems likely to threaten the scarce and specialized dipper *(Cinclus cinclus)*, which hunts insects underwater in hill streams right down to sea level.

Freshwater crayfish that form part of the diet of feral mink.

Whatever fashionable colors might have been produced by selective breeding in captivity, mink in the wild in Ireland seem predominantly black. The wide wake they leave when swimming in calm water and their bold, bounding progress through marsh and reed bed add to a distinctive image in the countryside. They are certainly less nervous than the pine marten, the native Irish mustelid closest to their size (about 24 inches, or 60 cm) and with, historically, an equally covetable pelt.

This beautiful animal, with chestnut fur and creamy bib, was discussed earlier, in Chapter 8. The rocky region of the Burren still provides its most notable modern refuge. Since its major decline in the mid-1900s, when poisoned carrion baits set by sheep farmers for foxes and stray dogs took a heavy toll of the near-omnivorous pine martens, the species' population has recovered and expanded. There are now pine martens in about half the counties of Ireland, mainly in mature woodland but also in the transient scrub of clear-felled conifer forest. They are not specially shy of humans, and have been seen rummaging for food in the refuse bins of golf clubs, a hint, perhaps, of the potential of the "tree cat" as a troublesome Irish "raccoon."

Half the size of a pine marten and appearing as a sudden, thrilling ripple on a dry-stone wall or road bank, the Irish stoat *(Mustela erminea hibernica)* is at once the smallest, most elusive of the island's mustelids and the one most likely to hold its ground to return, with impassive brightness, the gaze of the human observer. As a mammal, after all, it has probably had the longest continuous presence on the island, exceeding that of people by perhaps two thousand years. The problem set by such an early presence (namely what it would have eaten at that sparsely furnished period) is discussed in Chapter 4.

What is certain, however, is that the Irish stoat is now an endemic subspecies, shared only with the Isle of Man. In the north of Ireland, stoats are markedly smaller than in Britain, but the size increases as one goes south, so that a Munster stoat is nearly as big as one in Cornwall, across the Irish Sea. Along with a general reluctance to turn white in winter, the Irish stoat differs in another way from its British and Continental counterparts: the line where the chestnut-brown upper coat meets the white fur of the belly is wavy instead of straight.

The fact that most rural Irish people persist in calling the stoat a "weasel" must not complicate the story. Britain does have the weasel—not quite as small as North America's least weasel *(Mustela nivalis)* but substantially smaller than a stoat—and Ireland does not. The one certain diagnostic feature that marks the Irish stoat from a weasel on sight is the conspicuous tuft of black hair that has evolved at the tip of its tail, a device that may serve to confuse the aim of a swooping hawk.

The stoat is becoming more common in Ireland as rabbit populations develop resistance to the introduced myxomatosis virus and provide a more dependable food supply. Most human sightings are in open country, yet Sleeman's study of radio-tagged stoats charted a clear preference for woodland. It also found them spending two-thirds of their time resting in underground dens and much of their hunting time up trees, in pursuit of birds and nestlings. Stoats are a totally protected species, no longer persecuted for their traditional threat to unguarded farmyard poultry and free to enrich country lore with sometimes bizarre behavior. "There are few more entertaining mammals in Ireland," promised a recent Heritage Service handbook, "particularly when stoats are seen in a family party." James Fairley, on the other hand, while quoting engaging descriptions of stoats at play, added others that describe the more militant maneuvers of stoat families hunting together in autumn, including an attack on a woman and her collie dog in County Donegal in the 1960s. "A stoat pack," he warned solemnly, "is not to be trifled with."

SELECTED REFERENCES

Fairley, J. 1984. *An Irish Beast Book.* Belfast: Blackstaff.

———. 2001. *A Basket of Weasels.* Belfast: Author.

Hayden, T., and R. Harrington. 2000. *Exploring Irish Mammals.* Dublin: Town House.

Kruuk, H. 1989. *The Social Badger.* Oxford University Press.

———. 1995. *Wild Otters: Predation and Populations.* Oxford University Press.

Macdonald, D. 1995. *European Mammals: Evolution and Behaviour.* London: Harper-Collins.

Sleeman, P. 1989. *Stoats & Weasels, Polecats & Martens.* London: Whittet.

—— *Chapter 17* ——

Making Room for Nature

In the first decade of Irish independence, the brilliant goldfinches of the struggling Free State were still preyed on by people who boiled up holly bark to make bird lime, which they poured like glue over bushes and thistle-heads. They called the birds down with decoy finches and later dispatched them, caged, to England, along with kidnapped skylarks and linnets *(Carduelis cannabina)*.

The Protection of Wild Birds Act 1930 put all this beyond the law, much to the disquiet of a young Dublin politician Sean Lemass. "If the economic situation becomes better," he said in a Dail debate, "we can then afford to indulge in luxury legislation of this kind, but we must put the necessities of human beings before those of wild birds." Thirty years later, as Taoiseach, or prime minister, Lemass led Ireland into a new program of economic development and then into membership of the European Economic Community (now the European Union, EU). Words like "amenity" and "heritage" crept into Ireland's physical planning, and a new concern for wildlife species was elaborated in a steady flow of conservation directives from Europe.

But in the rural population, environmentalists are still widely regarded as interfering, city-based do-gooders, and nature conservation is still largely identified with an Anglo Irish culture. A political reflection of this may be Ireland's low ranking in international conservation league tables: only 1 percent of the national territory is given the strict protection of national parks and reserves, compared with an average of 12 percent in other developed countries.

Among the Republic's own initiatives, however, in 1981, was a natural heritage inventory of sites worth conservation. Their designation as Areas of Scientific Interest did little to promote popular understanding of the need for controls on development. As ecologist John Feehan has put it:

> The drawing of boundaries on maps and describing of biota inside the reserves is the easy part. The more difficult challenge is to convince those who own these precious pockets of land—those surviving patches where some whisper of the richness of this island as it was before our arrival still remains—that their tenure is not absolute, that they are guardians of a heritage that belongs to the whole community.

Such an ideal, however, has to contend with the fierce possessiveness of land.

A change of name, from Areas of Special Scientific Interest to Natural Heritage Areas, was an attempt to widen the appreciation of their value, but did little to change attitudes. From some 1,200 of these designated areas, a top-priority list of Special Areas of Conservation were chosen as the Republic's contribution to a European network of outstanding nature conservation sites known as Natura 2000. Farmers were often suspicious of new restrictions on land use, even with compensation, and Dúchas, the state heritage agency, adopted a conciliatory, informal style in local boundary negotiations in the hope of good long-term cooperation. But the mainly urban-based conservation organizations have been severely critical of the number and range of sites put forward in response to the EU Habitats Directive—some 360 by 2000—and Europe, too, showed deep impatience with the slow pace of progress in Ireland, as in other EU countries.

A substantial shadow list of additional sites proposed by the conservation nongovernmental organizations (NGOs) to the EU's environment commissioner showed special concern for river species such as otter and salmon and for habitats that would serve as "stepping stones" for wildlife between the major conservation zones. As one example, an inventory of fens carried out by the Irish Peatland Conservation Council found sixty-five new conservation-worthy fens, unlisted by Dúchas and under threat from drainage and landfill. Responding to the NGOs' appeals in 2002, the EU called for more protected sites in Ireland in thirteen kinds of threatened habitats, including orchid-rich grasslands and alkaline fens.

At that stage, the proposed Special Areas of Conservation covered almost 2,471,000 acres (1 million ha)—some 10 percent of the Republic's land and lake area—together with 4 percent of marine areas (including, for example, the whole of the Shannon estuary). Their acceptance by farmers has been eased considerably by extra cash grants offered in the Rural Environment Protection Scheme (REPS), devised by the EU in reform of its common agricultural policy. In a swing away from subsidizing intensive farming, which produced huge surpluses of food at great cost to the environment, REPS pays farmers to manage their land in a way that satisfies society at large. Management plans worked out with local advisers match livestock numbers to the land, reduce chemical fertilizers, protect streams and wildlife habitats, rehabilitate hedges and field walls, find corners for trees, and in general realize the popular (perhaps mainly urban) image of a clean, green, nature-friendly countryside. Extra incentives to promote organic farming have matched an unprecedented consumer demand for food grown or reared without synthetic chemicals, hormones, or genetic modification. At the end of 1999, some forty-three thousand Irish farmers had enlisted in REPS, most of them with holdings of fewer than 100 acres (40 ha).

It was headage payments on sheep that led to such destructive overstocking of western hills. On the overgrazed peatlands, where root mats are not too badly damaged, the surface vegetation stands some chance of repair. But a bog cut away for fuel or gardeners' peat mixes, drained and planted with conifers, or plowed up for grass is gone forever. Conservation of a range of bogs in the Republic has come gradually, in response to the EU Habitats Directive and lobbying by the Irish Peatland Conservation Council. By the year 2000, the remnants of intact raised bog had dwindled to fewer than 22,000 acres (9,000 ha), all zoned for protection. On many of these, traditional turf-cutters continued to harvest peat for their home fires, resisting all offers of compensation to phase out their activities.

The main inroads on the blanket bogs of the hills and Atlantic moorlands have come from mechanical peat-cutting and planting of conifers. Along with the new Special Areas of Conservation, some 74,000 acres (30,000 ha) of bog have already been safeguarded, mostly within national parks. An additional 25,000 acres (10,000 ha) of mountain and Atlantic bog, much of it grossly overgrazed by sheep or forested with conifers, falls within the new Mayo National Park in the Owenduff–Nephin Beg region of the county. The failure to curb overgrazing in this upland refuge of the red grouse earned Ireland censure in 2002 by the European Court of Justice. It was responding to complaints by the European Commission about environmental lapses by several EU-member states, but under a "first offense" clause, exacted no financial penalty. The commission, meanwhile, has continued to pursue the Republic for what is sees as inadequate protection of bird habitats, especially in coastal bays and estuaries.

As new programs emerge to control sheep numbers on the uplands, their very presence on the hills is coming under challenge. "Restoration ecology" projects initiated by Dúchas have been reaching into the past for alternative patterns of hill farming that could replenish and sustain a richer natural biodiversity. Replacement of cattle by sheep in the uplands is less than half a century old, and the sheep's selective grazing has had a different impact. On Mangerton Mountain in Killarney National Park, a decade of trials with small, black Kerry cattle (now a rare breed in national terms) has found them reduc-

Mountain slope stripped of vegetation by overstocking of sheep.

ing the coarse and dominant purple moor-grass and heavily trampling the bracken, while soft grasses and heather, suited to hares, red grouse, and deer, have begun to expand.

Other trials aim at restoring native hill woodland, including Scots pine, and vanished dwarf shrubs such as juniper and cowberry *(Vaccinium vitis-idea)* have been reared in Killarney's conservation nursery. The mix of light-weight grazing with reconstructed woodland now promises to transform the park's bare slopes and perhaps to offer both an economic and ecological model for the western uplands as a whole. The director of these projects, Rory Harrington, has taken his inspiration from the dense woodland vegetation surviving on the many lake islands of Kerry. He accepts that a planted restoration is resisted by many ecologists as "wildlife gardening," but argues that seed banks for many lost plants (such as *Vaccinium* species) have been devastated by thousands of years of burning and overgrazing.

Transformation of Ireland's landscape on a scale and at a rate unprecedented in its history is being brought about by commercial afforestation. Already, great tracts of densely planted conifers blanket many upland bogs, to the general dismay of naturalists. Now the state intends that forests shall cover one-fifth of the countryside by 2030. Grant schemes are shifting the initiative from state forestry to planting by farmers and corporate interests, and, despite a weighting of grants to favor broadleaved trees, the great bulk of new plantations, including those of Coillte, the state forestry company, will continue to be of short-rotation, fast-growing conifers from the Pacific Coast of North America.

New ecological forestry standards, developed in Europe after the Rio Earth Summit of 1992, have brought impressive pledges on biodiversity from Coillte and the Forestry Service, but the shadow of coniferization continues to creep across the island. The state-funded National Heritage Council has joined conservation NGOs in urging the planting of conifers and broadleaves in equal numbers and in mixed woodlands.

After generations of indifference, there is a fast-growing liberation of the ordinary human feeling for trees and the networks of life they support. A groundswell of voluntary planting of native species has been fostered particularly by the NGO called Crann (the Irish word for tree). The new century has also brought generous state funding to restore long-neglected fragments of native woodland and to plant new woods with native species. Some existing woods are in nature reserves, and others have been salvaged and rehabilitated by Coillte from old Anglo Irish estates planted with conifers half a century ago. These are among the People's Millennium Forests in which every Irish household has a (nominal) certificate to a tree. Still more grants have been directed to native woodland in private hands and often unfenced, heavily grazed, or choked with rhododendron.

All are potential reservoirs of biodiversity, preserving the ancient seed stock of native tree species and woodland plants such as bluebells, wood anemones *(Anemone nemorosa),* and wild garlic *(Allium ursinum),* which have flowered in woods of oak and hazel for many thousands of years. Only the most ecologically depleted woods will be replanted, with a mix of native trees, and some of these will be planned for low-intensity, close-to-nature forestry, with selective felling and a continuous cover. Others will be riverbank plantings of mixed native trees to become new wildlife corridors, enriching aquatic ecosystems in the progress.

Such advances have been eased by the recent change in Ireland's economic fortunes. But this also brought new and deeply significant expectations for the small-farm families who effectively manage so much of the conservation countryside. Well-paid work on construction was suddenly abundant in small towns and holiday resorts within daily commuting distance. In scenic small-farm areas, demand for sites for building bungalows has overcome traditional reluctance to part with land. For the first time in Irish history, a pattern of part-time farming has taken on solid reality, and the "green" concerns and restrictions of REPS have seemed less inhibiting or unreasonable.

The ecological implications will take time to appear. Smaller herds and flocks and less chemical fertilizer should bring a resurgence in plant and insect diversity (including, quite probably, a striking advance of bracken, thorn-scrub, and thistles). More farmers will plant part of their land with conifers, often at great cost to wetland and scrub habitats, or use higher grants to plant broadleaf trees in conservation areas. The sale of sites for housing has brought a suburban intrusion to many hitherto open moorland roads, and ribbons of bungalows reaching out from towns and villages have destroyed long stretches of hedgerow, field bank, and wayside wetlands. The creation of new gardens filled with trees, shrubs, and flowers (most of them exotic) may seem some compensation, even an enrichment, but they draw Ireland's biodiversity yet further away from stable and indigenous ecosystems.

As the population rises to service economic growth, new housing and road building will continue to fracture natural habitats and set barriers across the leafy or watery links between them. Wildlife needs room to travel, to meet and mate, swap genes within species, and recover from change and disaster. Tracts of the countryside preserved for nature are fast becoming islands in a sea of inhospitable, intensive farming and rural suburbia. The smaller a reserve, the fewer species it can sustain, and interchange between colonies gets harder all the time.

Ecological corridors and stepping stones are an important part of European conservation strategies, and an ambitious Pan-European Ecological Network was endorsed by all EU ministers of environment in 1995. In one tentative network sketched out for Ireland, an ingenious, if sprawling, long-distance corridor would link Connemara with the Mourne Mountains by way of the Burren, the Shannon callows, and surviving midland raised bogs; another would stretch up the northwest coast from Achill Island to Malin Head. But such regional corridors seem more appropriate to countries with wide-ranging mammals or deep-forest birds, and their conservation potential for Ireland is clearly limited.

In the island's more intimate landscapes, species to be conserved under the EU's Habitats Directive are often small and slow to disperse (bog mosses, for example, or tiny fenland snails) and the priority is to save original habitat on a scale that secures continuity. The island's most threatened butterfly, the marsh fritillary, is tied to damp meadows and hillsides where its food plant, the devil's-bit scabious, grows abundantly: connections between their small and isolated populations become vital corridors. For the lesser horseshoe bats of Clare, particular hedgerows link their roosting places with woodland feeding grounds.

Such closer focus seems more fitting to Ireland's ecological fabric. It also reinforces the importance of river and canal banks, railways, eskers, hedgerows, and road verges as local wildlife corridors and refuges threading an increasingly built-up countryside.

At the reelection of its Coalition Government in 2002, the Republic had still to absorb its National Biodiversity Plan, prepared under the United Nations Convention on Biological Diversity ratified by Ireland in 1996. With its priority lists of species and habitats to be protected, the plan may have unwelcome implications for farmers and other landowners and for the policies and practices of government departments usually not concerned with nature conservation.

Northern Ireland, as a region of the United Kingdom, published its first program of biodiversity action plans (for the Irish hare, chough, and curlew) in 2000. The program spelled out the logic of shared strategies in conservation across the island, a cross-border cooperation already shown in jointly prepared *Red Data Books* of threatened species and in deciding on Special Protection Areas for Birds.

The somewhat more orderly advance of conservation in Northern Ireland has been reflected in forty-five statutory nature reserves up to 1992 and a clear policy of substantial protection for remaining areas of raised and blanket bog and fens. Designation of Lough Neagh as an Area of Special Scientific Interest (ASSI), with almost 98,840 acres (40,000 ha) of water and wetland, made it the second largest ASSI in Ireland and Britain after the Wash, the great bird wetland of England's east coast. The two main tools of habitat care in Northern Ireland both return compensatory payments to farmers. Designation of Environmentally Sensitive Areas (as in the Mourne Mountains and the Antrim Glens) works broadly in the same way as the REPS system in the Republic; the more recent Countryside Management Scheme rewards farmers for taking special care of particular bird habitats (for example, in avoiding cutting field rushes on curlew and snipe territories).

In the medley of conservation acronyms current in the later twentieth century, the AONB, or Area of Outstanding Natural Beauty, was found only in Northern Ireland. This frank and protective admiration of landscape was extended to the glens and lofty headlands of the northeast coast and to the Mourne and Sperrin Mountains. It derived from the planning code of UK administration and a largely anglicized conservation culture.

The early years of designations such as AONB were marked in the north by strict control of building in the open countryside and attempts by planners to concentrate new housing in existing villages and towns. As in the South, such aspirations soon succumbed to local political pressures and were eventually fulfilled only in the "most sensitive" landscapes. The Republic's dramatic economic growth produced an urban housing crisis in which commuters reach farther and farther into the countryside, and the burden of new planning applications has often overwhelmed local authority resources.

As the new century began, the Government Exchequer, enriched with budget surpluses, prepared to fund new roads, tunnels, and other infrastructural works on an unprecedented scale. Millions in state funding were used for the first time to prime research and development in biotechnology, aquaculture, and other marine enterprises. Rural population seemed destined to rise to densities more familiar in England or Holland, and tourist housing in scenic areas was already overrunning some traditional villages and towns. Recreational use of mountains, lakes, and seashores was making a significant ecological impact for the first time, with tourist trampling of hill paths a particular problem for the national parks.

The implications of all this for wildlife and its habitats have made Irish ecologists doubly grateful for the exacting conservation standards and structures emanating from the European Commission. Only in the supranational framework of the EU could such an objective, scientific consensus have been achieved, and without its external authority (and the structural funding that often hinged on compliance), Ireland's natural heritage would have been consumed in compromises with development. Few impacts of Ireland's European membership have been as far-reaching as the insistence on concern for nature, or are linking so many separate agencies into a network of mutual consultation and support.

Outstanding among them has been the state-funded Heritage Council, whose advisory brief extends to "the totality of a landscape," from flora and fauna to archaeology and heritage buildings. Among its many valuable interventions has been commitment to a national biological records center for the Republic, similar to that established for Northern Ireland by the Ulster Museum in Belfast. A central data bank of species and their distribution is essential to conservation strategies and the monitoring of biodiversity.

In a country now transformed by full employment and more generous national resources, and one claiming to value its green image, the general aims of conservation should be more easily achieved. But regional inequalities persist, and areas that have shared least in the new prosperity also tend to be those of highest conservation value. The underpinning of small-farm livelihoods through schemes such as REPS seems essential to protecting the intimate weave of meadow, hedgerow, and stream-side habitats, along with the broader sweeps of mountain and blanket bog. The farmers, on the other hand, so ready to accept the mantle of "traditional guardians of the landscape," need to accept the right of the state to sustain natural habitats and biodiversity and the right of the taxpayer to reach and walk the hills and the seashore.

THE NATIONAL PARKS

Outside of great wilderness areas such as east Greenland or southern Chile, most national parks preserve landscape that bears, intact, some essence of the pristine, even if the human hand is not too far away. The brilliant meadow floor of America's Yosemite Valley was the creation of regular burnings by its original Ahwahneechee occupants, but that does not make it any less sublime or ecologically precious.

Ireland's parks are small by most national standards, given the mountain landscapes they protect. At Killarney in Kerry and Glenveagh in Donegal, their settings are redolent of former colonial wealth and romantic taste. But their crags, lakes, and waterfalls still have the power to prompt an exalted response to nature, and they provide an exciting arena for ecological restoration and education. Their first purpose is to conserve natural plant and animal communities and scenic landscapes, a statement of aims directly inherited from the act of 1916 that established the U.S. National Parks Service. It bears repetition in an Ireland that can covet wild landscape for other local tourist amenities such as golf or water-skiing.

The nucleus of Killarney National Park is a 10,625-acre (4,300 ha) family estate with a Victorian mansion—Muckross House, bought by an American, William Bowers Bourn, as a wedding present for his daughter, and subsequently presented to the state in 1932 as a memorial to her. The house and its gardens are now a museum of the lifestyle of the landed gentry, including traditional farms that present a working model of Irish farming in the 1930s, before tractors and electricity.

Now much extended, with its famous lakes and tracts of mountain assembled from other estates, the park broadly coincides with the native oakwoods and mountainous deer range described in Chapter 13. It sits in the foothills of a range of conical peaks known as Macgillycuddy's Reeks, with Ireland's highest mountain, Carrauntoohil, rising among them to 3,410 feet (1,039 m). This backdrop of dramatically glaciated sandstone descends to a valley floored with limestone, sculpted into promontories and islands in Killarney's Lough Leane.

Glenveagh National Park, in the Derryveagh Mountains at the northwest of County Donegal, also has at its heart a lakeside Victorian mansion, in this case a Gothic highland castle, surrounded by gardens, presented to the state in 1983. The estate had a nineteenth-century history of harrowing evictions of tenants. Its owner then, a land speculator named John George Adair, died while visiting his ranching interests in America, and his widow developed deer stalking at Glenveagh with animals imported from Britain. The American links continued through its next two owners, a Harvard professor of fine arts, Arthur Kingsley Porter, and an Irish American of similar interests, Henry McIlhenny, who sold the estate for creation of the national park and then gave the castle and gardens to go with it.

A herd of 450 red deer of mixed ancestry has been the main wildlife focus of the park, but the year 2000 also saw the beginning of a long-term project to reintroduce the golden eagle *(Aquila chrysaetos)* after almost a century's absence. In the Ireland of the mid-1800s one could have seen a dozen eagles in a day in the mountains of Kerry, and some fifty traditional breeding sites were scattered down the Atlantic seaboard from Donegal to Cork. The main work of extinction was carried through in the later 1900s, as gamekeepers and shepherds waged war on the birds with rifles, traps, and strychnine, and collectors vied for the last eggs and young. The last two pairs of native birds nested on the crags above Glenveagh in 1910 and on the remote cliffs of north Mayo in 1912. Ireland is the only country in the world to have lost its golden eagles in such recent times.

Nonbreeding birds from the Scottish population of eagles (now much recovered, to almost one thousand birds) have occasionally wandered to Ireland, but their appearance has been too sporadic to offer much promise of natural recolonization. The reintroduction program at Glenveagh compels a heavy investment of young chicks: seventy-five of them, removed from eyries in the Scottish highlands and released in the park over five summers. It also required a study that compared the numbers of grouse, hares, and rabbits living in the home ranges of Scottish eagles to the prey available in Donegal. Glenveagh passed the test: big stretches of heather regenerating within the park give cover to the prey species, and the original nesting crags are virtually unchanged. But even with full cooperation from the hill farmers, predictably suspicious of such an awesome predator, natural mortality will be high: as many as 80 percent of eagle fledglings do not survive to breeding age. A successful outcome to the Glenveagh venture, so long a dream of Irish naturalists, would be three or four pairs of eagles soaring on the thermals between Muckish Mountain and Glendowan, and their offspring prospecting new summits the length of the island.

The gleaming, quartzite peaks of the Connemara National Park include four of the famous Twelve Bens that cluster at the west of County Galway. This was a region severely overgrazed by sheep in the closing decades of the twentieth century, so that the waist-deep heather so abundant on many of the park's steep hillsides is a dramatic demonstration of the biomass more natural to the region. The heather offers winter grazing to two small breeding herds of native red deer, brought from Killarney to build up a second genetic reservoir. The original red deer of west Connacht had been reduced to a dozen by the 1830s and died out soon afterward. In 1984 the roaring of a stag in the Kylemore Valley brought back a sound that had not been heard there in 150 years.

Much of the 5,000 acres (2,000 ha) of the Connemara park was originally part of the Kylemore Abbey estate (its Victorian castle, now a girls' school, is perched between a lake and the towering cliffs of Dubh Cruagh), and its southern swathe of mountain was once owned by Richard ("Humanity Dick") Martin, who helped to found Britain's Royal Society for the Prevention of Cruelty to Animals during the early nineteenth century. The park, entered at the village of Letterfrack, has been a particular focus of research into postglacial landscape history and bog formation. Some of its peatland is 16 feet (5 m) deep, concealing stumps of pine trees 4,000 years old, and the information from pollen studies has made the displays in the visitor center a particularly effective presentation of the blanket bog ecosystem.

In Kerry, Donegal, and Connemara, the creation of parks was facilitated by the gift or acquisition of large tracts of land held in private estates; those elsewhere have been patiently pieced together at the margins of farming, sometimes using as core areas land originally purchased for state forestry. In the Wicklow Mountains National Park, most of the central mountain spine, with its summits of glacially rounded granite and broad slopes of blanket bog, is already state-owned, and flanking tracts of state forestry will revert to seminatural vegetation once the conifers have been felled. But future conservation will depend also on management of the uplands in partnership with neighboring farmers. In a model of grassroots community organization, the Wicklow Uplands Council represents more than thirty interest groups with a stake in the future of the mountains.

In County Clare, the core of the Burren National Park is centered on Mullaghmore Mountain, where the strange, shattered beauty of the Burren takes on, for many visitors, an almost metaphysical power of silence and solitude. The construction of an official interpretive center, with attendant parking lots, at this magical heartland was passionately resisted by conservation groups and ultimately abandoned.

The Mayo National Park is the youngest and most ambitious, taking in a great sweep of Atlantic blanket bog and the panoramic Nephin Beg mountain range. Its 25,000 acres (10,000 ha) hold the largest remaining area of uninhabited landscape in Ireland, with a compelling sense of postglacial wilderness. The Owenduff River, almost entirely within the park, is the last in western Europe that drains such a large and intact blanket bog system, dotted with pools, lakes, and streams. On the mountain range, overgrazing of the blanket bog has eroded the peat, in places, to bare rock, but the ancient Bangor Trail, way-marked from Bangor Erris to Lough Feagh, remains an exhilarating test of stamina. The Mayo park holds great promise of long-term restoration and species enrichment now that grazing is under control.

CONSERVATION AND THE SEA

After long indifference to the ocean, and slow recovery from colonial repression of the fishing industry, Ireland finds itself embarked on a novel encounter with the marine world. With money to invest in science and technology, the Republic has woken up to its pivotal ocean location, and some 77,000 square miles (200,000 km²) of continental shelf with its teeming marine life are presented as a new frontier of prosperity.

With stocks of traditional whitefish reaching critically low levels in European waters, Irish trawlers have joined a new fishing frenzy at the edge of the shelf to catch new, deep-water species unprotected by EU fishery quotas: among them, such long-lived and late-maturing species as orange roughy *(Hoplostethus atlanticus)* and bluemouth *(Helicolenus dactylopterus dactylopterus)*, whose population dynamics are still virtually unassessed. Bottom trawling in the Porcupine, Seabight, and Rockall troughs has been damaging spectacular mounded reefs of Atlantic cold-water corals, living seabed cities up to 9.4 miles

(15 km) across, photographed for the first time in the 1970s. These circular reefs are found along the continental slope from Portugal to Norway, and their rich biodiversity, with records of more than 860 species of animals, give them a role in the deep-sea ecosystem quite as important as that of the tropical corals. They have been linked by geologists with methane seeps from unexploited hydrocarbon reservoirs, so that the Pan-European Atlantic Coral Ecosystem Study, launched in 2000, has become of great interest to deep-sea oil and gas exploration companies. The outcome of this urgent multidisciplinary project is of close concern to Ireland, and will be crucial to conservation of the seabed and sustainable human use of its natural resources.

At the inshore margin of the continental shelf, the 4,660 miles (7,500 km) of Ireland's much-indented coastline would seem to offer plenty of room for even the trebling of coastal fish farming and shellfish mariculture to which the government is now committed. But siting operations to allow a proper coexistence with nature immediately changes the picture. The inshore waters with the richest natural ecosystems are also the most productive for mariculture, so that many of the bays, which became Special Areas of Conservation under the EU Habitats Directive are already actively farmed with salmon cages and shellfish culture.

How much mariculture is too much? How is unacceptable impact to be defined? Rule-of-thumb limits, to food waste, chemical and genetic pollution, competition for plankton, occupation of the intertidal seashore, seabed damage from dredging, only begin to engage with the real impact of such intensive monocultures, and the effectiveness of such limits will change from one local ecosystem to the next. What will be the impact on the predatory attentions of seals, cormorants, and herons, or even the ordinary feeding habitats of seashore birds, and how can this be reconciled with goals of conservation?

As an island people begins, at last, to look seaward, the entrepreneurial adventure is shared at a popular level through the education of subaqua television and new aquariums and a sudden, affluent explosion of pleasure in water-based recreation. One can even be glad that the Irish awakening has taken so long: that it is happening in greener times and within a wider European concern for ocean ecology. Unlike the exploitation of natural resources on land, each new step in marine development now depends heavily on the initiative and expertise of scientists; to a degree unique in history, they mediate human intrusion on the sea. The urge to exploit and the need to conserve can be un-easy partners in "sustainable development," but at least that ethic now promises a steadying keel to the formidable program of marine investment launched by the Irish state as the twenty-first century began.

ON THE OTHER HAND: AN EPILOGUE

Few places remain in the developed world where a naturalist can rest easy about the future of habitats and species or of landscapes that conjure the spirit of the wild. In my coastal cul-de-sac, below a mountain and encircled by rock, rough grassland, and sea, I can live in something of a dream. But any journey to the interior is a procession into change, and the rising glow of town and suburb make my dark, starry nights an almost eccentric privilege. It is easy to let concerns for nature heap up into a self-righteous gloom.

Ireland is still a beautiful island, still sparsely populated where it counts. Despite the many wounds, there is still enough peatland to offer not just the fabric and wildlife of this fascinating landform but its great airy passages of silence and peace. Most rivers and lakes are still sufficiently pristine, their banks so undisturbed that the otter lives almost everywhere it could wish. It still swims the west coast in daylight, sharing grassy shores with the chough, a dashing and cheerful crow dispossessed from much of Europe. The diversity of Ireland in everything from rocks to weather, the intimate, intricate mix of habitats, is itself a treasure in a world preoccupied with largeness of scale and homogenous use of land.

Reflecting a harsh agrarian past in which poverty lived close to the soil and its resources, the Irish view of nature has been essentially utilitarian: what's it good for? The conservation ethic that grants intrinsic value to the rest of the world's species is still mysterious and dubious to many of the island's land-owners, and it is no bad test of a naturalist's convictions, not to mention his or her coherence, to decide why a snail the size of a pinhead or a liverwort best enjoyed through a microscope should be worth the loss of some lucrative new development. The EU Habitats Directive, however necessary, has been a crash course in the loftiness of scientific consensus.

On the other hand, new schemes for planting and restoring native woods are the most promising means yet through which the Irish people can claim ownership of a simple, Arcadian enthusiasm for nature, however modern its genesis may be. The communal gathering of acorns by the sackful for the propagation of local, native oaks has an air of ritual and remembrance as well as ecological amends.

Tucked away in the rocky folds of my hillside are shreds and remnants of woods that were never planted: they grew up and hung on naturally at the edges of human activity. Behind the bluffs at the shore are ferny grottoes of oaks, hollies, and hazel; in a ravine at the foot of the mountain, the oaks lean out over waterfalls and are joined by aspens that shiver in the wind. All over Ireland, there are patches of woods with this ragged, leftover look, as if they were grudged a living. Their adoption and nourishment will confirm a widespread will to live on better terms with nature.

SELECTED REFERENCES

Biodiversity in Northern Ireland: Northern Ireland Species Action Plans: Irish Hare, Chough, Curlew. 2000. Belfast: Her Majesty's Stationery Office.

Curtis, T. G. F., and H. N. McGough, eds. 1988. *The Irish Red Data Book 1: Vascular Plants.* Dublin: Stationery Office.

Stapleton, L., M. Lehane, and P. Toner, eds. 2000. *Ireland's Environment: A Millennium Report.* Johnstown Castle, Ireland: Environmental Protection Agency.

Whilde, A. 1993. *Irish Red Data Book 2: Vertebrates—Threatened Mammals, Birds, Amphibians, and Fish in Ireland.* Belfast: Her Majesty's Stationery Office.

——— Appendix ———

Nature Reserves in Ireland

The reserves, administered by Dúchas, The Heritage Service, are listed under their counties, which are given in alphabetical order. They are also marked on the Discovery Series of the Ordnance Survey maps. Leaflets are available in local Tourist Information Centers.

COUNTY CLARE

Cahermurphy Wood, east of Lough Graney in the northeast of County Clare, is an oak-wood with birch and holly.

Dromore Wood, near Dromore lake, 6.2 miles (10 km) north of Ennis, is a woodland and wetland showing transition from aquatic to woodland communities and with interesting fauna.

Keelhilla, 1.2mile (2 km) southwest of Cappaghmore on the northeast corner of the Burren, is an example of woodland on limestone pavement.

Ballyteigue, 1.2 miles (2 km) west of Lisdoonvarna, is 16 acres (6.4 ha) of wet meadow on shale.

COUNTY CORK

Knockomagh Wood, adjoining Lough Hyne, is a mixed oak and birch wood with some introduced species.

Glengarriff Wood is an internationally important oak woodland, with birch and rowan, with a well-developed understory of holly, rich in ferns, mosses, and liverworts.

The Gearagh, 3.1 miles (5 km) south of Macroom, is a drowned forest, the remnants of a large and unique postglacial alluvial forest that was submerged to make a reservoir on the River Lee for a hydroelectric scheme. Regeneration of the oak, ash, and birch has occurred in the marshes and on the islands. There are interesting and well-documented dragonflies, rare aquatic species, and large numbers of waders and duck.

Kilcolman Bog, northeast of Buttevant, comprises two reserves. They are the site of a former lake, which is now marsh, fen, and ponds. They are important for wintering flocks of teal, wigeon, shoveller, whooper swans, and Greenland white-fronted geese.

Lough Hyne, just south of Skibbereen, is a deep sea lough joined to the Atlantic by a very narrow channel. It is a study area of international importance because of its great diversity and abundance of species, and the Skibbereen Heritage Centre has audiovisual displays and an aquarium.

Capel Island and *Knockadoon Head,* south of Youghal, are important for nesting and migrant seabirds. The cliff face of the headland is of geological interest, showing successive rock layers.

COUNTY DONEGAL

Dunally Wood, at Creeslough in the Dunally river valley, contains hazel, ash, and alder and a rich flora.

Rathmullan Wood, on the western shores of Lough Swilly, is a mature woodland of oak, beech, and birch.

Ballyarr Wood, 3.1 miles (5 km) west of Ramelton and north of Letterkenny, is an ecologically important oakwood with a well-developed understory of hazel, blackthorn, and holly.

Pettigo Plateau, 9.3 miles (15 km) southeast of Donegal town, is a typical example of Atlantic blanket bog surrounded by lakes. It also contains raised bog domes, and is used in winter by Greenland white-fronted geese.

Lough Barra Bog, at the head of the Gweebarra River, is a good example of western blanket bog.

Meenachullion Bog, south of Lough Barra and between Gubbin Hill and the Owenwee River, is a lowland blanket bog.

COUNTY DUBLIN

Rogerstown Estuary, between Donabate and Rush in north County Dublin, is an important feeding ground for Brent geese, waders, and many different species of duck.

North Bull Island, in Dublin Bay, is an internationally important site of grassland, mudflats, salt marsh, and sand dunes, running almost the length of the north side of the bay and only 5 miles (8 km) from Dublin's city center. Tens of thousands of waders and ducks feed and roost there, especially in winter, when several hundred Brent geese graze beside passing traffic. An interpretive center is run by BirdWatch Ireland.

Baldoyle Estuary, just north of Howth Head, is also important for Brent geese.

COUNTY GALWAY

Derryclare Wood, on the west shore of Lough Inagh in Connemara, is a seminatural oakwood with rich communities of lichens and invertebrates.

Leam West Bog, southeast of Maam Cross, is an internationally important area of blanket bog with diverse habitats.

Rosturra/Derrylahan Wood, Derrycrag Wood, and *Pollnaknockaun Wood* are all near Woodford, and their woodlands of oak and ash are remnants with continuity to ancient forest.

Ballynastaig Wood and *Coole/Garryland,* are both near Gort. Coole park was the home of Lady Gregory, and frequently visited by W. B. Yeats. The two reserves contain deciduous woods, lakes, turloughs, and a rich flora and fauna.

Richmond Esker, near Moylough, is one of the last remaining eskers with native woodland.

COUNTY KERRY

Uragh Wood, southwest of Kenmare on the southwest shore of Inchiquin Lough, is a fine Atlantic oakwood with birch, rowan, aspen, strawberry tree, and juniper.

Derrycunihy Wood, in Killarney Valley at the southern end of the Upper Lake, is a natural oak and holly wood, rich in ferns, mosses, and lichens, as well as invertebrates.

Tralee Bay is important for waterfowl, especially wintering Brent geese.

Castlemaine Harbour, south of the Dingle Peninsula, is an important site for wintering wildfowl, and a salt marsh rich in species. It has some of the earliest plant fossil remains in shale, called the Kiltorcan beds.

Puffin Island, north of St. Finan's Bay off the Iveragh Peninsula. This is an important nesting site for Manx shearwater, puffin, storm petrel, guillemot, razorbill and kittiwake.

Great Skellig and *Little Skellig* are islands southwest of County Kerry and noted for breeding sea birds. Little Skellig has an internationally important gannet colony.

Inishtearaght, one of the Blasket Islands off the Dingle Peninsula, which are all rich in seabirds, choughs, and gray seals.

Eirk Bog, on the floodplain of the Owenreagh River, Killarney, just north of Moll's Gap, contains intermediate stages between blanket bog, raised bog, and fen.

Cummeragh River Bog, northeast of Waterville and southwest of Derriana Lough, is a
  blanket bog of international importance at the head waters of the Cummeragh
  River.

Mount Brandon is on the Dingle Peninsula. Part of the mountain range contains blanket bog and heath, with a concentration of alpine and arctic-alpine plants.

Sheheree Bog, near Killarney, is a raised bog in an area of blanket bog, and one of the few
  areas where slender cottongrass is found. (Another is near Maam Cross in County
  Galway.)

Lough Yganavan and *Lough Nambrackdarrig,* both south of Castlemaine Harbour, are
  breeding sites for the natterjack toad.

COUNTY KILDARE

Pollardstown Fen, 2 miles (3 km) northwest of Newbridge, is the country's best-devel-
  oped fen. It is maintained by alkaline drainage water from the Curragh gravels,
  and supports the usual fen flora and fauna. It is of international importance.

COUNTY KILKENNY

Ballykeefe Wood, Garryricken Wood, and *Kyleadohir Wood,* all near Callan, are small
  woods of mainly oak.

Fiddtown Island, on the Suir River, 4.3 miles (7 km) is southeast of Carrig on Suir. A
  unique, low-lying river island, it is completely covered with reed swamp and with
  an interesting variety of willow species and their hybrids.

COUNTY LAOIS

Grantstown Wood and *Coolacurragh Wood,* at Durrow, are examples of wet woodland.

Timahoe Esker, near Timahoe, is one of the last eskers with native woodland.

Slieve Bloom Mountains, in Counties Laois and Offaly, have blanket bog best seen at
  the mountaintops, because the lower reaches have been afforested.

COUNTY MAYO

Old Head Wood, on the south shore of Clew Bay, 12.4 miles (20 km) west of Westport,
  is a rare example of Atlantic oakwood. It is mixed with introduced beech and
  sycamore, and is rich in lichens.

Knockmoyle/Sheskin is north of Bellacorick on the Ballina-Belmullet road. It is an iron-
  rich spring and fen in blanket bog, of international importance for rare fen plants.

Owenboy Bog, south of Eskeragh Bridge just east of Bellacorick, has low domes resembling raised bog. There is a network of mineral-rich flushes that have interesting mosses. Used in winter by Greenland white-fronted geese.

COUNTY OFFALY

Clara Bog, 1.2 miles (2 km) south of Clara, is the last large and relatively intact raised midland bog with a hummock-and-hollow system. It has diverse habitats and vegetation.

Mongan Bog, near Clonmacnoise, is a wet raised bog with a hummock-and-pool system. It is bordered by an esker, the Pilgrim's Road on the north side.

Raheenmore Bog is a good example of a raised bog in a basin situation with well-developed flora and fauna. It is actively growing, with hummocks and hollows.

COUNTY SLIGO

Ballygilgan Lissadell is a grassland that floods in winter and an important mainland site for barnacle geese.

Easkey Bog, south of Dromore West, offers blanket bog from lowland to mountain.

COUNTY TIPPERARY

Redwood Bog, on the Tipperary side of the Little Brosna River callows, is about 6 miles (10 km) northeast of Portumna Bridge. It is a wet and well-developed raised bog with pools and an intact dome and is crossed by several eskers of botanical interest.

COUNTY WESTMEATH

Scragh Bog, 4.3 miles (7 km) northwest of Mullingar, is a fen in transition to raised bog with several rare plants and mosses, and is of international importance.

COUNTY WEXFORD

Wexford Wildfowl Reserve, on reclaimed slob land north of Wexford Harbour, is a wintering site of international significance for thousands of waders, geese, ducks and swans.

The Raven, on the east coast north of Wexford town, is a dune system with interesting flora and invertebrate fauna. It is also a feeding area for migrant birds.

Ballyteigue Burrow, east of Kilmore Quay, is a series of dunes on a shingle spit, with rare dune and salt marsh plants.

COUNTY WICKLOW

Glen of the Down is just northwest of Delgany. The valley is a good example of a glacial overflow channel and has a well-developed community of flora and fauna.

Deputy's Pass, 3.7 miles (6 km) southwest of Wicklow town, is an oakwood that was originally coppiced.

Vale of Clara, 3.7 miles (6 km) northwest of Rathdrum, is a seminatural broadleaf woodland, one of the largest in the country.

Glendalough is an ecologically and geomorphologically important valley, with oak and conifer woods, two lakes, and interesting marsh and peat plant communities.

Knocksink Wood, on the banks and gorge of the Glencullin River northeast of Enniskerry, is a deciduous woodland planted with conifers with rich and diverse ground flora and insects. It also has springs that are protected under the EU Habitats Directive. There is an educational center at the location.

Glenealo Valley, west of Glendalough, is a valley of blanket bog and heath, surrounded by mountains.

National Nature Reserves in Northern Ireland

COUNTY ANTRIM

Breen is an oak and birch wood.

Straidkilly, 1.2 miles (2 km) northeast of Glenarm, protects a hazel wood on a steep slope.

Glenariff Waterfalls, 3 miles (5 km) inland from Waterfoot, is a deep, wooded gorge with spectacular waterfalls: Ess-na-crub, Ess-na-laragh, and Tears of the Mountain are the largest.

Lough Neagh–Randalstown Forest, 2 miles (3 km) south of Randalstown on the northern shore of Lough Neagh, west of Shane's Castle, is a mixed deciduous woodland with deer, red squirrels, and waterbirds.

Lough Neagh–Rea's Wood is a fen and deciduous woodland, rich in invertebrates.

Lough Neagh–islands are all important nesting sites for gulls, terns, and duck.

Giant's Causeway, 7.4 miles (12 km) east of Portrush and 9.3 miles (15 km) west of Ballycastle, is an internationally famous complex of basalt columns leading from the shore and disappearing under the sea.

Kebble has cliffs of basalt and chalk with thousands of nesting sea birds.

Portrush protects 2.5 acres (1 ha) of marine fossils in rocks of baked clay.

Swan Island, in Larne Lough, is an important tern colony.

Brackagh Moss, 2 miles (3 km) south of Portadown, is a cutaway bog showing layers of former fen peat with a diversity of plants and insects.

Slieveanorra, 6.2 miles (10 km) southwest of Cushendun, in the middle of the Glens of Antrim, is a bog at various stages of development with pools and mature peatland.

Lough Neagh–Farr's Bay is a fen formed by successive lowerings of the lake.

Belshaw's Quarry shows a profile of the geology of the south of the county.

COUNTY ARMAGH

Mullenakill and Annagariff, between Portadown and Lough Neagh, are part of Peatlands Country Park, designed to interpret peatland ecology. There are many rare plants in the area.

Lough Neagh–Oxford Island is a peninsula and series of bays on the southern shore of the lake, north of Lurgan. There are bird-watching hides and the Lough Neagh Discovery Centre.

COUNTY DERRY

Banagher Glen, 3.7 miles (6 km) south of Dungiven, is a mixed woodland with interesting mosses and liverworts.

Ballymaclary, 9.3 miles (15 km) northwest of Coleraine, is part of the large expanse of Magilligan Strand. It has sand dunes and wet slacks with interesting vegetation.

Magilligan Point, east of the entrance to Lough Foyle, has extensive sand dunes and wet slacks.

Roe Estuary, 6.2 miles (10 km) north of Limavady, has mud flats, salt marsh, and sand banks. It is a site for wintering and migrant waders and wildfowl.

Binevenagh, south of Magilligan, is a grassland on basalt with the Bishop's Road running along cliffs 788 feet (240 m) high.

COUNTY DOWN

Bohill is 2.5 acre (1 ha) of oak, holly, birch, and hazel woodland important for the holly blue butterfly.

Hollymount, 1.2 miles (2 km) east of Downpatrick, is an ancient woodland rich in lichens and some fen.

Rostrevor is an oakwood, 2.4 miles (1 km) northeast of Rostrevor on the southwest slopes of Slievemartin.

Ballyquintin Point, on the Ards Peninsula, 4.3 miles (7 km) south of Portaferry, has a rocky shore with raised shingle beach and some salt marsh.

Cloghy Rocks, on the east of the Ards Peninsula, is a beach with offshore rocks and interesting marine species.

Dorn, on the eastern shore of Strangford Lough, is an important site for marine biology.

Granagh Bay, at the eastern side of Strangford Narrows, is rich in marine life.

Killard, 3.7 miles (6 km) south of Strangford, has grassy dunes rich in plant species, particularly orchids.

Murlough, just north of Newcastle, is a notable sand dune system showing all stages of development, with rich plant communities that include the bee orchid. The sands are used in summer by breeding common seals.

North Strangford Lough is a stable mud flats and bird sanctuary for waders and water-fowl. At Castle Espie, a center run by the Wildfowl and Wetlands Trust, are bird hides and an education center.

Quoile Pondage Basin, northeast of Downpatrick, is a lake formed by the construction of a barrage across the estuary of the River Quoile, with a profusion of wetland plants and a high population of wintering wildfowl.

COUNTY FERMANAGH

Castle Archdale Islands, on the east shore of Lower Lough Erne, contains three islands: Inishmakill, Cleenishmeen and Cleenishgarve. There is a mixed, deciduous wood-land, rich in species.

Correl Glen, 3.7 miles (6 km) northwest of Derrygonnelly, is a waterfall and ancient natural woodlands.

Hanging Rock and Rosses, 1.2 miles (2 km) southeast of Belcoo near the south shore of Lower Lough Macnean, is a hanging cliff with an ash wood at its foot.

Marble Arch, 3.7 miles (6 km) east of Belcoo, is a moist ash wood in a steep-sided glen leading to the Marble Arch Cave.

Reilly Wood is an oakwood.

Gole Woods is scrub.

Lough Naman Bog, north of the road from Derrygonnelly to Garrison, is a large tract of blanket bog with well-developed hummock and hollow communities.

Castlecaldwell, 3.7 miles (6 km) east of Beleek, on the western shore of Lower Lough Erne, is fen and scrub in sheltered bays.

Ross Lough, 6.2 miles (10 km) northwest of Enniskillen, with fen and open water on the east end of the lough.

Crossmurrin, on the northern slopes of Cuilcagh Mountain, is an area of mixed grass-land on limestone.

COUNTY TYRONE

Boorin conserves heathland on glacial till.

Killeter, on the border with County Donegal and 10 miles (16 km) west of Castlederg has two areas of mountain, blanket bog with diverse mosses.

Meenadoan, west of Lough Bradan Forest and northeast of Pettigo, is a small bog with interesting plant succession.

The Murrins, 10 miles (16 km) northeast of Omagh, is an area of mountain blanket bog, including some wet heathland.

Index